G. GARNER
LSW 405

MATRIX ALGEBRA FOR

THE BIOLOGICAL SCIENCES

(Including Applications in Statistics)

SERIES ON QUANTITATIVE METHODS FOR BIOLOGISTS AND MEDICAL SCIENTISTS

Wilfrid J. Dixon, Editor

Basic Statistics: A Primer for the Biomedical Sciences
Olive Jean Dunn

Matrix Algebra for the Biological Sciences (Including Applications in Statistics)
S. R. Searle

MATRIX ALGEBRA FOR THE BIOLOGICAL SCIENCES

(Including Applications in Statistics)

S. R. SEARLE

Associate Professor of Biological Statistics
Biometrics Unit, Department of Plant Breeding,
N.Y. State College of Agriculture,
Cornell University, Ithaca, N.Y.

John Wiley & Sons, Inc., New York · London · Sydney

10 9 8 7 6 5 4

Library of Congress Catalog Card Number: 66-11528
Printed in the United States of America

FOREWORD TO THE SERIES

The biological and medical sciences are in a period of explosive growth whose rate has been increased by the diversification of experimental methods, procedures, and concepts. Biological systems are multi-dimensional, and individual differences are great; therefore quantitative methods in biology will necessarily vary greatly depending on the environment. All biologists know that the adaptation of techniques from other fields is not simple, usually indirect, and almost always impossible without radical revision, extension, or reformulation. The simulation of physiological systems requires a mathematical base and in most cases large electronic computers. But as theoretical concepts in biology mature, mathematical methods and models become more effective. The basic experimental methods originating in the biological sciences may now be joined by the methods of the physical sciences and made technically feasible by the rapid advances in electronics and computers.

It is the goal of this series to provide the various biological research workers expositions of the modern powerful analytical tools now available. Some books will be of interest to specialized groups; others will provide a broader survey.

WILFRID J. DIXON

PREFACE

Algebra, in its simplest connotation, is a shorthand for the language of mathematics. Matrix algebra is also a shorthand, one that provides concise facilities for dealing with many numbers simultaneously. While in no sense new, matrix algebra has heretofore been set in a mathematical environment sufficiently abstruse to make it difficult for a person with only meagre training in mathematics to assimilate it. Now, however, many such people are coming to find a knowledge of matrices useful in their work, especially those in the biological sciences where the use of mathematics is greatly increasing. It is to these people that I address this book.

I have attempted throughout to keep the mathematical discussion relatively informal, in the belief that it will appeal to the reader who may not enjoy the more formal manner of precise mathematical writing. The basic prerequisite for using the book is high-school algebra. Differential calculus is used on only a few pages, and these can be omitted if desired; nothing will be lost insofar as a general understanding of matrix algebra is concerned. Proofs and demonstrations of most of the theory are given, for without them the presentation would be lifeless. But in every chapter the theoretical development is profusely illustrated both with elementary numerical examples and with illustrations taken from biology and statistics. (Suitability of the illustrative material is discussed at the end of Chapter 1 and the topics used are listed in the index under the headings "Biological Illustrations" and "Statistical Illustrations".) In addition, two entire chapters are devoted to uses of matrix algebra that are important in the analysis of biological (and other) data, namely the statistical analyses relating to regression and linear models. Only a brief outline of statistical theory (based on examples) is given in these chapters, with the result that they may be most valued by those with some knowledge of statistical methods and their derivation. However, there is such increasing use of these methods in the biological sciences that it seems appropriate to end the book in this manner.

The mainstream of the text is the first seven chapters, beginning with one on introductory concepts that includes a discussion of subscript and summation notation. This is followed by four chapters that deal with elementary matrix operations, determinants, inverse matrices, and rank and linear independence. The sixth chapter considers the problem of solving simultaneous linear equations (especially those not of full rank) in which the concept of a generalized inverse is introduced, and is shown to be of great value. Chapter 7 is concerned with latent roots and vectors, and a plethora of miscellaneous topics are presented in Chapter 8; and Chapters 9 and 10 describe the matrix algebra of regression and linear models. All chapters except the last two end with exercises.

Occasional sections and paragraphs can be omitted at a first reading, especially by those whose experience in mathematics is somewhat limited. These portions of the book are marked with an asterisk and printed in small type and, generally speaking, contain material subsidiary to the main flow of the text—material that may be a little more advanced in mathematical presentation than the general level otherwise maintained.

Chapters, and sections within chapters, are numbered with Arabic numerals 1, 2, 3, ... References to sections are made by means of the decimal system, with an appropriate Section number shown in the heading of each page; e.g. Section 1.3 is Section 3 of Chapter 1, with [1.3] being part of the heading on page 5. Numbered equations have the numbers (1), (2), . . . within each chapter. Those of one chapter are seldom referred to from another, but when they are the chapter reference is explicit; but a reference such as "equation (3)", or more simply "(3)", relates to the equation numbered (3) in the chapter containing the reference.

I am greatly indebted to D. S. Robson for his encouraging me to write this book and for his continuing assistance and advice at all stages of the work. Many others also provided sterling aid in reading draft manuscripts and suggesting improvements, notably L. N. Balaam, W. T. Federer, M. R. Mickey, Jr., B. L. Raktoe, H. E. Schaffer, E. C. Townsend and N. S. Urquhart. Valuable assistance in preparatory work was given by Mrs. D. VanOrder; particular thanks go to Mrs. A. T. Holman for her endless patience and her magnificent work at the typewriter. To all these people I am extremely grateful.

S. R. SEARLE

Ithaca, N.Y.
December 1965

CONTENTS

1. *Introduction* *1*

1. The scope of matrix algebra, 1
2. General description of a matrix, 3
3. Subscript notation, 4
4. Summation notation, 6
5. Dot notation, 10
6. Definition of a matrix, 11
7. Vectors and scalars, 13
8. General notation, 14
9. Illustrative examples, 15
10. Exercises, 16

    References, 17

2. *Elementary matrix operations* *18*

1. Addition, 18
2. Scalar multiplication, 20
3. Subtraction, 20
4. Equality and the null matrix, 22
5. Multiplication, 22
6. Linear transformations, 33
7. The laws of algebra, 35
8. The transpose of a matrix, 38
9. Quadratic forms, 42
10. Variance-covariance matrices, 46
11. Partitioning of matrices, 47
12. Multiplication of partitioned matrices, 49
13. Exercises, 50

    References, 55

3. Determinants *56*

1. Simple evaluation, 56
2. Formal definition, 61
3. Elementary expansions, 63
4. Addition and subtraction of determinants, 70
5. Diagonal expansion, 71
6. The Laplace expansion, 74
7. Multiplication of determinants, 75
8. Conclusion, 77
9. Exercises, 77

   References, 79

4. The inverse of a matrix *80*

1. Introduction, 80
2. Products equal to *I*, 84
3. The cofactors of a determinant, 86
4. Derivation of the inverse, 88
5. Conditions for existence of inverse, 91
6. Properties of the inverse, 91
7. Left and right inverses, 92
8. Some uses of inverses, 93
9. Inverting matrices on high-speed computers, 98
10. Exercises, 105

    References, 107

5. Rank and linear independence *108*

1. Solving linear equations, 108
2. Linear independence, 110
3. Linear dependence of vectors, 112
4. Linear dependence and determinants, 113
5. Sets of linearly independent vectors, 114
6. Rank, 117
7. Elementary operators, 121
8. Rank and the elementary operators, 123
9. Finding the rank of a matrix, 123
10. Equivalence, 125
11. Reduction to equivalent canonical form, 126
12. Congruent reduction of symmetric matrices, 129
13. The rank of a product matrix, 132
14. Exercises, 133

    Reference, 135

6. *Linear equations and generalized inverses* *136*

1. Equations having many solutions, 136
2. Consistent equations, 137
3. More or fewer equations than unknowns, 143
4. Generalized inverse matrices, 144
5. Solving linear equations using generalized inverses, 147
6. Rectangular matrices, 158
7. Exercises, 160
References, 162

7. *Latent roots and vectors* *163*

1. Age distribution vectors, 163
2. Derivation of latent roots and vectors, 165
3. Latent roots all different, 167
4. Multiple latent roots, 174
5. Some properties of latent roots, 177
6. Dominant latent roots, 181
7. Factorizing the characteristic equation, 186
8. Symmetric matrices, 188
9. Exercises, 193
References, 195

8. *Miscellanea* *196*

1. Orthogonal matrices, 196
2. Matrices having all elements equal, 197
3. Idempotent matrices, 199
4. Nilpotent matrices, 202
5. A vector of differential operators, 203
6. Jacobians, 209
7. Inversion by partitioning, 210
8. Matrix functions, 212
9. Direct sums, 213
10. Direct products, 215
11. Exercises, 220
References, 223

9. *The matrix algebra of regression analysis* *225*

1. General description, 225
2. Estimation, 228
3. The case of k x-variables, 230
4. The mathematical model, 231
5. Unbiasedness and variances, 232
6. Predicted y- values, 232
7. Estimating the error variance, 233
8. Deviations from means, 233
9. Analysis of variance, 241
10. Multiple correlation, 243
11. Tests of significance, 243
12. Fitting variables one at a time, 247
13. Inverting the matrix of all variables, 249
14. Summary of calculations, 252

References, 253

10. *Some matrix algebra of linear statistical models* *254*

1. General description, 254
2. The normal equations, 257
3. Solving the normal equations, 259
4. Expected values and variances, 260
5. Estimating the error variance, 260
6. Estimable functions, 262
7. Examples, 264
8. Analysis of variance, 273
9. Regression as part of a linear model, 284
10. Summary of calculations, 288

References, 288

Book list *290*

Index *291*

CHAPTER 1

INTRODUCTION

1. THE SCOPE OF MATRIX ALGEBRA

The biological sciences are today in the process of changing from being primarily descriptive to being very much quantitative. As a result, biologists find themselves confronted more and more with large amounts of numerical data, measurements of one form or another gathered from their laboratories, field experiments and surveys. But the mere collecting and recording of data achieve nothing; having been collected, they must be investigated to see what information may be contained concerning the biological problem at hand. Any investigation so made, being quantitative in nature, will involve mathematics in some way even if only the calculating of percentages or averages. Frequently, however, biologists have to subject their data to more complex calculations, requiring procedures that involve mathematical details beyond their general experience. In order to carry out the mathematics the biologist in this situation must either learn the procedures himself, or at least learn something of the language of mathematics, that he may communicate satisfactorily with the mathematician whose aid he enlists. In either case there will be numerous mathematical tools that it will be valuable for him to understand. Matrices and their algebra are one of these tools.

A matrix is simply a rectangular array of numbers set out in rows and columns. As such it is a device frequently used in organizing the presentation of numerical data that they may be handled with ease mathematically. Herein lies the value of the matrix: the facility it provides of condensing into a small set of symbols a wealth of mathematical manipulations. As a result, matrices greatly assist in problems of data analysis such as those that currently face biologists through the increasing growth of quantitative methods in their work. The same is true of other fields of science, of course, for so often when mathematical analysis is needed matrices are found useful both in organizing the calculations involved and in clarifying an understanding of them.

Numerous examples are cited throughout this book of using matrix algebra in situations involving mathematical analysis of biological data. One arises in population dynamics where, in studying a living population, we want to investigate the distribution of individuals according to their age. If n_t represents a series of values describing the age distribution at time t, then the relationships between successive age distributions can often be described as $n_t = Mn_{t-1}$ for some appropriate matrix M. An equation such as this is, as we shall see, simple enough in the context of matrix algebra, yet it has contributed greatly to the study of stable age distributions and logistic growth, yielding results that might have been obtained only with great difficulty without the use of matrix algebra. Some of the methods involved are mentioned in Chapter 7. A similar equation occurs in the study of inbreeding where, for any particular mating system, one is interested in relating the frequencies of mating types in one generation with those in another. As shown in Chapter 2, the relationships involved can be expressed as a set of linear equations but, with a suitable knowledge of matrix terminology, they can also be put in the simple and succinct form $f_n = Af_{n-1}$. The symbols f_n and f_{n-1} represent an array of frequencies at the nth and $(n - 1)$th generations respectively and A (a matrix) represents the relationships between them. From this can be derived the result $f_n = A^n f_0$, where f_0 represents a set of initial frequencies. Relative to manipulating sets of simultaneous equations, the brevity of this symbolism is readily apparent. Another instance of brevity, and one that also illustrates the universal nature of matrix expressions, occurs in the technique of regression analysis which is often applicable to biological data. This involves calculating values $b_0, b_1, \ldots, b_k$ for the equation

$$y = b_0 + b_1x_1 + b_2x_2 + \cdots + b_kx_k$$

where we have a whole series of observations on the variable y and on each of the k variables $x_1, x_2, \ldots, x_k$. With proper definition the b's are obtained by solving the equations $X'Xb = X'y$ as $b = (X'X)^{-1}X'y$, where X and y, both matrices, represent all the observations in the x and y variables respectively, and b represents the series of b's. Again, the brevity of such expressions is evident, as is their universality too when one realizes that they apply no matter how many x-variables there are nor how numerous the observations on each. Once the necessary matrix algebra is assimilated [for example the meanings attached to X, X' and $(X'X)^{-1}$], the method of expressing the results is the same for all manner of situations. Chapter 9 deals with this topic in detail and Chapter 10 is concerned with extensions of it.

The manipulative skill required for using matrices in these and numerous other situations of applying mathematics to biology is neither great nor excessively demanding, whereas in terms of brevity, simplicity and clarity the value of matrix algebra is often very appreciable. Furthermore, the almost universal nature of matrix expressions has great appeal, for so often can the same results be applied, with only minor changes, to situations involving both small amounts of data and to those involving extremely large amounts. Matrix algebra is therefore a vehicle by which mathematical procedures for many problems, both large and small, can be described independently of the size of the problem. Size affects not the understanding of the procedures, only the amount of calculating involved which in turn determines time and cost, factors which in today's world of high-speed computers, are rapidly diminishing. The presence of such computers is additional reason for advancing the value of matrix algebra. If a set of calculations can be expressed in terms of matrices the problem of making easy and efficient use of a high-speed computer for carrying them out is then often simplified. Matrix algebra and high-speed computers therefore go hand in hand to some extent.

2. GENERAL DESCRIPTION OF A MATRIX

The aspect of a matrix which makes it an aid in organizing data provides a convenient starting point for its description. We begin with an example and subsequently proceed to a formal definition.

Illustration. Wallace (1959) reports the percentage of sterile cultures found in successive generations of several different experimental populations of *drosophila melanogaster*. In doing so he finds it convenient to summarize results in tabular form, as extracted in Table 1.

TABLE 1. PERCENTAGE OF STERILE CULTURES AMONG DIFFERENT POPULATIONS IN SUCCESSIVE GENERATIONS

Generation	Population		
	1	2	3
1	18	17	11
2	19	13	6
3	6	14	9
4	9	11	4

Suppose now that the array of numbers in the table is extracted and written simply as

$$\begin{bmatrix} 18 & 17 & 11 \\ 19 & 13 & 6 \\ 6 & 14 & 9 \\ 9 & 11 & 4 \end{bmatrix},$$

where the position of an entry in this array determines its meaning; for example, the entry in the second row and third column, 6, represents the percentage of sterile cultures observed in the second generation of the third population. In this way a row represents the percentage of sterile cultures seen in the same generation of different populations, and a column represents the percentage of sterile cultures seen in the same population at different generations. For example, the first row shows the percentage of sterile cultures observed in the first generation of the three populations, and the first column shows the percentage of sterile cultures seen in the four generations of the first population. Such an array of numbers is called a *matrix*.

The first thing to notice about any matrix is that it is a rectangular (or square) array, with the entries therein set out in rows and columns. The elements of each row usually have something in common, as do those of each column. The individual entries in the array are called the *elements* or *terms* of the matrix; in general they can be numbers of any sort (real or complex, rational and irrational) or even functions of one or more variables. For our purposes we will usually think of them as being real numbers, positive, negative or zero.

Matrix algebra, being the algebra of matrices, is the algebra of arrays such as we just have described, each treated as an entity and denoted by a single symbol. The algebra involves the elements within the arrays, but it is the handling of the arrays themselves, as separate entities, that constitutes matrix algebra.

Before proceeding to a formal definition of a matrix we present some mathematical notation that is widely used in matrix algebra. (Readers familiar with subscript and summation notation can readily omit the next three sections.)

3. SUBSCRIPT NOTATION

Algebra is arithmetic with letters of the alphabet representing numbers. Thus the first two rows of the matrix given previously, namely

$$\begin{bmatrix} 18 & 17 & 11 \\ 19 & 13 & 6 \end{bmatrix}$$

might be written as $\begin{bmatrix} a & b & c \\ d & e & f \end{bmatrix}$

where a, b, c, d, e and f represent the numbers 18, 17, 11, 19, 13 and 6 respectively. To this matrix, the array thought of as a self-contained entity, might be given the name A. We could then write

$$A = \begin{bmatrix} a & b & c \\ d & e & f \end{bmatrix}.$$

More generally, A in this case might stand for any matrix of two rows and three columns.

Since a matrix described in this manner using the letters a through z could have only twenty-six elements, a less restrictive notation is needed. Instead of specifying each number by a letter we use one letter for a whole series of numbers, modified by attaching to it a series of integers to provide the necessary identification. Thus the matrix A could be written as

$$A = \begin{bmatrix} a_1 & a_2 & a_3 \\ b_1 & b_2 & b_3 \end{bmatrix}.$$

The integers 1, 2 and 3 are called *subscripts* and the elements of the first row, for example, are read as "*a*, one", "*a*, two" and "*a*, three", or as "*a*, sub one", "*a*, sub two" and "*a*, sub three". In this instance the subscripts describe the column in which each element belongs and all elements in the same column have a common subscript; for example, the second column elements have subscript 2.

The principle just described can be applied to the rows and columns of a matrix simultaneously, using a single letter of the alphabet for all elements. Two subscripts are used for each element and written alongside one another, the first specifying the row the element is in and the second the column. Thus A could be written as

$$A = \begin{bmatrix} a_{11} & a_{12} & a_{13} \\ a_{21} & a_{22} & a_{23} \end{bmatrix}.$$

The elements are read as "*a*, one, one", "*a*, one, two" or as "*a*, sub one, one", "*a*, sub one, two" and so on. In this way an element's subscripts uniquely locate its position in the matrix. For, just as we have used the letter a to denote the value of an element of a matrix, so we can use other letters to denote other things; for example we might refer to a row as "row i", and were i equal to 4 "row i" would be the fourth row of the matrix. In the same way we can use j for referring to columns, and "column j" or the "jth column" is, for $j = 3$, the third column. By this means we refer to the element in the ith row and jth column of a matrix as a_{ij}, using i and j as subscripts to the letter a to denote the row and column the element is in, just as a_{12}, for example, is in the first row and second column. In general then, a_{ij} is in the ith row and jth column, e.g. a_{ij} for $i = 2$ and $j = 3$ is a_{23}, the element in the second row and third column. No comma is placed between the two subscripts unless it is needed to avoid confusion. For example, were there twelve columns in a matrix, the element in row 1 and column 12 could be written as $a_{1,12}$ to distinguish it from $a_{11,2}$, the element in the eleventh row and second column.

This notation provides ready opportunity to detail the elements of a row, a column or even a complete matrix, in a very compact manner. Thus the first row of A,

$$a_{11} \quad a_{12} \quad a_{13}$$

can be described as

$$a_{1j} \qquad \text{for} \quad j = 1, 2, 3.$$

By so doing, we mean that the elements of the row are a_{1j} for j taking in turn each one of the integer values 1, 2 and 3. Columns can be described in like fashion. The whole matrix is accounted for by enclosing a_{ij} in curly brackets and writing

$$A = \{a_{ij}\} \qquad \text{for} \quad i = 1, 2 \qquad \text{and} \quad j = 1, 2, 3.$$

This notation completely specifies the elements of the matrix, its name and its size. It is used in a later paragraph to specify

$$A = \{a_{ij}\} \qquad \text{for} \quad i = 1, 2, \ldots, r, \qquad \text{and} \quad j = 1, 2, \ldots, c$$

as the general matrix of r rows and c columns, the dots indicating that i takes all integer values from 1 through r and j takes all integer values from 1 through c.

4. SUMMATION NOTATION

The most frequently used arithmetic operation is that of adding numbers together. Subscript notation provides a means of describing this operation in a very succinct manner. Suppose we wish to add five numbers represented as a_1, a_2, a_3, a_4 and a_5. Their sum is clearly

$$a_1 + a_2 + a_3 + a_4 + a_5.$$

It can be put in words as

"the sum of all values of a_i for $i = 1, 2, \ldots, 5$."

The phrase "the sum of all values of" is customarily represented by the capital form of the Greek letter sigma, namely $\sum$. Accordingly the sum of the a's is written as

$$\sum a_i \quad \text{for} \quad i = 1, 2, \ldots, 5.$$

A further abbreviation is $\sum_{i=1}^{i=5} a_i$, where the phrase "for $i = 1, 2, \ldots, 5$" has been replaced by the "$i = 1$" and "$i = 5$" below and above the $\sum$ sign. This indicates that the summation is for all integer values of i from 1 through 5. Thus

$$\sum_{i=1}^{i=5} a_i = a_1 + a_2 + a_3 + a_4 + a_5.$$

Many variations on this summation notation are to be found. Frequently the "$i = 5$" above the $\sum$ is reduced to just 5, so that a commonly seen form is

$$\sum_{i=1}^{5} a_i = a_1 + a_2 + a_3 + a_4 + a_5.$$

Similarly

$$\sum_{i=1}^{3} x_i = x_1 + x_2 + x_3$$

and

$$\sum_{i=1}^{n} x_i = x_1 + x_2 + x_3 + \cdots + x_{n-2} + x_{n-1} + x_n,$$

the sum of n x's. Notice the algebraic expressions for subscripts, $n - 2$ and $n - 1$. This is permissible, and occasionally quite extensive expressions are used this way. Sometimes the terms above and below the $\sum$ sign are omitted when they are clearly obvious from the context, and sometimes they are used as subscripts and superscripts to the $\sum$:

$$\textstyle\sum_1^4 y_i = y_1 + y_2 + y_3 + y_4.$$

Another variant of the notation is that the lower limit of i need not necessarily be the value 1: with $i = 3$ written below the $\sum$ sign we have

$$\sum_{i=3}^{7} y_i = y_3 + y_4 + y_5 + y_6 + y_7.$$

Furthermore, although subscripts in a summation are usually consecutive integers, omissions can be accounted for:

$$\sum_{\substack{i=3 \\ i \neq 4}}^{7} y_i = y_3 + y_5 + y_6 + y_7.$$

Summation notation has so far been described in terms of simple sums, but it also encompasses sums of squares and sums of products, and indeed sums of any series of expressions that can be identified by subscript notation. Thus the sum of squares $c_1^2 + c_2^2 + c_3^2 + c_4^2$ can be written as

$$\sum_{i=1}^{4} c_i^2 = c_1^2 + c_2^2 + c_3^2 + c_4^2.$$

Similarly, if we have two series of numbers p_1, p_2, p_3 and q_1, q_2, q_3 the sum of their products term by term, $p_1q_1 + p_2q_2 + p_3q_3$, is written

$$\sum_{i=1}^{3} p_iq_i = p_1q_1 + p_2q_2 + p_3q_3.$$

The notation we have described applies equally well to cases involving two subscripts: in summing over either subscript the other remains unchanged. For example,

$$\sum_{j=1}^{3} a_{1j} = a_{11} + a_{12} + a_{13} \qquad \text{and} \qquad \sum_{i=1}^{2} a_{i2} = a_{12} + a_{22}$$

are specific instances of the general results

$$\sum_{j=1}^{3} a_{ij} = a_{i1} + a_{i2} + a_{i3} \qquad \text{and} \qquad \sum_{i=1}^{2} a_{ij} = a_{1j} + a_{2j}.$$

Summing each of the latter leads to the self-explanatory operation of double summation:

$$\begin{aligned}\sum_{i=1}^{2}\left(\sum_{j=1}^{3} a_{ij}\right) &= \sum_{i=1}^{2} (a_{i1} + a_{i2} + a_{i3}) \\ &= \sum_{i=1}^{2} a_{i1} + \sum_{i=1}^{2} a_{i2} + \sum_{i=1}^{2} a_{i3} \\ &= a_{11} + a_{21} + a_{12} + a_{22} + a_{13} + a_{23}.\end{aligned}$$

Similarly

$$\begin{aligned}\sum_{j=1}^{3}\left(\sum_{i=1}^{2} a_{ij}\right) &= \sum_{j=1}^{3} (a_{1j} + a_{2j}) \\ &= \sum_{j=1}^{3} a_{1j} + \sum_{j=1}^{3} a_{2j} \\ &= a_{11} + a_{12} + a_{13} + a_{21} + a_{22} + a_{23},\end{aligned}$$

which is seen at once to be the same as $\sum_{i=1}^{2}\left(\sum_{j=1}^{3} a_{ij}\right)$. Removing the brackets gives the important result

$$\sum_{i=1}^{2}\sum_{j=1}^{3} a_{ij} = \sum_{j=1}^{3}\sum_{i=1}^{2} a_{ij},$$

thus illustrating the general principle that the order of summation in double summation is of no consequence. In general

$$\sum_{i=1}^{m}\sum_{j=1}^{n} a_{ij} = \sum_{j=1}^{n}\sum_{i=1}^{m} a_{ij}.$$

In terms of a matrix of m rows and n columns the left-hand side is the sum of row totals and the right-hand side is the sum of column totals, both sums equaling the total of all elements.

Sums of squares and products can be written in a similar fashion:

$$\sum_{j=1}^{4} a_{ij}^2 = a_{i1}^2 + a_{i2}^2 + a_{i3}^2 + a_{i4}^2,$$

$$\sum_{i=1}^{3} a_{ij}b_{ij} = a_{1j}b_{1j} + a_{2j}b_{2j} + a_{3j}b_{3j},$$

and

$$\sum_{j=1}^{3} a_{1j}b_{j1} = a_{11}b_{11} + a_{12}b_{21} + a_{13}b_{31}.$$

Some readers may question the use of j in b_{j1} in this last expression since until now we have used i as the first subscript. There is, however, nothing authoritarian about i and j as subscripts, any letters of the alphabet may be so used. Certainly i and j are commonly found in this role, both in this book and elsewhere, but they are by no means the only letters used as subscripts; for example,

$$\sum_{i=1}^{2}\sum_{j=1}^{3} a_{ij} = \sum_{p=1}^{2}\sum_{q=1}^{3} a_{pq} = \sum_{k=1}^{2}\sum_{t=1}^{3} a_{kt}$$

$$= a_{11} + a_{12} + a_{13} + a_{21} + a_{22} + a_{23}.$$

The expression $\sum_{j=1}^{3} a_{1j}b_{j1}$ developed earlier is an example of the more general form

$$\sum_{j=1}^{n} a_{ij}b_{jk} = a_{i1}b_{1k} + a_{i2}b_{2k} + a_{i3}b_{3k} + \cdots + a_{in}b_{nk},$$

an expression used extensively in multiplying two matrices together. It is discussed in detail in the next chapter (Section 2.5), but to anticipate for a moment we may note that if A is the matrix

$$A = \begin{bmatrix} a_{11} & a_{12} & a_{13} \\ a_{21} & a_{22} & a_{23} \end{bmatrix} \quad \text{and } B \text{ is} \quad B = \begin{bmatrix} b_{11} & b_{12} & b_{13} \\ b_{21} & b_{22} & b_{23} \\ b_{31} & b_{32} & b_{33} \end{bmatrix},$$

then
$$\sum_{j=1}^{3} a_{1j}b_{j1} = a_{11}b_{11} + a_{12}b_{21} + a_{13}b_{31}$$

is the sum of the term by term products of the elements of the first row of A and the first column of B. Similarly

$$\sum_{j=1}^{3} a_{ij}b_{jk} = a_{i1}b_{1k} + a_{i2}b_{2k} + a_{i3}b_{3k}$$

is the sum of the term by term products of the elements of the ith row of A and the kth column of B.

Another attribute of summation operations may be noted. It involves summation of terms that do not have subscripts; for example,

$$\sum_{i=1}^{4} x = x + x + x + x = 4x$$

and

$$\sum_{i=1}^{3} ky_i = ky_1 + ky_2 + ky_3 = k\left(\sum_{i=1}^{3} y_i\right).$$

The generalizations of these results are easily derived:

$$\sum_{i=1}^{n} x = nx \qquad \text{and} \qquad \sum_{i=1}^{n} ky_i = k\left(\sum_{i=1}^{n} y_i\right).$$

It is appropriate to mention at this point a multiplication procedure analogous to $\sum$. It is denoted by the symbol pi, Π, and involves multiplying together all the terms to which it applies. Thus, whereas

$$\sum_{i=1}^{5} b_i = b_1 + b_2 + b_3 + b_4 + b_5,$$

$$\prod_{i=1}^{5} b_i = b_1b_2b_3b_4b_5,$$

the product of all the b's. Operationally, Π is equivalent to $\sum$ except that it denotes multiplication instead of addition.

5. DOT NOTATION

A further abbreviation used extensively is

$$\sum_{i=1}^{m} a_{ij} = a_{.j}$$

The dot subscript in place of i denotes that summation has taken place over the i subscript. Since the notation $a_{.j}$ shows no indication of the limits of i over which summation has occurred, it is used only when these limits are clear from the context of its use. In line with $a_{.j}$ we also have

$$a_{i.} = \sum_{j=1}^{n} a_{ij}$$

and

$$a_{..} = \sum_{i=1}^{m} a_{i.} = \sum_{j=1}^{n} a_{.j} = \sum_{i=1}^{m} \sum_{j=1}^{n} a_{ij}$$

6. DEFINITION OF A MATRIX

A matrix is a rectangular (or square) array of numbers arranged in rows and columns. The rows are of equal length, as are the columns. In terms of the notation of Section 3 we will let a_{ij} denote the element in the ith row and jth column of a matrix A, and if A has r rows and c columns it is written as

$$A = \begin{bmatrix} a_{11} & a_{12} & a_{13} & \dots & a_{1j} & \dots & a_{1c} \\ a_{21} & a_{22} & a_{23} & \dots & a_{2j} & \dots & a_{2c} \\ \cdot & & & & \cdot & & \cdot \\ \cdot & & & & \cdot & & \cdot \\ \cdot & & & & \cdot & & \cdot \\ a_{i1} & a_{i2} & a_{i3} & \dots & a_{ij} & \dots & a_{ic} \\ \cdot & & & & \cdot & & \cdot \\ \cdot & & & & \cdot & & \cdot \\ \cdot & & & & \cdot & & \cdot \\ a_{r1} & a_{r2} & a_{r3} & \dots & a_{rj} & \dots & a_{rc} \end{bmatrix}.$$

The three dots indicate, in the first row for example, that the elements a_{11}, a_{12} and a_{13} continue in sequence up to a_{1c}; likewise in the first column the elements a_{11}, a_{21}, continue in sequence up to a_{i1} and on up to a_{r1}. The use of three dots to represent omitted values of a long sequence in this manner is standard and will be used extensively. This form of writing a matrix clearly specifies its terms, and also its size, namely the number of rows and columns. An alternative and briefer form is

$$A = \{a_{ij}\}, \qquad \text{for} \quad i = 1, 2, \dots, r, \qquad \text{and} \quad j = 1, 2, \dots, c,$$

the curly brackets indicating that a_{ij} is a typical element, the limits of i and j being r and c respectively.

The element a_{ij} is sometimes called the ijth *element*, the first subscript referring to the row the element is in and the second to the column. Thus a_{23} is the element in the second row and third column. The size of the matrix, i.e. the number of rows and columns, is referred to as its *order* (or sometimes as its *dimensions*). Thus A with r rows and c columns has order $r \times c$ (read as "r by c") and, to emphasize its dimensions, the matrix can be written $A_{r \times c}$. When $r = c$, the number of rows equals the number of columns, A is square and is referred to as a *square matrix* and, provided there is no ambiguity, is described as being of order r. In this case the elements $a_{11}, a_{22}, a_{33}, \ldots, a_{rr}$ are referred to as the *diagonal elements* of the matrix, and the sum of them is the *trace* of the matrix; that is, when A is square the trace of A equals $\sum_{i=1}^{r} a_{ii}$. In every instance the first term in the first row of a matrix, a_{11} in this case, is called the *leading term* of the matrix.

A simple example of a matrix, one of order 2×3, is

$$A_{2 \times 3} = \begin{bmatrix} 4 & 0 & -3 \\ -7 & 2.73 & 1 \end{bmatrix}.$$

Notice that zero is legitimate as an element, that the elements need not all have the same sign, and that integers and decimal numbers can both be elements of the same matrix. When all the non-diagonal elements of a square matrix are zero, the matrix is described as a *diagonal matrix*; for example,

$$A = \begin{bmatrix} 3 & 0 & 0 \\ 0 & -17 & 0 \\ 0 & 0 & 99 \end{bmatrix}$$

is a diagonal matrix. Another variant of a square matrix is that in which all elements above (or below) the diagonal are zero; for example,

$$B = \begin{bmatrix} 1 & 5 & 13 \\ 0 & -2 & 9 \\ 0 & 0 & 7 \end{bmatrix} \quad \text{and} \quad C = \begin{bmatrix} 2 & 0 & 0 \\ 8 & 3 & 0 \\ 1 & -1 & 2 \end{bmatrix}.$$

Such forms are usually given the name *triangular matrix*, and in particular B is called an *upper triangular matrix* and C is a *lower triangular matrix*, these names being used for obvious reasons.

Illustration. Matrices in biology often arise as matrices of transition probabilities. This use is illustrated, for example, by the work of Neyman et al. (1956), who attempt to give mathematical explanation for the apparent phenonemon of a state of equilibrium existing in the population of a particular species of flour beetle. In doing so they suppose "that the life cycle of a beetle can be divided into a finite number of different states", which can be numbered 1, 2, 3, ..., m. "When an egg is laid this egg is considered as a beetle in state 1; state 2 denotes a particular phase in the life of the egg" and so on, with state m representing the final phase in the life of an adult beetle. Various postulates are then adopted about the probabilities of beetles surviving from one state to another. These may be summarized in a matrix

$$P = \begin{bmatrix} p_{11} & p_{12} & p_{13} & \cdots & p_{1m} \\ p_{21} & p_{22} & p_{23} & \cdots & p_{2m} \\ \cdot & \cdot & \cdot & \cdot & \cdot \\ \cdot & \cdot & \cdot & \cdot & \cdot \\ \cdot & \cdot & \cdot & \cdot & \cdot \\ p_{m1} & p_{m2} & p_{m3} & \cdots & p_{mm} \end{bmatrix}$$

$$= \{p_{ij}\} \qquad \text{for} \quad i, j = 1, 2, \ldots, m,$$

where p_{ij} is the probability that a beetle in state i will, within some given unit of time, transfer to state j. Assumptions about the values of the p's, together with the application of statistical theory and matrix algebra, lead to a variety of conclusions about the general condition of the beetle population. The use of a matrix as an entity for handling an array of transition probabilities is one we shall return to on numerous occasions, for it occurs again and again in the study of natural populations. Such a matrix is often called a *probability transition matrix*.

7. VECTORS AND SCALARS

A matrix consisting of only a single column is called a *column vector*; for example,

$$x = \begin{bmatrix} 3 \\ -2 \\ 0 \\ 1 \end{bmatrix}$$

is a column vector of order 4. Likewise a matrix that is just a single row is a *row vector*.

$$y' = [4 \quad 6 \quad -7]$$

is a row vector of order 3.

A single number such as 2, 6.4, or -4 is called a *scalar*; the elements of a matrix are (usually) scalars although, as shall be discussed later, a matrix can be expressed as a matrix of smaller matrices. In some situations it is convenient to think of a scalar as a matrix of order 1×1; for example, the trace (see Section 1.6) of a scalar has meaning only if the scalar is considered as a square matrix of order 1, and in this sense the trace of a scalar is obviously the scalar itself.

8. GENERAL NOTATION

A well-recognized notation that shall be used here is that of denoting matrices by upper case letters and their elements by the lower case counterparts with appropriate subscripts. Vectors are denoted by lower case letters, usually from the end of the alphabet, using the prime superscript to distinguish a row vector from a column vector. Thus x is a column vector and x' is a row vector. The symbol λ is often used for a scalar.

Throughout this book the notation for displaying a matrix is

$$A = \begin{bmatrix} 1 & 4 & 6 \\ 0 & 2 & 3 \end{bmatrix},$$

enclosing the array of elements in square brackets. A variety of forms can be found in the literature, some of which are

$$\begin{pmatrix} 1 & 4 & 6 \\ 0 & 2 & 3 \end{pmatrix}, \quad \begin{Bmatrix} 1 & 4 & 6 \\ 0 & 2 & 3 \end{Bmatrix} \quad \text{and} \quad \begin{Vmatrix} 1 & 4 & 6 \\ 0 & 2 & 3 \end{Vmatrix}.$$

Single vertical lines are seldom used, since they are usually reserved for defining a determinant (see Chapter 3).

Another valuable notation that has already been used is

$$A = \{a_{ij}\} \qquad \text{for} \quad i = 1, 2, \ldots, r \qquad \text{and} \quad j = 1, 2, \ldots, c.$$

The curly brackets indicate that a_{ij} is a typical term of the matrix A for all pairs of values of i and j from unity up to the limits shown, in this case r and c; i.e. A is a matrix of r rows and c columns. This is by no means a universal notation and several variants of it can be found in the literature. One hastens to add that there is nothing sacrosanct about the letter A for denoting a matrix, any letter may be used. Despite this, there is repeated use of A in this book, in preference to a variety of other letters. Lower case letters are widely used to denote vectors, but not universally so. Nor is the procedure of using a prime superscript for distinguishing a row vector from a column vector to be considered absolute. In some texts matrices and vectors are found printed in bold face type to distinguish them from scalars denoted by the same letters, but this is not done here, since most of the book deals with matrices and vectors and there is little likelihood of confusion. Whenever a letter represents a scalar this fact is usually stated.

9. ILLUSTRATIVE EXAMPLES

A text on matrix algebra designed for pure mathematicians would deal with many topics that do not appear in this book because they have little connection with problems in biology. The text for mathematicians might also have few numerical illustrations. This will not be the case here, however, and indeed we have already used a simple illustration to introduce the concept of a matrix. In doing so we have capitalized on the great advantage of numerical illustrations, namely, that to many people arithmetic involving actual numbers is easier to follow than long expositions in algebra. For this reason we will, wherever possible, illustrate mathematical results with numerical examples.

The mathematical discussion in the ensuing chapters is relatively informal. Proofs and demonstrations of general results are not omitted, however, for to do so would result in a lifeless, "cookbook" presentation. Once a general mathematical result has been established it is usually illustrated by a numerical example, with additional examples available as exercises at the end of most chapters. Only a small proportion of these is algebraic; most are numerical.

Obtaining suitable illustrative material has not been easy, primarily because many useful applications of matrix algebra in biology (and in other areas too) are those based on relatively advanced techniques of the algebra. To prepare the reader for these techniques requires considerable preliminary development for which few good opportunities exist for using numerical illustrations drawn from the real world. The situation is something akin to one that might arise in studying chemistry. The potency of nitric acid is easily illustrated by its effect on a copper penny. The value of the same acid in a complex chemical reaction might also be easily illustrated once the details of the reaction are understood, but for each of these details the real world may hold few illustrations of use or interest.

In making use of numerical examples to illustrate mathematical procedures in matrix algebra we find ourselves with three types of illustration. The first (which has already been used) is to take real data and in a manner that may be somewhat artificial adapt the data to the concept of a matrix or some procedure related thereto. This is analogous to the nitric acid and the penny—useful for illustrating fundamentals. It is a procedure used repeatedly in Chapter 2. The second type of illustration is found at the other end of the scale. There, when we have developed results of sufficient mathematical complexity, we can readily find illustrations taken from real life for which matrix algebra is helpful in providing an answer. We see a number of such illustrations in the final chapters of this book. In our analogy they are the counterpart of the complex chemical reaction. Before reaching this advanced stage, however, there are many intermediate mathematical steps for which few, if any, real-world problems can be used as illustrations. For this part of the work we rely heavily on a third type of illustration, namely one which is nothing more than numbers that have been "pulled out of a hat" for the sole purpose of illustrating a mathematical procedure. Such numbers have no meaning and no use other than the very good one of assisting the reader in understanding the procedures being demonstrated. These in turn lead up to the more advanced techniques that will later be found useful in solving real-life problems. Numerical illustrations created in this manner are labeled "Example"; those based on situations from real life are labeled "Illustration".

10. EXERCISES

1. For $a_{11} = 17 \quad a_{12} = 31 \quad a_{13} = 26 \quad a_{14} = 11$
   $a_{21} = 19 \quad a_{22} = 27 \quad a_{23} = 16 \quad a_{24} = 14$
   $a_{31} = 21 \quad a_{32} = 23 \quad a_{33} = 15 \quad a_{34} = 16$

   show that

   i. $a_{1.} = 85$, $a_{2.} = 76$, and $a_{3.} = 75$,
   ii. $a_{.1} = 57$, $a_{.2} = 81$, $a_{.3} = 57$, and $a_{.4} = 41$,
   iii. $a_{..} = 236$.

2. For the above series of values show that

   i. $\sum_{i=1}^{3} a_{ii} = 59$,

   ii. $\sum_{\substack{j=1 \\ j \neq 2}}^{4} a_{ij} = 54, 49$ and 52 for $i = 1, 2$ and 3 respectively,

   iii. $\sum_{\substack{i=1 \\ i \neq 2}}^{3} \sum_{\substack{j=1 \\ j \neq 3}}^{4} a_{ij} = 119$.

3. Prove the following identities and demonstrate their validity from the above numerical example.

   i. $\sum_{i=1}^{m} \sum_{j=1}^{n} a_{ij}^2 = \sum_{j=1}^{n} \sum_{i=1}^{m} a_{ij}^2$

   ii. $\sum_{i=1}^{m} (\sum_{j=1}^{n} a_{ij})^2 = \sum_{i=1}^{m} a_{i.}^2$

   iii. $\sum_{j=1}^{n} \sum_{k=1}^{n} a_{ij} a_{hk} = a_{i.} a_{h.}$, for $i \neq h$

iv. $$a_{..}^2 = \sum_i a_{i.}^2 + 2 \sum_{i=1}^{n-1} \sum_{h=i+1}^{n} a_{i.}a_{h.}$$

$$= \sum_i a_{i.}^2 + 2 \sum_{i=1}^{n-1} \sum_{h>i} a_{i.}a_{h.}$$

$$= \sum_i a_{i.}^2 + \sum_{i=1}^{n} \sum_{\substack{h=1 \\ h\neq i}}^{n} a_{i.}a_{h.}$$

v. $$\sum_{\substack{i=1 \\ i\neq p}}^{m} \sum_{\substack{j=1 \\ j\neq q}}^{n} a_{ij} = a_{..} - a_{p.} - a_{.q} + a_{pq}$$

vi. $$\sum_{i=1}^{m} \sum_{j=1}^{n} (a_{ij} - 1) = a_{..} - mn$$

vii. $$\sum_{j=1}^{n} \sum_{\substack{k=1 \\ k\neq j}}^{n} a_{ij}a_{ik} = a_{i.}^2 - \sum_{j=1}^{n} a_{ij}^2$$

viii. $$\left(\sum_{j=1}^{n} a_{ij}\right)^2 = \sum_{j=1}^{n} a_{ij}^2 + 2\sum_{j=1}^{n-1} \sum_{p=j+1}^{n} a_{ij}a_{ip}$$

$$= \sum_{j=1}^{n} a_{ij}^2 + 2\sum_{j=1}^{n-1} \sum_{p>j}^{n} a_{ij}a_{ip}$$

$$= \sum_{j=1}^{n} a_{ij}^2 + \sum_{j=1}^{n} \sum_{\substack{p=1 \\ p\neq j}}^{n} a_{ij}a_{ip}$$

ix. $$\sum_{i=1}^{m} \sum_{j=1}^{n} 4a_{ij} = 4a_{..}$$

4. i. Show that, for the values of a_{ij} given in Exercise 1, trace(A) = 59 where $A = \{a_{ij}\}$ for $i, j = 1, 2, 3$.
   ii. What is the value of trace(B) for $B = \{a_{i,j+1}\}$ for $i, j = 1, 2, 3$?
   iii. Is there a value for trace(M) when

$$M = \{a_{ij}\} \text{ for } i = 1, 2, 3 \quad \text{and} \quad j = 1, 2, \ldots, 4?$$

5. Show that

i. $$\sum_{i=3}^{5} 3^i = 351,$$

ii. $$\sum_{k=2}^{7} 2^k = 252,$$

iii. $$\sum_{r=1}^{5} r = 15,$$

iv. $$\sum_{\substack{s=1 \\ s\neq 2}}^{6} s(s+1) = 106,$$

v. $$\prod_{i=1}^{4} 2^i = 1024.$$

REFERENCES

Neyman, Jerzy, Thomas Park and Elizabeth L. Scott (1956). Struggle for existence. The *Tribolium* model: biological and statistical aspects. *Proceedings Third Berkeley Symposium on Mathematical Statistics and Probability*, Vol. IV, Edited by Jerzy Neyman, University of California Press, Berkeley.

Wallace, Bruce (1959). Studies of the relative fitness of experimental populations of *Drosophila melanogaster*. *Am. Naturalist*, **93**, 295–314.

CHAPTER 2

ELEMENTARY MATRIX OPERATIONS

This chapter describes the simple arithmetic operations of addition, subtraction and multiplication as they apply to matrices. Division, being somewhat more complicated for matrices than it is in customary arithmetic, is not considered in this chapter but is given one to itself, Chapter 4. Other operations peculiar to matrix algebra are discussed in the present chapter, however, as are certain specific forms of a matrix; for example, the counterparts of zero and unity in ordinary arithmetic. Numerical illustrations adapted from a variety of sources are used wherever possible.

1. ADDITION

We begin with the operation of addition, introducing it by means of an illustration.

Illustration. The distribution of moose in different vegetative areas of Montana is discussed by Peek (1962). Part of his data for 1958 consists of Table 1.

TABLE 1. NUMBER OF MOOSE SEEN, 1958

Sex	Vegetative Area		
	1	2	3
Bulls	98	24	42
Cows	39	15	22
Cows (with calves)	22	15	17

Let us write the body of the table as a 3 × 3 matrix

$$A = \begin{bmatrix} 98 & 24 & 42 \\ 39 & 15 & 22 \\ 22 & 15 & 17 \end{bmatrix},$$

so that with the same frame of reference the data for 1959 can also be written as a matrix:

$$B = \begin{bmatrix} 55 & 19 & 44 \\ 43 & 53 & 38 \\ 11 & 40 & 20 \end{bmatrix}.$$

Then, over the two years, the total number of bulls seen in vegetative area 1 is the sum of the elements in the first row and first column of each matrix, namely 98 + 55 = 153; and the total number of cows with calves seen in area 2 is 15 + 40 = 55. In this way the matrix of all such sums

$$\begin{bmatrix} 98+55 & 24+19 & 42+44 \\ 39+43 & 15+53 & 22+38 \\ 22+11 & 15+40 & 17+20 \end{bmatrix} = \begin{bmatrix} 153 & 43 & 86 \\ 82 & 68 & 60 \\ 33 & 55 & 37 \end{bmatrix}$$

represents the numbers of moose seen over the two years. This is the matrix sum A plus B; it is the matrix formed by adding the matrices A and B element by element. Hence if we write $A = \{a_{ij}\}$ and $B = \{b_{ij}\}$ for $i = 1, 2, \ldots, r$ and $j = 1, 2, \ldots, c$ the matrix representing the sum of A and B is

$$A + B = \{a_{ij} + b_{ij}\} \qquad \text{for} \quad i = 1, 2, \ldots, r \qquad \text{and} \quad j = 1, 2, \ldots, c;$$

i.e. the sum of two matrices is the matrix of sums, element by element.

It is evident from this definition that matrix addition can take place only when the matrices involved are of the same order; i.e. two matrices can be added only if they have the same number of rows and the same number of columns. Thus it is meaningless to attempt the addition of

$$\begin{bmatrix} 1 & 2 \\ 6 & -4 \end{bmatrix} \quad \text{and} \quad \begin{bmatrix} 3 & 9 \\ -6 & 1 \\ 4 & 3 \end{bmatrix}.$$

Matrices that have the same order and can therefore be added together are said to be *conformable for addition*. For example,

$$\begin{bmatrix} 1 & 2 & 3 \\ 6 & -4 & 5 \end{bmatrix} \quad \text{and} \quad \begin{bmatrix} -3 & 1 & 9 \\ 2 & 4 & 6 \end{bmatrix}$$

are conformable for addition and their sum is $\begin{bmatrix} -2 & 3 & 12 \\ 8 & 0 & 11 \end{bmatrix}$.

2. SCALAR MULTIPLICATION

We have just described matrix addition. A simple use of it shows that

$$\begin{aligned} A + A &= \{a_{ij}\} + \{a_{ij}\} \\ &= \{2a_{ij}\} \\ &= 2A. \end{aligned}$$

Extending this to the case where λ is a positive integer, we have

$$\lambda A = A + A + A + \dots + A,$$

there being λ A's in the sum on the right. Carrying out these matrix additions gives

$$\lambda A = \{\lambda a_{ij}\}, \qquad \text{for} \quad i = 1, 2, \dots, r \qquad \text{and} \quad j = 1, 2, \dots, c.$$

This result, extended for λ being any scalar, is the definition of *scalar multiplication*. Thus the matrix A multiplied by the scalar λ is the matrix A with every element multiplied by λ. For example,

$$3\begin{bmatrix} 1 & -7 \\ 3 & 5 \end{bmatrix} = \begin{bmatrix} 3 & -21 \\ 9 & 15 \end{bmatrix},$$

and

$$2.71\begin{bmatrix} 1 & -7 \\ 3 & 5 \end{bmatrix} = \begin{bmatrix} 2.71 & -18.97 \\ 8.13 & 13.55 \end{bmatrix}.$$

3. SUBTRACTION

Illustration. Kamar (1962) reports the mean weight (in grams) of four breeds of duck as shown in Table 2:

TABLE 2. MEAN WEIGHT OF DUCKS AT THREE MONTHS OF AGE

Sex	Breed			
	1	2	3	4
Male	910	1275	1210	1304
Female	860	967	667	1048

If the entries in the table are represented by the matrix

$$A = \begin{bmatrix} 910 & 1275 & 1210 & 1304 \\ 860 & 967 & 667 & 1048 \end{bmatrix},$$

the weights of the same breeds as reported at six months of age can be represented by

$$B = \begin{bmatrix} 2050 & 1340 & 1344 & 1384 \\ 1380 & 1058 & 1011 & 1189 \end{bmatrix}.$$

Now the gain in weight from three to six months of age for males of breed 1 is clearly $2050 - 910 = 1140$, and that for females of breed 4 is $1189 - 1048 = 141$. Hence the matrix

$$\begin{bmatrix} 2050 - 910 & 1340 - 1275 & 1344 - 1210 & 1384 - 1304 \\ 1380 - 860 & 1058 - 967 & 1011 - 667 & 1189 - 1048 \end{bmatrix}$$

$$= \begin{bmatrix} 1140 & 65 & 134 & 80 \\ 520 & 91 & 344 & 141 \end{bmatrix}$$

represents the weight gains from three to six months of age for both sexes of all four breeds. This is the matrix operation of subtraction. Thus if $A = \{a_{ij}\}$ and $B = \{b_{ij}\}$ for $i = 1, 2, \ldots, r$ and $j = 1, 2, \ldots, c$, subtraction in matrix algebra is defined as

$$A - B = \{a_{ij} - b_{ij}\} \qquad \text{for} \quad i = 1, 2, \ldots, r \text{ and} \quad j = 1, 2, \ldots, c;$$

i.e. the difference between two matrices is the matrix of differences element by element. We can prove this more formally as a direct outcome of the rules of addition and scalar multiplication. Since

$$\begin{aligned} -B &= (-1)B, \\ A - B &= A + (-1)B \\ &= \{a_{ij}\} + (-1)\{b_{ij}\} \\ &= \{a_{ij}\} + \{-b_{ij}\} \\ &= \{a_{ij} - b_{ij}\}. \end{aligned}$$

Example.

$$\begin{bmatrix} 3 & 6 \\ 8 & 2 \\ 4 & 1 \end{bmatrix} - \begin{bmatrix} 1 & 1 \\ 0 & -3 \\ 2 & -5 \end{bmatrix} = \begin{bmatrix} 2 & 5 \\ 8 & 5 \\ 2 & 6 \end{bmatrix}.$$

As with addition, only matrices that are of the same order can be subtracted from one another. Thus matrices that are conformable for addition are also conformable for subtraction, and vice versa.

4. EQUALITY AND THE NULL MATRIX

Two matrices are said to be equal when they are identical element by element. Thus $A = B$ when $\{a_{ij}\} = \{b_{ij}\}$, meaning that $a_{ij} = b_{ij}$ for $i = 1, 2, \ldots, r$ and $j = 1, 2, \ldots, c$. If

$$A = \begin{bmatrix} 2 & 6 & -4 \\ 3 & 0 & 1 \end{bmatrix}, \quad B = \begin{bmatrix} 2 & 6 & -4 \\ 3 & 0 & 1 \end{bmatrix} \quad \text{and} \quad C = \begin{bmatrix} 2 & 6 & -4 \\ 2 & 0 & 1 \end{bmatrix},$$

A is equal to B but not equal to C. It is also apparent that equality of two matrices has no meaning unless they are of the same order.

Combining the ideas of subtraction and equality leads to the definition of zero in matrix algebra. For, when

$$A = B,$$
$$a_{ij} = b_{ij}$$

$$\text{for} \quad i = 1, 2, \ldots, r \quad \text{and} \quad j = 1, 2, \ldots, c, \quad \text{and so}$$

$$A - B = \{a_{ij} - b_{ij}\} = 0.$$

The matrix on the right is a matrix of zeros, i.e. every element is zero. Such a matrix is called a *null matrix*; it is the zero of matrix algebra and as such is referred to by some writers as a *zero matrix*. It is, of course, not a unique zero because for a matrix of any order there is a null matrix of the same order. For example, null matrices of order 2×4 and 3×3 are respectively

$$\begin{bmatrix} 0 & 0 & 0 & 0 \\ 0 & 0 & 0 & 0 \end{bmatrix} \quad \text{and} \quad \begin{bmatrix} 0 & 0 & 0 \\ 0 & 0 & 0 \\ 0 & 0 & 0 \end{bmatrix}.$$

5. MULTIPLICATION

Before describing the multiplication of matrices brief discussion is given to two simpler operations involving vectors. Once again, numerical illustrations are used to introduce general methods.

a. A product of vectors

Illustration. A two-year study of ewe fertility reported by Tallis (1962) concerns, in part, 92 ewes that had not lambed in 1952. In 1953, 58 of them did not lamb, 26 had a single lamb and 8 had twins. The total number of lambs dropped by these ewes in 1953 is therefore $58(0) + 26(1) + 8(2) =$

42. Suppose that the numbers of ewes having 0, 1 and 2 lambs respectively are written as a row vector

$$a' = [58 \quad 26 \quad 8],$$

and the numbers of lambs per ewe are written as a column vector

$$x = \begin{bmatrix} 0 \\ 1 \\ 2 \end{bmatrix}.$$

Then the total number of lambs born to these ewes, 42, is the sum of products of the elements of a' each multiplied by the corresponding element of x. This is the definition of the product $a'x$. It is written as

$$a'x = [58 \quad 26 \quad 8] \begin{bmatrix} 0 \\ 1 \\ 2 \end{bmatrix}$$

without any multiplication symbol between the vectors, and is calculated as

$$a'x = 58(0) + 26(1) + 8(2) = 42.$$

This example illustrates the general procedure for obtaining $a'x$: multiply each element of the row vector a' by the corresponding element of the column vector x and add the products. The sum is $a'x$. Thus if

$$a' = [a_1 \quad a_2 \quad \ldots \quad a_n]$$

and

$$x = \begin{bmatrix} x_1 \\ x_2 \\ \cdot \\ \cdot \\ \cdot \\ x_n \end{bmatrix}$$

their product $a'x$ is defined as

$$a'x = a_1x_1 + a_2x_2 + \cdots + a_nx_n$$
$$= \sum_{i=1}^{n} a_ix_i.$$

The definition applies only when a' and x have the same order; if they are not of the same order the product $a'x$ is undefined and does not exist.

b. A matrix-vector product

Illustration. The preceding illustration is only part of the study reported by Tallis (1962). More complete data are shown in Table 3:

TABLE 3. CLASSIFICATION OF EWES ACCORDING TO NUMBER OF LAMBS BORN IN EACH OF TWO YEARS, 1952 AND 1953

Number of Lambs in 1952	Number of Lambs in 1953		
	0	1	2
	Number of Ewes		
None	58	26	8
Single	52	58	12
Twins	1	3	9

The ewes having no lambs in 1952 had a total of $58(0) + 26(1) + 8(2) = 42$ in 1953, and similar calculations can be made for the ewes bearing a single lamb in 1952 and for those having twins; they are summarized in Table 4:

TABLE 4. TOTAL NUMBERS OF LAMBS BORN

Ewes Classified According to Number of Lambs Born in 1952	Total Number of Lambs Born in 1953
None	$58(0) + 26(1) + 8(2) = 42$
Single	$52(0) + 58(1) + 12(2) = 82$
Twins	$1(0) + 3(1) + 9(2) = 21$

Let us write the entries of Table 3 as a matrix

$$A = \begin{bmatrix} 58 & 26 & 8 \\ 52 & 58 & 12 \\ 1 & 3 & 9 \end{bmatrix},$$

and the results of Table 4 as a vector $\begin{bmatrix} 42 \\ 82 \\ 21 \end{bmatrix}$. As can be seen from Table 4 the elements of this vector are derived in exactly the same way as the product $a'x$ developed earlier, using the successive rows of A as the vector a'. The result is the product Ax; that is, Ax is obtained by repetitions of the product $a'x$ using the rows of A successively for a' and writing the results as a column vector. Thus

$$Ax = \begin{bmatrix} 58 & 26 & 8 \\ 52 & 58 & 12 \\ 1 & 3 & 9 \end{bmatrix}\begin{bmatrix} 0 \\ 1 \\ 2 \end{bmatrix}$$

$$= \begin{bmatrix} 58(0) + 26(1) + 8(2) \\ 52(0) + 58(1) + 12(2) \\ 1(0) + 3(1) + 9(2) \end{bmatrix} = \begin{bmatrix} 42 \\ 82 \\ 21 \end{bmatrix}.$$

General subscript notation for this example is as follows.

$$A = \begin{bmatrix} a_{11} & a_{12} & a_{13} \\ a_{21} & a_{22} & a_{23} \\ a_{31} & a_{32} & a_{33} \end{bmatrix}, \quad \text{and} \quad x = \begin{bmatrix} x_1 \\ x_2 \\ x_3 \end{bmatrix},$$

and

$$Ax = \begin{bmatrix} a_{11}x_1 + a_{12}x_2 + a_{13}x_3 \\ \\ a_{21}x_1 + a_{22}x_2 + a_{23}x_3 \\ \\ a_{31}x_1 + a_{32}x_2 + a_{33}x_3 \end{bmatrix} = \begin{bmatrix} \sum_{k=1}^{3} a_{1k}x_k \\ \sum_{k=1}^{3} a_{2k}x_k \\ \sum_{k=1}^{3} a_{3k}x_k \end{bmatrix}.$$

We see at once that the first element of Ax is the sum of products of the a_{ij}'s of the first row of A with the elements of x, and likewise for the other elements of Ax. Generalization is evident. The product Ax of a matrix A and a column vector x is a column vector whose ith term is the sum of products of the elements of the ith row of A each multiplied by the corresponding element of x. Both from this definition and from the example it is clear that Ax is defined only when the number of elements in the rows of A (i.e. number of columns) is the same as the number of elements in the column vector x. And when this occurs Ax is a column vector having the same number of elements as there are rows in A. Therefore, when A has r rows and c columns and x is of order c, Ax is a column vector of order r; its ith element is $\sum_{k=1}^{c} a_{ik}x_k$ for $i = 1, 2, \ldots, r$. More formally we have

$$A = \{a_{ij}\} \quad \text{for} \quad i = 1, 2, \ldots, r \quad \text{and} \quad j = 1, 2, \ldots, c;$$
$$x = \{x_j\} \quad \text{for} \quad j = 1, 2, \ldots, c;$$

and

$$Ax = \{\sum_{j=1}^{c} a_{ij}x_j\} \quad \text{for} \quad i = 1, 2, \ldots, r.$$

Example.

$$\begin{bmatrix} 4 & 2 & 1 & 3 \\ 2 & 0 & -4 & 7 \end{bmatrix}\begin{bmatrix} 1 \\ 0 \\ -1 \\ 3 \end{bmatrix} = \begin{bmatrix} 4(1) + 2(0) + & 1(-1) + 3(3) \\ 2(1) + 0(0) + & -4(-1) + 7(3) \end{bmatrix} = \begin{bmatrix} 12 \\ 27 \end{bmatrix}.$$

Illustration. Typical of many uses of the matrix-vector product is one that occurs in the study of inbreeding, where what is known as the generation matrix is used to relate the frequencies of mating types in one generation to those in another. Kempthorne (1957, page 108 et seq.), for example, gives the result earlier stated by Fisher (1949) that if, after one generation of full-sib mating $f^{(1)}$ represents the vector of frequencies of the seven distinct "types of mating possible in this situation", then $f^{(1)} = Af^{(0)}$, where $f^{(0)}$ is the initial vector of frequencies and where A is the matrix

$$A = \begin{bmatrix} 1 & \frac{2}{16} & 0 & \frac{1}{4} & 0 & \frac{1}{16} & 0 \\ 0 & \frac{4}{16} & 1 & \frac{1}{4} & \frac{1}{2} & \frac{3}{16} & \frac{4}{16} \\ 0 & \frac{2}{16} & 0 & 0 & 0 & 0 & 0 \\ 0 & \frac{8}{16} & 0 & \frac{2}{4} & 0 & \frac{4}{16} & 0 \\ 0 & 0 & 0 & 0 & 0 & \frac{2}{16} & 0 \\ 0 & 0 & 0 & 0 & \frac{1}{2} & \frac{6}{16} & \frac{8}{16} \\ 0 & 0 & 0 & 0 & 0 & 0 & \frac{4}{16} \end{bmatrix}.$$

This simply means that if $f_i^{(1)}$ is the frequency of the ith type of mating after a generation of full-sib mating (i.e. the ith element of the vector $f^{(1)}$) and if $f_i^{(0)}$ is the corresponding initial frequency (ith element of $f^{(0)}$), then, for example,

$$f_1^{(1)} = f_1^{(0)} + \tfrac{2}{16}f_2^{(0)} + \tfrac{1}{4}f_4^{(0)} + \tfrac{1}{16}f_6^{(0)};$$

and as another example

$$f_4^{(1)} = \tfrac{8}{16}f_2^{(0)} + \tfrac{2}{4}f_4^{(0)} + \tfrac{4}{16}f_6^{(0)}.$$

These and five other similar equations are implied in the vector equation $f^{(1)} = Af^{(0)}$. The matrix A which represents the relationships between the two sets of frequencies is known in this context as the generation matrix.

c. A product of two matrices

Multiplying two matrices can be explained as a simple repetitive extension of multiplying a matrix by a vector. To obtain the product of two matrices A and B think of the matrix B as being a series of column vectors. Then the product AB is the matrix obtained by setting alongside one another each of the product vectors of A with the columns of B. We demonstrate the procedure using a simple numerical example. If

$$A = \begin{bmatrix} 1 & 0 & 2 \\ 3 & 1 & 1 \\ 1 & 2 & 1 \\ -1 & 3 & 2 \end{bmatrix}, \quad \text{and} \quad B = \begin{bmatrix} 1 & 2 \\ 0 & 1 \\ 0 & -1 \end{bmatrix},$$

we think of B as being composed of the two column vectors

$$x = \begin{bmatrix} 1 \\ 0 \\ 0 \end{bmatrix} \quad \text{and} \quad w = \begin{bmatrix} 2 \\ 1 \\ -1 \end{bmatrix}.$$

The products of A with each of the columns of B are

$$Ax = \begin{bmatrix} 1(1) + 0(0) + 2(0) \\ 3(1) + 1(0) + 1(0) \\ 1(1) + 2(0) + 1(0) \\ -1(1) + 3(0) + 2(0) \end{bmatrix} = \begin{bmatrix} 1 \\ 3 \\ 1 \\ -1 \end{bmatrix}$$

and

$$Aw = \begin{bmatrix} 1(2) + 0(1) + 2(-1) \\ 3(2) + 1(1) + 1(-1) \\ 1(2) + 2(1) + 1(-1) \\ -1(2) + 3(1) + 2(-1) \end{bmatrix} = \begin{bmatrix} 0 \\ 6 \\ 3 \\ -1 \end{bmatrix}.$$

Setting these vectors alongside each other yields the matrix product AB:

$$AB = \begin{bmatrix} 1 & 0 \\ 3 & 6 \\ 1 & 3 \\ -1 & -1 \end{bmatrix},$$

and the complete operation is written as

$$AB = \begin{bmatrix} 1 & 0 & 2 \\ 3 & 1 & 1 \\ 1 & 2 & 1 \\ -1 & 3 & 2 \end{bmatrix} \begin{bmatrix} 1 & 2 \\ 0 & 1 \\ 0 & -1 \end{bmatrix} = \begin{bmatrix} 1 & 0 \\ 3 & 6 \\ 1 & 3 \\ -1 & -1 \end{bmatrix}.$$

The result of this process is that in the product AB the ith element of the first column is the sum of products of the elements of the ith row of A with those of the first column of B, and the ith element in the second column of AB is the sum of products of the elements of the ith row of A with those of the second column of B.

We can generalize at once: the ijth element (the element in the ith row and jth column) of the product AB of two matrices A and B is the sum of products (element by element) of the elements of the ith row of A with the corresponding elements of the jth column of B. Thus if the ith row of A is

$[a_{i1} \quad a_{i2} \quad \ldots \quad a_{ic}]$ and the jth column of B is $\begin{bmatrix} b_{1j} \\ b_{2j} \\ \cdot \\ \cdot \\ \cdot \\ b_{cj} \end{bmatrix}$, the ijth element of AB is

$$a_{i1}b_{1j} + a_{i2}b_{2j} + \cdots + a_{ic}b_{cj} = \sum_{k=1}^{c} a_{ik}b_{kj}.$$

To obtain this expression think of moving from element to element along the ith row of A and simultaneously down the jth column of B, summing the products of corresponding elements. The resulting sum is the ijth element of AB.

Once again this is a definition that holds only if a certain condition is met, in this case if the jth column of B (and hence all columns) has the same number of elements as does the ith row of A (and hence all rows). Since the number of elements in a column of a matrix is the number of rows in the matrix (and the number of elements in a row is the number of columns) this means that there must be exactly as many rows in B as there are columns in A. Thus the matrix product AB is defined only if the number of columns of A equals the number of rows of B. It might also be noted, as in the numerical example given previously, that the product AB has the same number of rows as A and the same number of columns as B. This is true in general.

The vitally important consequences of the definition of matrix multiplication are therefore as follows. The product AB of two matrices A and B exists only if there is the same number of columns in A as there are rows in B; the matrices are then said to be *conformable for multiplication for the product* AB, and AB has the same number of rows as A and the same number of columns as B. Hence if A is $r \times c$ and B is $c \times s$, with

$$A = \{a_{ij}\} \qquad \text{for} \quad i = 1, 2, \ldots, r, \quad \text{and} \quad j = 1, 2, \ldots, c,$$

and

$$B = \{b_{ij}\} \qquad \text{for} \quad i = 1, 2, \ldots, c, \quad \text{and} \quad j = 1, 2, \ldots, s,$$

the product AB is $r \times s$, and its ijth element is $\sum_{k=1}^{c} a_{ik}b_{kj}$. Hence we write

$$AB = \left\{\sum_{k=1}^{c} a_{ik}b_{kj}\right\} \qquad \text{for} \quad i = 1, 2, \ldots, r, \qquad \text{and} \quad j = 1, 2, \ldots, s.$$

The expression for the ijth element is the sum of products of the elements of the ith row of A with those of the jth column of B, and as well as being the ijth element of AB is sometimes called the *inner product* of the ith row of A and the jth column of B. The matrix product AB is therefore obtained by calculating the inner product of each row of A with each column of B, the inner product of the ith row of A with the jth column of B being the ijth element of AB.

Example.

For $\quad A = \begin{bmatrix} 1 & 0 & 2 \\ -1 & 4 & 3 \end{bmatrix}$ and $B = \begin{bmatrix} 0 & 6 & 1 & 5 \\ 1 & 1 & 2 & 7 \\ 2 & 4 & 4 & 3 \end{bmatrix}$

the element in the first row and first column of the product AB is based on the first row of A and the first column of B and is

$$1(0) + 0(1) + 2(2) = 4;$$

that in the first row and second column is

$$1(6) + 0(1) + 2(4) = 14;$$

and the element of AB in the second row and third column is

$$-1(1) + 4(2) + 3(4) = 19.$$

In this way AB is obtained as

$$AB = \begin{bmatrix} 4 & 14 & 9 & 11 \\ 10 & 10 & 19 & 32 \end{bmatrix}.$$

The reader should satisfy himself of the validity of this result.

d. Existence of matrix products

As noted in Section 1.6, subscript notation can be used to denote the order of a matrix; viz $A_{r \times c}$ is a matrix of order r by c (r rows and c columns). A product AB can be written in this notation as

$$A_{r \times c} B_{c \times s} = P_{r \times s},$$

a form which provides opportunity both for checking the conformability of A and B and for ascertaining the order of their product. Repeated use of this also simplifies determining the order of a matrix derived by multiplying several matrices together. Adjacent subscripts (which must be equal for conformability) "cancel out", leaving the first and last subscripts as the order of the product. For example, the product

$$A_{r \times c} B_{c \times s} C_{s \times t} D_{t \times u}$$

is a matrix of order $r \times u$.

This notation also demonstrates what is by now readily apparent from the definition of matrix multiplication, namely that the product BA does not necessarily exist, even if AB does. For BA can be written as $B_{c \times s}A_{r \times c}$, which we see at once is a legitimate product only if $s = r$. Otherwise BA is not defined. There are therefore three situations regarding the product of two matrices A and B. If A is of order $r \times c$,

(i) AB exists only if B has c rows,
(ii) BA exists only if B has r columns,
and
(iii) AB and BA both exist only if B is $c \times r$.

A corollary to (iii) is that A^2 exists only when A is square. Another corollary is that both AB and BA always exist when A and B are square and of the same order. But as shall be shown subsequently, the two products are not necessarily equal. Their inequality will be discussed when considering the commutative law of multiplication, but meanwhile we simply state that they are not in general equal. As a means of distinguishing between them, AB is described as A *post-multiplied* by B, and BA as A *pre-multiplied* by B.

Illustration. Kemeny et al. (1957) mention an interesting example of matrix multiplication relating to the pecking order of poultry. Suppose we have three chickens, 1, 2, 3 whose pecking order is such that chicken 1 dominates chicken 2, chicken 2 dominates chicken 3, and chicken 1 dominates chicken 3. These relationships can be expressed in a matrix $A = \{a_{ij}\}$ where $a_{ii} = 0$, and $a_{ij} = 1$ and $a_{ji} = 0$ if chicken i dominates chicken j. In the example

$$A = \begin{bmatrix} 0 & 1 & 1 \\ 0 & 0 & 1 \\ 0 & 0 & 0 \end{bmatrix}.$$

This is called the matrix of one-stage dominances, and likewise A^2 is the matrix of two-stage dominances and in this case

$$A^2 = \begin{bmatrix} 0 & 0 & 1 \\ 0 & 0 & 0 \\ 0 & 0 & 0 \end{bmatrix}.$$

To evolve a ranking of the chickens the sum of A and A^2 is formed,

$$S = A + A^2 = \begin{bmatrix} 0 & 1 & 2 \\ 0 & 0 & 1 \\ 0 & 0 & 0 \end{bmatrix},$$

and the row sums of this matrix, called the powers of the individuals, are used for comparing the chickens in terms of the general pecking order within the flock. Thus chicken number one has power of three; that of number two is one and number three has zero power. The ranking of the chickens according to pecking order is therefore 1, 2, 3. These principles are readily extended to comparisons involving any number of chickens, and interestingly enough these same dominance matrices can also be used for ranking participants in athletic events in which each participant competes against all others.

e. Vector products

The topics discussed in parts **a** and **b** of this section are special cases of the general matrix product $A_{r\times c}B_{c\times s} = P_{r\times s}$. Thus in **a** we considered the case for $r = 1$ and $s = 1$, which means that $A_{r\times c}$ becomes $A_{1\times c}$, a row vector, and $B_{c\times s}$ becomes $B_{c\times 1}$, a column vector. Retaining the convention that a' is a row vector, further emphasized by writing its order as subscripts $a'_{1\times c}$, we then have

$$a'_{1\times c}b_{c\times 1} = p_{1\times 1},$$

a scalar. Conversely

$$b_{c\times 1}a'_{1\times c} = P_{c\times c},$$

a square matrix. In **b** we considered a column vector pre-multiplied by a matrix; that is,

$$A_{r\times c}b_{c\times 1} = p_{r\times 1},$$

the product being a column vector; similarly a row vector post-multiplied by a matrix is

$$a'_{1\times c}B_{c\times r} = p'_{1\times r},$$

a row vector.

In words these four results are as follows:

(i) A row vector post-multiplied by a column vector is a scalar.
(ii) A column vector post-multiplied by a row vector is a matrix.
(iii) A matrix post-multiplied by a column vector is a column vector.
(iv) A row vector post-multiplied by a matrix is a row vector.

Examples. Given

$$A = \begin{bmatrix} 1 & 2 \\ 3 & 4 \end{bmatrix}, \quad B = \begin{bmatrix} 6 & 3 & 1 \\ -1 & 2 & 5 \end{bmatrix}, \quad x' = [1 \quad 5], \quad y = \begin{bmatrix} 3 \\ 1 \end{bmatrix},$$

the following results are true:

$$AB = \begin{bmatrix} 1(6) + 2(-1) & 1(3) + 2(2) & 1(1) + 2(5) \\ 3(6) + 4(-1) & 3(3) + 4(2) & 3(1) + 4(5) \end{bmatrix} = \begin{bmatrix} 4 & 7 & 11 \\ 14 & 17 & 23 \end{bmatrix},$$

$$x'B = [1 \quad 13 \quad 26], \qquad Ay = \begin{bmatrix} 5 \\ 13 \end{bmatrix},$$

$$x'Ay = 70, \qquad x'y = 8 \qquad \text{and} \quad yx' = \begin{bmatrix} 3 & 15 \\ 1 & 5 \end{bmatrix}.$$

f. Products with diagonal matrices

A diagonal matrix was defined in Section 1.6 as a square matrix which has all non-diagonal elements zero. (Some but not all of the diagonal elements may also be zero.) Such a matrix is sometimes called a *quasi-scalar* matrix.

The procedure for multiplying matrices results in multiplication by diagonal matrices being particularly easy: pre-multiplication of a matrix A by a diagonal matrix D gives a matrix whose rows are those of A multiplied by the respective diagonal elements of D. Thus with

$$D = \begin{bmatrix} 1.3 & 0 \\ 0 & 2.1 \end{bmatrix} \quad \text{and} \quad A = \begin{bmatrix} 2 & -1 & 7 \\ -1 & 0 & 1 \end{bmatrix},$$

$$DA = \begin{bmatrix} 2.6 & -1.3 & 9.1 \\ -2.1 & 0 & 2.1 \end{bmatrix}.$$

Post-multiplication leads in a similar way to the columns being multiplied by the respective diagonal elements of D, for example, with

$$A = \begin{bmatrix} 2 & 1 \\ 0 & -5 \\ 12 & 7 \end{bmatrix} \quad \text{and} \quad D = \begin{bmatrix} -7 & 0 \\ 0 & 4 \end{bmatrix}, \quad AD = \begin{bmatrix} -14 & 4 \\ 0 & -20 \\ -84 & 28 \end{bmatrix}.$$

g. The trace of a product

The trace of a matrix is defined in Section 1.6. Thus if $A = \{a_{ij}\}$ for $i, j = 1, 2, \ldots, n$, the trace of A is $\text{tr}(A) = \sum_{i=1}^{n} a_{ii}$, the sum of the diagonal elements. We now show that for a product matrix $\text{tr}(AB) = \text{tr}(BA)$ and hence $\text{tr}(ABC) = \text{tr}(BCA) = \text{tr}(CAB)$. Note that $\text{tr}(AB)$ exists only if AB is square, which occurs only when A is $r \times c$ and B is $c \times r$. Then if $AB = \{(ab)_{ij}\}$,

$$\begin{aligned} \text{tr}(AB) &= \sum_{i=1}^{r} (ab)_{ii} = \sum_{i=1}^{r} \left(\sum_{j=1}^{c} a_{ij} b_{ji} \right) \\ &= \sum_{j=1}^{c} \left(\sum_{i=1}^{r} b_{ji} a_{ij} \right) = \sum_{j=1}^{r} (ba)_{jj} \\ &= \text{tr}(BA). \end{aligned}$$

Extension to products of three or more matrices is obvious.

h. Term-by-term product

Occasional reference may be found to the "term-by-term" or "element-by-element" product of two matrices. This is simply the matrix formed from two matrices of similar order by multiplying corresponding elements;

i.e. for $A = \{a_{ij}\}$ and $B = \{b_{ij}\}$, the term-by-term product of A and B is $\{a_{ij}b_{ij}\}$. For example, the term-by-term product of

$$A = \begin{bmatrix} 2 & 3 & 6 \\ 1 & -1 & 0 \end{bmatrix} \quad \text{and} \quad \begin{bmatrix} 1 & 2 & 7 \\ 3 & 8 & 4 \end{bmatrix} \quad \text{is} \quad \begin{bmatrix} 2 & 6 & 42 \\ 3 & -8 & 0 \end{bmatrix}.$$

6. LINEAR TRANSFORMATIONS

We have just seen that the product Ax of a matrix A and a vector x is itself a vector. Suppose this product vector is called y. Then $y = Ax$. By the nature of the multiplying procedure we at once appreciate that the elements of y are linear combinations[1] of those of x; i.e. by the multiplication process the vector x has been transformed into the vector y. The matrix A in this situation is said to represent the *linear transformation* of x into y. It is the operational means by which the elements of x are transformed into elements of y. For example,

$$\begin{bmatrix} y_1 \\ y_2 \end{bmatrix} = y = Ax = \begin{bmatrix} a_{11} & a_{12} & a_{13} \\ a_{21} & a_{22} & a_{23} \end{bmatrix} \begin{bmatrix} x_1 \\ x_2 \\ x_3 \end{bmatrix}$$

gives

$$y_1 = a_{11}x_1 + a_{12}x_2 + a_{13}x_3$$

and

$$y_2 = a_{21}x_1 + a_{22}x_2 + a_{23}x_3.$$

This idea of a linear transformation of one set of variables into another (that is, of one vector into another) arises in a variety of circumstances. An example drawn from genetics will serve as an illustration.

Illustration. When considering a single locus on a chromosome, one at which there are only two alleles G and g say, the three possible genotypes are GG, Gg and gg. (G and g are symbols representing genes; they are not matrices or vectors.) Gene effects relative to these genotypes can be defined (Anderson and Kempthorne, 1954, for example) in terms of a mean μ, a measure of gene substitution α, and a measure of dominance δ, such that

$$\begin{aligned} \mu &= \tfrac{1}{4}GG + \tfrac{1}{2}Gg + \tfrac{1}{4}gg \\ \alpha &= \tfrac{1}{4}GG \qquad\quad - \tfrac{1}{4}gg \\ \delta &= -\tfrac{1}{4}GG + \tfrac{1}{2}Gg - \tfrac{1}{4}gg. \end{aligned} \tag{1}$$

[1] See page 110 for definition of 'linear combination'.

Using matrix notation these equations are

$$\begin{bmatrix} \mu \\ \alpha \\ \delta \end{bmatrix} = \begin{bmatrix} \frac{1}{4} & \frac{1}{2} & \frac{1}{4} \\ \frac{1}{4} & 0 & -\frac{1}{4} \\ -\frac{1}{4} & \frac{1}{2} & -\frac{1}{4} \end{bmatrix} \begin{bmatrix} GG \\ Gg \\ gg \end{bmatrix}. \tag{2}$$

Here we clearly see the manner in which the vector of genotypes is transformed into the vector of gene effects, namely, by pre-multiplication by the matrix $\begin{bmatrix} \frac{1}{4} & \frac{1}{2} & \frac{1}{4} \\ \frac{1}{4} & 0 & -\frac{1}{4} \\ -\frac{1}{4} & \frac{1}{2} & -\frac{1}{4} \end{bmatrix}$. As a consequence the matrix is said to represent the linear transformation of the vector $\begin{bmatrix} GG \\ Gg \\ gg \end{bmatrix}$ into the vector $\begin{bmatrix} \mu \\ \alpha \\ \delta \end{bmatrix}$.

A special use of the general linear transformation $y = Ax$ is that characteristics of y can often be derived very easily from those of x. To continue the illustration, a different set of definitions used for the genetic effects by Hayman and Mather (1955) are m, a and d where

$$\begin{aligned} GG &= m + a \\ Gg &= m + d \\ gg &= m - a. \end{aligned} \tag{3}$$

Suppose that we wish to find the relationship of μ, α and δ to m, a and d. It could readily be derived by substituting into equations (1) the values for GG, Gg and gg given by equations (3). However, writing (3) in matrix notation as

$$\begin{bmatrix} GG \\ Gg \\ gg \end{bmatrix} = \begin{bmatrix} 1 & 1 & 0 \\ 1 & 0 & 1 \\ 1 & -1 & 0 \end{bmatrix} \begin{bmatrix} m \\ a \\ d \end{bmatrix} \tag{4}$$

enables this substitution to be carried out very simply, for using (4) in (2) gives

$$\begin{bmatrix} \mu \\ \alpha \\ \delta \end{bmatrix} = \begin{bmatrix} \frac{1}{4} & \frac{1}{2} & \frac{1}{4} \\ \frac{1}{4} & 0 & -\frac{1}{4} \\ -\frac{1}{4} & \frac{1}{2} & -\frac{1}{4} \end{bmatrix} \begin{bmatrix} 1 & 1 & 0 \\ 1 & 0 & 1 \\ 1 & -1 & 0 \end{bmatrix} \begin{bmatrix} m \\ a \\ d \end{bmatrix} \tag{5}$$

and carrying out the matrix multiplication on the right-hand side leads to the result

$$\begin{bmatrix} \mu \\ \alpha \\ \delta \end{bmatrix} = \begin{bmatrix} 1 & 0 & \frac{1}{2} \\ 0 & \frac{1}{2} & 0 \\ 0 & 0 & \frac{1}{2} \end{bmatrix} \begin{bmatrix} m \\ a \\ d \end{bmatrix} = \begin{bmatrix} m + \frac{1}{2}d \\ \frac{1}{2}a \\ \frac{1}{2}d \end{bmatrix}. \tag{6}$$

Hence $\mu = m + \frac{1}{2}d$, $\alpha = \frac{1}{2}a$ and $\delta = \frac{1}{2}d$. The effort involved in this derivation is less than that of making the direct algebraic substitutions of equations (3) into (1), and would be considerably less in a situation concerned with more variables than the three of this illustration. Notice further that the major effort required, that of reducing (5) to (6), is solely arithmetical—effort that is nowadays speedily circumvented by high-speed computers, even when dealing with large numbers of variables.

The illustration just given exemplifies the general result that if $y = Ax$ and $x = Bw$ then $y = ABw$. And provided the products Ax and Bw exist this is true for any vectors x, y and w and any matrices A and B. On all occasions the algebra of the substitutions implicit in these expressions is written in the same way as just shown, and the primary calculation involved is nothing more than matrix multiplication.

Illustration. An application of the foregoing occurs in the illustration given earlier (Section 5b) concerning inbreeding. There we had two vectors $f^{(0)}$ and $f^{(1)}$ related by the equation $f^{(1)} = Af^{(0)}$. By the same arguments the equation $f^{(2)} = Af^{(1)}$ also holds true, and therefore $f^{(2)} = AAf^{(0)} = A^2f^{(0)}$; and in general $f^{(n)} = A^nf^{(0)}$, $f^{(n)}$ being the vector of frequencies of different types of mating after n generations of inbreeding.

Two practical problems arise here: that of computing A^n, the nth power of A, and that of finding the limiting value of A^n, if any, for infinitely large values of n. We return to these problems in later chapters.

7. THE LAWS OF ALGEBRA

We now give formal consideration to the associative, commutative and distributive laws of algebra as they relate to the addition and multiplication of matrices.

(a) The addition of matrices is associative provided the matrices are conformable for addition. For, if A, B and C have the same order,

$$\begin{aligned}(A + B) + C &= \{a_{ij} + b_{ij}\} + \{c_{ij}\}\\ &= \{a_{ij} + b_{ij} + c_{ij}\}\\ &= A + B + C.\end{aligned}$$

Also,

$$\begin{aligned}\{a_{ij} + b_{ij} + c_{ij}\} &= \{a_{ij}\} + \{b_{ij} + c_{ij}\}\\ &= A + (B + C),\end{aligned}$$

so proving the associative law of addition.

(b) The associative law is also true for multiplication provided the matrices are conformable for multiplication. For, if A is $p \times q$, B is $q \times r$ and C is $r \times s$, then

$$\begin{aligned}(AB)C &= \left\{\sum_{k=1}^{r}\left(\sum_{j=1}^{q} a_{ij}b_{jk}\right)c_{kh}\right\}\\ &= \left\{\sum_{k=1}^{r}\sum_{j=1}^{q} a_{ij}b_{jk}c_{kh}\right\}\\ &= \left\{\sum_{j=1}^{q} a_{ij}\left(\sum_{k=1}^{r} b_{jk}c_{kh}\right)\right\}\\ &= A(BC)\end{aligned}$$

and we write $(AB)C = A(BC) = ABC$.

(c) The distributive law holds true also, for example,

$$A(B + C) = AB + AC$$

provided both B and C are conformable for addition (necessarily of the same order) and A and B are conformable for multiplication (and hence A and C also). If A is $p \times q$ and B and C are both $q \times r$

$$\begin{aligned}A(B + C) &= \left\{\sum_{j=1}^{q} a_{ij}(b_{jk} + c_{jk})\right\}\\ &= \left\{\sum_{j=1}^{q} a_{ij}b_{jk} + \sum_{j=1}^{q} a_{ij}c_{jk}\right\} = AB + AC.\end{aligned}$$

(d) Addition of matrices is commutative provided the matrices are conformable. If A and B are of the same order

$$A + B = \{a_{ij} + b_{ij}\} = \{b_{ij} + a_{ij}\} = B + A.$$

(e) Multiplication of matrices is not in general commutative. As seen earlier, there are two possible products that can be derived from matrices A and B, AB and BA, and if A is of order $r \times c$ both products exist only if B is of order $c \times r$. AB is then square, of order r, and BA is also square, of order c. Possible equality of AB and BA can therefore be considered only where $r = c$, in which case A and B are both square and have the same order r. The products are then

$$AB = \left\{\sum_{k=1}^{r} a_{ik}b_{kj}\right\} \quad \text{and} \quad BA = \left\{\sum_{k=1}^{r} b_{ik}a_{kj}\right\}$$

$$\text{for} \quad i \text{ and j} = 1, 2, \ldots, r.$$

It is seen at once that the ijth elements of these products do not necessarily have even a single term in common in their sums of products, let alone are they equal. It is therefore apparent that even when AB and BA both exist and are of the same order they are not in general equal.

But this does not mean that AB and BA are never equal, for they can be in particular cases. For example,

$$\begin{bmatrix} 3 & 2 \\ 2 & 3 \end{bmatrix}\begin{bmatrix} -1 & 2 \\ 2 & -1 \end{bmatrix} = \begin{bmatrix} 1 & 4 \\ 4 & 1 \end{bmatrix} = \begin{bmatrix} -1 & 2 \\ 2 & -1 \end{bmatrix}\begin{bmatrix} 3 & 2 \\ 2 & 3 \end{bmatrix},$$

whereas the general non-commutative property of matrix multiplication is illustrated by the example

$$\begin{bmatrix} 1 & 2 \\ 3 & 4 \end{bmatrix}\begin{bmatrix} 0 & -1 \\ 1 & -1 \end{bmatrix} = \begin{bmatrix} 2 & -3 \\ 4 & -7 \end{bmatrix} \neq \begin{bmatrix} 0 & -1 \\ 1 & -1 \end{bmatrix}\begin{bmatrix} 1 & 2 \\ 3 & 4 \end{bmatrix} = \begin{bmatrix} -3 & -4 \\ -2 & -2 \end{bmatrix}.$$

Multiplication of matrices is commutative in two instances of special importance: (i) multiplication involving a null matrix, that is, if A_p is a square matrix of order p, and if 0_p is likewise with all elements zero, then

$$0_pA_p = A_p0_p = 0_p;$$

(ii) multiplication involving a diagonal matrix having all diagonal elements equal to unity. Such a matrix is called an *identity matrix*, or occasionally a *unit matrix*, and is usually denoted by the letter I, sometimes with a subscript for the order. Thus

$$I_3 = \begin{bmatrix} 1 & 0 & 0 \\ 0 & 1 & 0 \\ 0 & 0 & 1 \end{bmatrix}.$$

In general, if A_p is of order p, then

$$I_pA_p = A_pI_p = A_p.$$

Just as a null matrix is a zero of matrix algebra, so is an identity matrix a "one" or unit of the algebra. For example,

$$\begin{bmatrix} 1 & 0 \\ 0 & 1 \end{bmatrix}\begin{bmatrix} 1 & 2 & -4 \\ 9 & 7 & 2 \end{bmatrix} = \begin{bmatrix} 1 & 2 & -4 \\ 9 & 7 & 2 \end{bmatrix}.$$

A matrix of the form λI where λ is a scalar is sometimes called a *scalar matrix*. Thus $4I = \begin{bmatrix} 4 & 0 \\ 0 & 4 \end{bmatrix}$ is a scalar matrix.

We may note in passing that for any matrix $A_{r\times s}$, pre- or post-multiplication by a null matrix of appropriate order results in a null matrix. Thus, if $0_{c\times r}$ is a null matrix of order $c \times r$, $0_{c\times r}A_{r\times s} = 0_{c\times s}$. Likewise

$$A_{r\times s}0_{s\times p} = 0_{r\times p}.$$

For example,

$$[0 \quad 0]\begin{bmatrix} 1 & 2 & -4 \\ 9 & 7 & 2 \end{bmatrix} = [0 \quad 0 \quad 0].$$

Another situation in which the commutative law of multiplication does hold is when a matrix is multiplied by a scalar, λ say, for

$$\lambda A = \{\lambda a_{ij}\} = \{a_{ij}\lambda\} = A\lambda.$$

It is clear that multiplication by the scalar λ is equivalent to multiplication by the scalar matrix λI.

This discussion of the laws of algebra emphasizes the need for considering the conformability of matrices before attempting the operations of addition, subtraction and multiplication. In practice this must always be so—conformability has always to be borne in mind, even if not mentioned explicitly. Although when the product AB is used we seldom write "AB, where A and B are conformable for multiplication," conformability is assumed, and should be demonstrable at all times.

We have now defined a matrix and considered all the arithmetic processes except division. This subject occupies Chapter 4, but before proceeding to it we deal with two procedures of matrix algebra that do not arise in scalar algebra.

8. THE TRANSPOSE OF A MATRIX

Illustration. In first defining a matrix (Section 1.2) we considered the percentage of sterile cultures of drosophila arising in four different generations of three different populations. The percentages were arrayed as a 4×3 matrix

$$A = \begin{bmatrix} 18 & 17 & 11 \\ 19 & 13 & 6 \\ 6 & 14 & 9 \\ 9 & 11 & 4 \end{bmatrix}$$

where rows represent percentages for the generations and columns represent those for the populations. This choice of rows is quite arbitrary as far as presentation of the percentages is concerned, and we could well interchange rows and columns with no loss of information, having rows for populations and columns for generations. The array of the percentages would then be

$$B = \begin{bmatrix} 18 & 19 & 6 & 9 \\ 17 & 13 & 14 & 11 \\ 11 & 6 & 9 & 4 \end{bmatrix}.$$

Although the elements of this matrix are the same as those of the first the definition of matrix equality shows at once that the two matrices are not the same—in fact, they are not even of the same order. They are related through the rows of one being the columns of the other, and whereas the first has order 3×4 the order of the second is 4×3. When matrices are related in this fashion each is said to be the *transpose* of the other; for example, B is said to be the transpose of A, and A is the transpose of B.

A formal description is as follows. The transpose of a matrix A is the matrix whose columns are the rows of A, with order retained, from first to last. The transpose is written as A'. An obvious consequence is that the rows of A' are the same as the columns of A, and if A is $r \times c$, the order of A' is $c \times r$. If a_{ij} is the term in the ith row and the jth column of A it is also the term in the jth row and ith column of A'. Hence if

$$A = \{a_{ij}\}$$

its transpose is

$$A' = \{a_{ij}\}' = \{a_{ji}\}$$

and on defining $A' = \{a'_{ij}\}$ we have $a'_{ij} = a_{ji}$, for $i = 1, 2, \ldots, c$, and $j = 1, 2, \ldots, r$. We shall now discuss several properties and consequences of this operation.

a. A reflexive operation

The transpose operation is reflexive; that is, the transpose of a transposed matrix is the matrix itself, i.e. $(A')' = A$.

Proof: $$(A')' = \{a'_{ij}\}' = \{a_{ji}\}' = \{a'_{ji}\} = \{a_{ij}\} = A.$$

b. Vectors

The transpose of a row vector is a column vector and vice versa. For example, the transpose of

$$x = \begin{bmatrix} 1 \\ 6 \\ 4 \end{bmatrix} \quad \text{is} \quad x' = [1 \quad 6 \quad 4].$$

This explains the notation already introduced, of denoting a row vector by a superscript prime. It indicates that a row vector is the transpose of the column vector of the same elements and distinguishes it from that column vector.

c. Products

The transpose of a product matrix is the product of the transposed matrices taken in reverse order, i.e. $(AB)' = B'A'$.

Proof: Let $$AB = C = \{c_{ij}\} = \{\sum_k a_{ik}b_{kj}\}.$$

Then

$$\begin{aligned}(AB)' = C' &= \{c'_{ij}\} \\ &= \{c_{ji}\} \\ &= \{\sum_k a_{jk}b_{ki}\}\end{aligned}$$

$$= \{\sum_k a'_{kj}b'_{ik}\}$$

$$= \{\sum_k b'_{ik}a'_{kj}\}$$

$$= B'A'.$$

Consideration of order and conformability for multiplication demonstrates the necessity of this result. If A is $r \times s$ and B is $s \times t$ the product $P = AB$ is $r \times t$; i.e., $A_{r\times s}B_{s\times t} = P_{r\times t}$. But A' is $s \times r$ and B' is $t \times s$ and the only product to be derived from these is $B'_{t\times s}A'_{s\times r} = Q_{t\times r}$ say. That $Q = B'A'$ is the transpose of $P = AB$ is also apparent from the definition of multiplication: the ijth term of Q is the inner product of the ith row of B' and the jth column of A', which in turn is the inner product of the ith column of B and the jth row of A, and this by the definition of multiplication is the jith term of P. Hence $Q = P'$, or $B'A' = (AB)'$.

Example.

$$AB = \begin{bmatrix} 1 & 0 & -1 \\ 2 & -1 & 3 \end{bmatrix}\begin{bmatrix} 1 & 1 & 1 \\ 0 & 2 & 4 \\ 3 & 0 & 7 \end{bmatrix} = \begin{bmatrix} -2 & 1 & -6 \\ 11 & 0 & 19 \end{bmatrix}$$

$$B'A' = \begin{bmatrix} 1 & 0 & 3 \\ 1 & 2 & 0 \\ 1 & 4 & 7 \end{bmatrix}\begin{bmatrix} 1 & 2 \\ 0 & -1 \\ -1 & 3 \end{bmatrix} = \begin{bmatrix} -2 & 11 \\ 1 & 0 \\ -6 & 19 \end{bmatrix} = (AB)'$$

An obvious extension of the result for the transpose of the product of two matrices is $(ABCD)' = D'C'B'A'$. The proof is left as an exercise for the reader.

d. Symmetric matrices

Illustration. Dorn and Burdick (1962) report mean wing lengths of three mutant types of drosophila crossed with themselves and with each other. Part of their data can be represented in a table as follows:

TABLE 5. MEAN WING LENGTH

Type	Type 1	Type 2	Type 3
1	1.59	1.69	2.13
2	1.69	1.31	1.72
3	2.13	1.72	1.85

The entries in the first row are the mean wing lengths of type 1 crossed to itself, to type 2 and to type 3. The entries in the second and third rows have similar meaning. Consider the matrix of entries in this table, namely

$$\begin{bmatrix} 1.59 & 1.69 & 2.13 \\ 1.69 & 1.31 & 1.72 \\ 2.13 & 1.72 & 1.85 \end{bmatrix}.$$

In the context of crossbreeding drosophila the crossing of type 1 with type 2 is, for example, identical to crossing type 2 with type 1; therefore the elements in the matrix form a symmetric pattern around the diagonal. Thus 1.69 appears both as the element in the first row and second column and as the element in the second row and first column. This symmetry is true for all off-diagonal elements, and as a result the elements of any row of the matrix are the same as those of the corresponding column. Consequently the transpose of the matrix is the same as the matrix itself. Such a matrix is called a *symmetric matrix*; it is, of course, square.

A square matrix is defined as symmetric when it equals its transpose, that is, A is symmetric when $A = A'$, in which case if it is of order r, $a_{ij} = a_{ji}$, for $i, j = 1, 2, \ldots, r$.

Three results concerning symmetry are worth mentioning.

(i) The product of two symmetric matrices is not generally symmetric. If $A = A'$ and $B = B'$, A and B are symmetric but the transpose of their product is

$$(AB)' = B'A' = BA$$

and in general BA is not the same as AB; i.e., AB is not symmetric.

Example.

With $\quad A = \begin{bmatrix} 1 & 2 \\ 2 & 3 \end{bmatrix} \quad$ and $\quad B = \begin{bmatrix} 3 & 7 \\ 7 & 6 \end{bmatrix}$,

$$(AB)' = \begin{bmatrix} 17 & 19 \\ 27 & 32 \end{bmatrix}' = \begin{bmatrix} 17 & 27 \\ 19 & 32 \end{bmatrix} = BA \neq AB.$$

(ii) The product of a matrix with its transpose is symmetric. Thus,

$$(AA')' = (A')'A' = AA'$$

and

$$(A'A)' = A'(A')' = A'A.$$

It is hardly necessary to point out that although both products AA' and $A'A$ are symmetric, they are not necessarily equal.

Example.

If $$A = \begin{bmatrix} 1 & 0 & 1 \\ 2 & -1 & 3 \\ 0 & 0 & 4 \end{bmatrix},$$

$$AA' = \begin{bmatrix} 1 & 0 & 1 \\ 2 & -1 & 3 \\ 0 & 0 & 4 \end{bmatrix}\begin{bmatrix} 1 & 2 & 0 \\ 0 & -1 & 0 \\ 1 & 3 & 4 \end{bmatrix} = \begin{bmatrix} 2 & 5 & 4 \\ 5 & 14 & 12 \\ 4 & 12 & 16 \end{bmatrix},$$

and

$$A'A = \begin{bmatrix} 1 & 2 & 0 \\ 0 & -1 & 0 \\ 1 & 3 & 4 \end{bmatrix}\begin{bmatrix} 1 & 0 & 1 \\ 2 & -1 & 3 \\ 0 & 0 & 4 \end{bmatrix} = \begin{bmatrix} 5 & -2 & 7 \\ -2 & 1 & -3 \\ 7 & -3 & 26 \end{bmatrix}.$$

As noted, both AA' and $A'A$ are symmetric, but they are not equal.

(iii) A row vector post-multiplied by a column vector of the same order yields a scalar and is therefore symmetric, i.e., $x'y = y'x = (x'y)'$.

Example. If $x' = [1 \quad 2 \quad 3]$ and $y' = [4 \quad 3 \quad 7]$, $x'y = 31 = y'x$.

9. QUADRATIC FORMS

With vectors

$$x' = [x_1 \quad x_2 \quad x_3] \qquad \text{and} \quad y' = [y_1 \quad y_2 \quad y_3]$$

where the subscripted x's and y's are scalars, let us consider the product $x'Ay$ where A is some 3×3 matrix. For example,

$$x'Ay = [x_1 \quad x_2 \quad x_3]\begin{bmatrix} 1 & 2 & 3 \\ 4 & 7 & 6 \\ 2 & -2 & 5 \end{bmatrix}\begin{bmatrix} y_1 \\ y_2 \\ y_3 \end{bmatrix}$$

$$= [x_1 + 4x_2 + 2x_3 \quad 2x_1 + 7x_2 - 2x_3 \quad 3x_1 + 6x_2 + 5x_3]\begin{bmatrix} y_1 \\ y_2 \\ y_3 \end{bmatrix}$$

$$= x_1y_1 + 4x_2y_1 + 2x_3y_1 + 2x_1y_2 + 7x_2y_2 - 2x_3y_2 + 3x_1y_3 + 6x_2y_3 + 5x_3y_3.$$

This is a second degree function of the first degree in each of the x's and y's. It is called a *bilinear form.*

Suppose, though, that y is replaced by x. Then we have

$$x'Ax = x_1^2 + 4x_2x_1 + 2x_3x_1 + 2x_1x_2 + 7x_2^2 - 2x_3x_2 + 3x_1x_3 + 6x_2x_3 + 5x_3^2, \quad (7)$$

which simplifies to

$$x'Ax = x_1^2 + x_1x_2(4 + 2) + x_1x_3(2 + 3) + 7x_2^2 + x_2x_3(-2 + 6) + 5x_3^2 \quad (8)$$

$$= x_1^2 + 7x_2^2 + 5x_3^2 + 6x_1x_2 + 5x_1x_3 + 4x_2x_3. \quad (9)$$

This is a quadratic function of the x's, and is referred to as a *quadratic form*. Two properties are apparent from its development. First, in equation (7) we see that $x'Ax$ is the sum of products of all possible pairs of the x_i's, each multiplied by an element of A; thus the second term in equation (7), $4x_2x_1$, is x_2x_1 multiplied by the element of A in the second row and first column. Second, in simplifying (7) to (9) we see in (8) that the coefficient of x_1x_2, for example, is the sum of two elements in A: the one in the first column and second row plus that in the second column and first row. These results are true generally.

If x is a vector of order n with elements x_i for $i = 1, 2, \ldots, n$, and if A is a square matrix of order n with elements a_{ij} for $i, j = 1, 2, \ldots, n$, then

$$\begin{aligned} x'Ax &= [\sum_i x_ia_{i1} \quad \sum_i x_ia_{i2} \quad \ldots \quad \sum_i x_ia_{in}]x \\ &= \sum_{j=1}^{n} (\sum_i x_ia_{ij})x_j \\ &= \sum_i \sum_j x_ix_ja_{ij} \quad \text{similar to equation (7),} \\ &= \sum_i x_i^2a_{ii} + \sum\sum_{i \neq j} x_ix_ja_{ij} \end{aligned}$$

and as in (8) and (9) this is

$$x'Ax = \sum_i x_i^2a_{ii} + \sum\sum_{j>i} x_ix_j(a_{ij} + a_{ji}). \quad (10)$$

Thus $x'Ax$ is the sum of squares of the elements of x, each square multiplied by the corresponding diagonal element of A, plus the sum of products of the elements of x, each product multiplied by the sum of the corresponding elements of A; i.e., the product of the ith and jth element of x is multiplied by $(a_{ij} + a_{ji})$.

Returning to the example, notice that just as (9) was derived from (8) so also can (9) be written as

$$x'Ax = x_1^2 + 7x_2^2 + 5x_3^2 + x_1x_2(1 + 5) + x_1x_3(1 + 4) + x_2x_3(0 + 4).$$

In this way we see that

$$x'Ax = x'\begin{bmatrix} 1 & 2 & 3 \\ 4 & 7 & 6 \\ 2 & -2 & 5 \end{bmatrix}x \quad \text{is the same as} \quad x'Bx = x'\begin{bmatrix} 1 & 1 & 1 \\ 5 & 7 & 0 \\ 4 & 4 & 5 \end{bmatrix}x,$$

where B is different from A. Note that the quadratic form is unchanged although the associated matrix is not the same. In fact there is no unique matrix A for which any particular quadratic form can be expressed as $x'Ax$. Many matrices can be so used. They must each have the same diagonal elements, and in each of them the sum of each pair of symmetrically placed off-diagonal elements a_{ij} and a_{ji} must be the same; for example, equation (9) can also be expressed as

$$x'Ax = x'\begin{bmatrix} 1 & 2342 & -789 \\ -2336 & 7 & 1.37 \\ 794 & 2.63 & 5 \end{bmatrix} x. \tag{11}$$

In particular, if we write

$$x'Ax = x_1^2 + 7x_2^2 + 5x_3^2 + x_1x_2(3 + 3) + x_1x_3(2.5 + 2.5) + x_2x_3(2 + 2)$$

we see that it can be expressed as

$$x'Ax = x'\begin{bmatrix} 1 & 3 & 2.5 \\ 3 & 7 & 2 \\ 2.5 & 2 & 5 \end{bmatrix} x \tag{12}$$

where A is now a symmetric matrix. As such it is unique; that is to say, for any quadratic form there is a unique symmetric matrix A for which the quadratic form can be expressed as $x'Ax$. It can be found in any particular case by rewriting the quadratic $x'Ax$ where A is not symmetric as $x'[\frac{1}{2}(A + A')]x$, for $\frac{1}{2}(A + A')$ is symmetric. For example, if A is the matrix used in equation (11), it is easily observed that $\frac{1}{2}(A + A')$ is the symmetric matrix used in (12).

Taking A as symmetric with $a_{ij} = a_{ji}$, we see from equation (10) that the quadratic form $x'Ax$ can be expressed as

$x'Ax = \sum_i x_i^2 a_{ii} + 2 \sum\sum_{j>i} x_i x_j a_{ij}$. For example,

$$x'Ax = x'\begin{bmatrix} a_{11} & a_{12} & a_{13} \\ a_{12} & a_{22} & a_{23} \\ a_{13} & a_{23} & a_{33} \end{bmatrix} x$$

$$= a_{11}x_1^2 + a_{22}x_2^2 + a_{33}x_3^2 + 2(a_{12}x_1x_2 + a_{13}x_1x_3 + a_{23}x_2x_3).$$

Illustration. For data from an experiment or a survey we often want to calculate a mean and a variance. Suppose n observations are represented by the vector $x' = [x_1 \;\; x_2 \;\; x_3 \;\; \dots \;\; x_n]$. Then the mean of the observations is $\bar{x} = \sum_{i=1}^{n} x_i \Big/ n$ and the sample variance is based on the sum of squares

$$SS = \sum_{i=1}^{n} (x_i - \bar{x})^2 = \sum_{i=1}^{n} x_i^2 - n\bar{x}^2.$$

In matrix notation it is easily seen that

$$\sum_{i=1}^{n} x_i^2 = [x_1 \quad x_2 \quad \ldots \quad x_n]\begin{bmatrix} x_1 \\ x_2 \\ \cdot \\ \cdot \\ \cdot \\ x_n \end{bmatrix} = x'x,$$

and

$$n\bar{x}^2 = \bar{x}\sum x_i$$

$$= [\bar{x} \quad \bar{x} \quad \ldots \quad \bar{x}]\begin{bmatrix} x_1 \\ x_2 \\ \cdot \\ \cdot \\ \cdot \\ x_n \end{bmatrix} = x'\begin{bmatrix} \frac{1}{n} & \frac{1}{n} & \cdots & \frac{1}{n} \\ \cdot & & & \cdot \\ \cdot & & & \cdot \\ \cdot & & & \cdot \\ \frac{1}{n} & \frac{1}{n} & \cdots & \frac{1}{n} \end{bmatrix} x.$$

Hence

$$SS = x'x - x'U_n x = x'(I - U_n)x$$

where U_n is the square matrix of order n whose every element equals $1/n$, and where $(I - U_n)$ is a symmetric matrix of order n, with every diagonal element equal to $(n - 1)/n$ and all off-diagonal elements equal to $-1/n$. We return to this example in Chapter 8.

We have just seen a simple illustration of how an expression that arises frequently in data analysis can be put in matrix form. It has been presented in detail to illustrate the use of the arithmetic operations of matrix algebra. The reader who has some familiarity with statistics can readily visualize extensions to analysis of variance procedures generally, and can envisage the simplicity of notation that may thereby be achieved.

A quadratic form $x'Ax$ which is positive for all values of x other than $x = 0$ is said to be a *positive definite* quadratic form. And if $x'Ax$ is positive or zero for all x other than $x = 0$ (i.e. $x'Ax = 0$ for some $x \neq 0$) then it is a *positive semi-definite* quadratic form. When $x'Ax$ is positive (semi-) definite the associated matrix A is called positive (semi-) definite. In the case of the foregoing illustration, SS, as a sum of squares, is never negative and therefore $x'(I - U_n)x$ is positive semi-definite. This property is utilized at the end of Chapter 7.

*10. VARIANCE-COVARIANCE MATRICES

Matrices important in statistical work are those whose elements are the variances of and covariances between a set of variables. Developing these matrices involves many of the topics just discussed.

Suppose $x_1, x_2, \ldots, x_n$ represent a set of n random variables with means $\mu_1, \mu_2, \ldots, \mu_n$, variances $\sigma_1^2, \sigma_2^2, \ldots, \sigma_n^2$ and covariances $\sigma_{12}, \sigma_{13}, \ldots, \sigma_{1n}, \sigma_{23}, \ldots, \sigma_{2n}, \ldots, \sigma_{n-1,n}$. If the random variables are represented by the vector $x' = [x_1\ x_2 \ldots x_n]$ and their means by $\mu' = [\mu_1\ \mu_2 \ldots \mu_n]$ then using E to denote expected value we have $E(x') = \mu'$, or $E(x) = \mu$. Furthermore, the variances and covariances can be written in matrix form as

$$V = \begin{bmatrix} \sigma_1^2 & \sigma_{12} & \ldots & \sigma_{1n} \\ \sigma_{12} & \sigma_2^2 & \ldots & \sigma_{2n} \\ \cdot & & & \cdot \\ \cdot & & & \cdot \\ \cdot & & & \cdot \\ \sigma_{1n} & \sigma_{2n} & \ldots & \sigma_n^2 \end{bmatrix}.$$

This matrix is symmetric: $V' = V$. Its ith diagonal term is the variance of x_i, and its ijth term, for $i \neq j$, is the covariance between x_i and x_j. The matrix is known as the *variance-covariance matrix* of the x's. From the basic definitions of variance and covariance, namely

$$\sigma_i^2 = E(x_i - \mu_i)^2$$

and

$$\sigma_{ij} = E(x_i - \mu_i)(x_j - \mu_j) \qquad \text{for} \quad i \neq j,$$

it is easily seen that

$$V = E(x - \mu)(x' - \mu').$$

If the means are zero, $\mu = 0$, and this becomes

$$V = E(xx').$$

These forms are often utilized in multivariate analysis where they offer a most useful shorthand.

The variance-covariance matrix of linear functions of the x's is easily obtained. Let us suppose that instead of the x's we are interested in k variables $y_1, y_2, \ldots, y_k$, each of which is a linear function of some or all of the x's. Then the linear transformation between the x's and y's can be represented by a vector equation $y = Tx$ where T is $k \times n$. Immediately we see that the vector of the means of the y variables is

$$E(y) = E(Tx) = TE(x) = T\mu.$$

Likewise the variance-covariance matrix of y is readily shown to be

$$E[y - E(y)][y - E(y)]' = TVT'.$$

Notice that if we are interested in just a single linear function of the x's, y is a scalar and T will be a row vector, t'. We then have $\text{var}(y) = t'Vt$ and, since $\text{var}(y)$ is a variance, it is always positive and therefore so is $t'Vt$. V is therefore a positive semi-definite matrix. This is true for any variance-covariance matrix.

As a final comment on these matrices, notice the generality of the result TVT'. It is the variance-covariance matrix of any set of linear combinations $y = Tx$. Algebraically it is a simple result and it applies to any number of linear combinations of any number of variables. This is just the kind of generality that makes matrix algebra so useful.

11. PARTITIONING OF MATRICES

Illustration. A study of blowfly populations by MacLeod and Donelly (1962) includes counts of flies caught in traps laid out in sets of four, each set forming a square with the traps identified as N, E, S and W. Fly counts in one such set of traps counted over five consecutive days, are reported as

$$\begin{bmatrix} 18 & 3 & 7 & 9 \\ 15 & 19 & 19 & 7 \\ 9 & 12 & 15 & 11 \\ 6 & 4 & 6 & 9 \\ 27 & 15 & 12 & 14 \end{bmatrix} = A_1, \qquad \text{say,}$$

where rows represent days and columns represent traps. One study consisted of four such sets of traps observed for five days in one area, which we shall call A, and of another four sets of traps observed for four days in another area, B. Thus the fly counts for one set of traps in area B were

$$\begin{bmatrix} 0 & 3 & 2 & 4 \\ 8 & 11 & 13 & 9 \\ 16 & 61 & 20 & 33 \\ 14 & 14 & 24 & 7 \end{bmatrix} = B_1, \qquad \text{say,}$$

again with rows representing days and columns representing the traps N, E, S and W. If A_2, A_3, A_4 and B_2, B_3, B_4 represent matrices of similar counts of the other three sets of traps in each of the two areas, the complete array of the data could be presented as a combined matrix

$$C = \begin{bmatrix} A_1 & A_2 & A_3 & A_4 \\ B_1 & B_2 & B_3 & B_4 \end{bmatrix}.$$

Although C is a matrix of nine rows and sixteen columns it is expressed here as consisting of eight matrices, each of four columns, all the A matrices having five rows and the B matrices having four rows. In this form C is referred to as a *partitioned* matrix, and the A's and B's as *submatrices* of C.

Example. Consider the matrix

$$B = \begin{bmatrix} 1 & 6 & 8 & 9 & 3 & 8 \\ 2 & 4 & 1 & 6 & 1 & 1 \\ 4 & 3 & 6 & 1 & 2 & 1 \\ 9 & 1 & 4 & 6 & 8 & 7 \\ 6 & 8 & 1 & 4 & 3 & 2 \end{bmatrix}.$$

On defining

$$B_{11} = \begin{bmatrix} 1 & 6 & 8 & 9 \\ 2 & 4 & 1 & 6 \\ 4 & 3 & 6 & 1 \end{bmatrix}, \qquad B_{12} = \begin{bmatrix} 3 & 8 \\ 1 & 1 \\ 2 & 1 \end{bmatrix},$$

$$B_{21} = \begin{bmatrix} 9 & 1 & 4 & 6 \\ 6 & 8 & 1 & 4 \end{bmatrix} \quad \text{and} \quad B_{22} = \begin{bmatrix} 8 & 7 \\ 3 & 2 \end{bmatrix},$$

B can be written in partitioned form as

$$\begin{bmatrix} B_{11} & B_{12} \\ B_{21} & B_{22} \end{bmatrix}.$$

B_{11}, B_{12}, B_{21} and B_{22} are the sub-matrices of B.

Note in this example that B_{11} and B_{21} have the same number of columns and so do B_{12} and B_{22}. Likewise B_{11} and B_{12} have the same number of rows, as do B_{21} and B_{22} also. This is the usual method of partitioning, as expressed in the general case for an $r \times c$ matrix:

$$A_{r \times c} = \begin{bmatrix} K_{p \times q} & L_{p \times (c-q)} \\ M_{(r-p) \times q} & N_{(r-p) \times (c-q)} \end{bmatrix}$$

where K, L, M and N are the sub-matrices with their orders shown as subscripts.

Partitioning is not restricted to dividing a matrix into just four sub-matrices; it can be divided into numerous rows and columns of matrices. Thus in the previous example if

$$B_{01} = \begin{bmatrix} 1 & 6 & 8 & 9 \\ 2 & 4 & 1 & 6 \end{bmatrix}, \qquad B_{02} = \begin{bmatrix} 3 & 8 \\ 1 & 1 \end{bmatrix},$$

$$B_{11}^* = [4 \quad 3 \quad 6 \quad 1] \quad \text{and} \quad B_{12}^* = [2 \quad 1]$$

with B_{21} and B_{22} as already given, B can be written in partitioned form as

$$B = \begin{bmatrix} B_{01} & B_{02} \\ B_{11}^* & B_{12}^* \\ B_{21} & B_{22} \end{bmatrix}.$$

In general a matrix A, of order $p \times q$, can be partitioned in r rows and c columns of sub-matrices as

$$A = \begin{bmatrix} A_{11} & A_{12} & \cdots & A_{1c} \\ A_{21} & A_{22} & \cdots & A_{2c} \\ \cdot & \cdot & & \cdot \\ \cdot & \cdot & & \cdot \\ \cdot & \cdot & & \cdot \\ A_{r1} & A_{r2} & \cdots & A_{rc} \end{bmatrix}$$

where A_{ij} is the sub-matrix in the ith row and jth column of sub-matrices. If the ith row of sub-matrices has p_i rows of elements and the jth column of sub-matrices has q_j columns, then A_{ij} has order $p_i \times q_j$, where

$$\sum_{i=1}^{r} p_i = p \quad \text{and} \quad \sum_{j=1}^{c} q_j = q$$

12. MULTIPLICATION OF PARTITIONED MATRICES

The greatest use of partitioning is in matrix multiplication, for if two matrices A and B are partitioned so that their sub-matrices are appropriately conformable for multiplication, the product AB can be expressed in partitioned form having sub-matrices that are functions of the sub-matrices of A and B. For example, if

$$A = \begin{bmatrix} A_{11} & A_{12} \\ A_{21} & A_{22} \end{bmatrix} \quad \text{and} \quad B = \begin{bmatrix} B_{11} \\ B_{21} \end{bmatrix},$$

$$AB = \begin{bmatrix} A_{11} & A_{12} \\ A_{21} & A_{22} \end{bmatrix} \begin{bmatrix} B_{11} \\ B_{21} \end{bmatrix} = \begin{bmatrix} A_{11}B_{11} + A_{12}B_{21} \\ A_{21}B_{11} + A_{22}B_{21} \end{bmatrix},$$

provided the products $A_{11}B_{11}$, $A_{12}B_{21}$, $A_{21}B_{11}$ and $A_{22}B_{21}$ exist, and provided that $A_{11}B_{11}$ and $A_{12}B_{21}$ are conformable for addition as must be $A_{21}B_{11}$ and $A_{22}B_{21}$. This implies that A_{11} (and A_{21}) must have the same number of columns as B_{11} has rows, and A_{12} (and A_{22}) must have the same number of columns as B_{21} has rows. We see at once that when two matrices are appropriately partitioned the sub-matrices of their product are obtained by treating the sub-matrices of each of them as elements in a normal matrix product, and the individual elements of the product are derived in the usual way from the products of the sub-matrices. If

$$X = \begin{bmatrix} a_{11} & a_{12} & b_{11} & b_{12} & b_{13} \\ a_{21} & a_{22} & b_{21} & b_{22} & b_{23} \\ c_{11} & c_{12} & d_{11} & d_{12} & d_{13} \end{bmatrix}$$

is partitioned in an obvious way as

$$X = \begin{bmatrix} A & B \\ C & D \end{bmatrix}, \quad \text{and likewise} \quad Y = \begin{bmatrix} p_{11} & p_{12} \\ p_{21} & p_{22} \\ q_{11} & q_{12} \\ q_{21} & q_{22} \\ q_{31} & q_{32} \end{bmatrix} \quad \text{as} \quad Y = \begin{bmatrix} P \\ Q \end{bmatrix},$$

then $XY = \begin{bmatrix} AP + BQ \\ CP + DQ \end{bmatrix}$

$$= \begin{bmatrix} \sum_{j=1}^{2} a_{1j}p_{j1} + \sum_{j=1}^{3} b_{1j}q_{j1} & \sum_{j=1}^{2} a_{1j}p_{j2} + \sum_{j=1}^{3} b_{1j}q_{j2} \\ \sum_{j=1}^{2} a_{2j}p_{j1} + \sum_{j=1}^{3} b_{2j}q_{j1} & \sum_{j=1}^{2} a_{2j}p_{j2} + \sum_{j=1}^{3} b_{2j}q_{j2} \\ \sum_{j=1}^{2} c_{1j}p_{j1} + \sum_{j=1}^{3} d_{1j}q_{j1} & \sum_{j=1}^{2} c_{1j}p_{j2} + \sum_{j=1}^{3} d_{1j}q_{j2} \end{bmatrix}.$$

*In general, if A is $p \times q$ and is partitioned as

$$A_{p\times q} = \{A_{ij}(p_i \times q_j)\}, \quad \text{for} \quad i = 1, 2, \ldots, r \quad \text{and} \quad j = 1, 2, \ldots, c$$

with $\sum_{i=1}^{r} p_i = p$ and $\sum_{j=1}^{c} q_j = q$, where $p_i \times q_j$ is the order of the sub-matrix A_{ij}, and likewise if

$$B_{q\times s} = \{B_{jk}(q_j \times s_k)\}, \quad \text{for} \quad j = 1, 2, \ldots, c \quad \text{and} \quad k = 1, 2, \ldots, d,$$

with $\sum_{j=1}^{c} q_j = q$ and $\sum_{k=1}^{d} s_k = s$, then

$$(AB)_{p\times s} = \{\sum_{j=1}^{c} A_{ij}B_{jk}(p_i \times s_k)\}, \quad \text{for} \quad i = 1, 2, \ldots, r \quad \text{and} \quad k = 1, 2, \ldots, d.$$

Illustration. Feller (1957, page 355) gives an example of a matrix of transition probabilities that can be partitioned as

$$P = \begin{bmatrix} A & 0 & 0 \\ 0 & B & 0 \\ U_1 & V_1 & T \end{bmatrix},$$

where each of the sub-matrices is also a matrix of transition probabilities. It is readily shown that

$$P^n = \begin{bmatrix} A^n & 0 & 0 \\ 0 & B^n & 0 \\ U_n & V_n & T^n \end{bmatrix},$$

where

$$U_n = U_1A^{n-1} + TU_{n-1} \quad \text{and} \quad V_n = V_1B^{n-1} + TV_{n-1}.$$

13. EXERCISES

1. Using rows for season of the year (fall, spring, summer) and columns for species of trout (brown, brook, rainbow), the numbers of trout reported by Gard (1961) as having crossed beaver dams are

$$\begin{bmatrix} 3 & 0 & 0 \\ 8 & 3 & 1 \\ 3 & 0 & 0 \end{bmatrix} \text{ upstream} \quad \text{and} \quad \begin{bmatrix} 1 & 1 & 0 \\ 9 & 1 & 4 \\ 3 & 0 & 0 \end{bmatrix} \text{ downstream.}$$

Show that the number of trout crossing beaver dams in either direction is

$$\begin{bmatrix} 4 & 1 & 0 \\ 17 & 4 & 5 \\ 6 & 0 & 0 \end{bmatrix}.$$

2. The effect of light on the body weight of male turkeys is reported by Shoffner et al. (1962) for three different lighting intensities and two different experiments as average weight (in pounds) at three months of age as

$$\begin{bmatrix} 9.8 & 10.8 \\ 10.4 & 11.8 \\ 11.0 & 11.4 \end{bmatrix} \quad \text{and as} \quad \begin{bmatrix} 20.2 & 22.6 \\ 20.6 & 22.4 \\ 21.8 & 22.1 \end{bmatrix}$$ at six months of age. Show that the matrix of average weight gains is

$$\begin{bmatrix} 10.4 & 11.8 \\ 10.2 & 10.6 \\ 10.8 & 10.7 \end{bmatrix}.$$

3. Show that

(*a*) $$\begin{bmatrix} 6 & 3 \\ 0 & 7 \\ -5 & 1 \end{bmatrix} + \begin{bmatrix} 3 & 8 \\ 2 & -1 \\ 6 & -4 \end{bmatrix} = \begin{bmatrix} 12 & 17 \\ -11 & 2 \\ 3 & 9 \end{bmatrix} - \begin{bmatrix} 3 & 6 \\ -13 & -4 \\ 2 & 12 \end{bmatrix}$$

(*b*) $$4\begin{bmatrix} 3 & 8 \\ 1 & 9 \end{bmatrix} - 3\begin{bmatrix} 1 & -2 \\ 0 & 1 \end{bmatrix} = 2\begin{bmatrix} 2 & -1 \\ 7 & 4 \end{bmatrix} + 5\begin{bmatrix} 1 & 8 \\ -2 & 5 \end{bmatrix}$$

(*c*) $$\begin{bmatrix} 3 & 8 & 4 \\ 8 & 7 & -1 \\ 4 & -1 & 2 \end{bmatrix} + \begin{bmatrix} 1 & -1 & 3 \\ -1 & 2 & 4 \\ 3 & 4 & 6 \end{bmatrix}$$ is symmetric.

4. The mean daily intakes of hay, silage and grain for two of the rations fed to cows in an experiment reported by Bishop et al. (1963) are given as

Ration	Mean Weight Of Feed (Pounds)		
	Hay	Silage	Grain
1	16.2	29.2	11.7
2	4.8	14.8	29.3

The percentage of crude protein in these feeds, in the three different barns where the animals were fed, was

Feed	Per Cent Crude Protein		
	Barn 1	Barn 2	Barn 3
Hay	12.7	13.9	10.1
Silage	2.0	2.1	1.8
Grain	17.2	17.2	17.2

Show that the average daily intake of crude protein in the three barns was

Ration	Mean Daily Intake of Crude Protein (Pounds)		
	Barn 1	Barn 2	Barn 3
1	4.6538	4.8774	4.1742
2	5.9452	6.0176	5.7908

5. For $A = \begin{bmatrix} 3 & 6 \\ 2 & 1 \end{bmatrix}$, $B = \begin{bmatrix} 1 & 0 & 3 & 2 \\ 0 & -1 & -1 & 1 \end{bmatrix}$, $x = \begin{bmatrix} 1 \\ 1 \\ 0 \\ -1 \end{bmatrix}$ and $y = \begin{bmatrix} 1 \\ -1 \end{bmatrix}$

show that

(*a*) $$AB = \begin{bmatrix} 3 & -6 & 3 & 12 \\ 2 & -1 & 5 & 5 \end{bmatrix}, \quad \text{and} \quad A'B = \begin{bmatrix} 3 & -2 & 7 & 8 \\ 6 & -1 & 17 & 13 \end{bmatrix},$$

(*b*) $$(A + A')B = \begin{bmatrix} 6 & 8 \\ 8 & 2 \end{bmatrix} B = \begin{bmatrix} 6 & -8 & 10 & 20 \\ 8 & -2 & 22 & 18 \end{bmatrix} = AB + A'B,$$

(*c*) $$BB' = \begin{bmatrix} 14 & -1 \\ -1 & 3 \end{bmatrix} \quad \text{and} \quad B'B = \begin{bmatrix} 1 & 0 & 3 & 2 \\ 0 & 1 & 1 & -1 \\ 3 & 1 & 10 & 5 \\ 2 & -1 & 5 & 5 \end{bmatrix},$$

(*d*) trace BB' = trace $B'B = 17$,

(*e*) $$Bx = \begin{bmatrix} -1 \\ -2 \end{bmatrix}, \quad B'Bx = \begin{bmatrix} -1 \\ 2 \\ -1 \\ -4 \end{bmatrix}, \quad x'B'Bx = 5 = (Bx)'Bx,$$

(*f*) $$Ay = \begin{bmatrix} -3 \\ 1 \end{bmatrix}, \quad A'Ay = \begin{bmatrix} -7 \\ -17 \end{bmatrix}, \quad y'A'Ay = 10 = (Ay)'Ay,$$

(*g*) $A^2 - 4A - 9I = 0$,

(*h*) $$\tfrac{1}{9}A \begin{bmatrix} -1 & 6 \\ 2 & -3 \end{bmatrix} = \tfrac{1}{9} \begin{bmatrix} -1 & 6 \\ 2 & -3 \end{bmatrix} A = \begin{bmatrix} 1 & 0 \\ 0 & 1 \end{bmatrix}.$$

6. Confirm:

(*a*) $$\begin{bmatrix} 1 & 1 \\ 1 & 1 \end{bmatrix}\begin{bmatrix} 2 & 3 \\ 3 & 2 \end{bmatrix} = \begin{bmatrix} 2 & 3 \\ 3 & 2 \end{bmatrix}\begin{bmatrix} 1 & 1 \\ 1 & 1 \end{bmatrix},$$

(*b*) that if $A = \frac{1}{3}\begin{bmatrix} 1 & 1 & 1 \\ 1 & 1 & 1 \\ 1 & 1 & 1 \end{bmatrix}, \quad A^2 = A,$

(*c*) that if $B = \begin{bmatrix} 1/\sqrt{6} & 1/\sqrt{3} & 1/\sqrt{2} \\ -2/\sqrt{6} & 1/\sqrt{3} & 0 \\ 1/\sqrt{6} & 1/\sqrt{3} & -1/\sqrt{2} \end{bmatrix}, \quad BB' = B'B = I_3,$

(*d*) that if $C = \begin{bmatrix} 6 & -4 \\ 9 & -6 \end{bmatrix}, \quad C^2 = 0,$

(*e*) $$\tfrac{1}{9}\begin{bmatrix} 4 & -5 & -1 \\ 1 & 1 & 2 \\ 4 & 4 & -1 \end{bmatrix}\begin{bmatrix} 1 & 1 & 1 \\ -1 & 0 & 1 \\ 0 & 4 & -1 \end{bmatrix} = I_3.$$

7. For $A = \begin{bmatrix} 3 & 0 & 3 & 2 \\ 0 & -1 & 5 & 0 \\ -1 & 6 & 0 & 8 \end{bmatrix}$ and $B = \begin{bmatrix} -1 & 0 & 8 & 3 \\ 0 & -2 & -7 & -1 \\ -1 & 3 & 0 & 2 \end{bmatrix}$

find AA', BB', AB' and BA', and show that

$$\begin{aligned} (A+B)(A+B)' &= (A+B)(A'+B') \\ &= AA' + BA' + AB' + BB' \\ &= AA' + (AB')' + AB' + BB' \\ &= \begin{bmatrix} 150 & -27 & 46 \\ -27 & 14 & -37 \\ 46 & -37 & 185 \end{bmatrix}. \end{aligned}$$

8. For $X = \begin{bmatrix} 1 & 2 & 3 \\ 0 & -1 & -2 \\ -1 & 0 & 7 \end{bmatrix}$ and $Y = \begin{bmatrix} 6 & 0 & 0 \\ -3 & 4 & 0 \\ 0 & -5 & 2 \end{bmatrix}$ find X^2, Y^2, XY

and YX, and show that

$$(X+Y)^2 = X^2 + XY + YX + Y^2 = \begin{bmatrix} 40 & 5 & 44 \\ -28 & 13 & -33 \\ -1 & -62 & 88 \end{bmatrix}.$$

9. For $A_1 = \begin{bmatrix} 1 & 0 & 3 \\ 0 & 2 & -1 \end{bmatrix}, \quad A_2 = \begin{bmatrix} 0 & 1 \\ 1 & 1 \end{bmatrix}, \quad B_1 = \begin{bmatrix} -1 & 2 & 0 \\ 0 & 3 & 4 \\ -1 & 0 & 0 \end{bmatrix}$

$$B_2 = \begin{bmatrix} 6 & 0 & -1 \\ 2 & 1 & 6 \end{bmatrix}, \quad b_1 = \begin{bmatrix} 0 \\ -1 \\ 2 \end{bmatrix}, \quad b_2 = \begin{bmatrix} 3 \\ -1 \end{bmatrix},$$

show that if $X = [A_1 \quad A_2]$ and $Y = \begin{bmatrix} B_1 & b_1 \\ B_2 & b_2 \end{bmatrix}$,

$$XY = [A_1B_1 + A_2B_2 \quad A_1b_1 + A_2b_2] = \begin{bmatrix} -2 & 3 & 6 & 5 \\ 9 & 7 & 13 & -2 \end{bmatrix}.$$

10. Given $A = \begin{bmatrix} 1 & 0 & 2 \\ 0 & 1 & 1 \\ 2 & 0 & 2 \end{bmatrix}$, $\quad B = \begin{bmatrix} 1 & 3 & 0 \\ 0 & 4 & -1 \\ 2 & 3 & 0 \end{bmatrix}$, $\quad X = \begin{bmatrix} 6 & 5 & 7 \\ 2 & 2 & 4 \\ 3 & 3 & 6 \end{bmatrix}$

show that $AX = BX$ even though $A \neq B$.

11. For $\quad A = \{a_{ij}\} \quad$ for $\quad i, j = 1, 2, \ldots, r,$
and $\quad D = \{d_{ij}\}, \quad d_{ij} = 0 \quad$ for $\quad i \neq j, \quad$ for $\quad i, j = 1, 2, \ldots, r,$
show that $AD = \{d_{jj}a_{ij}\}$ and $DA = \{d_{ii}a_{ij}\}$ for $i, j = 1, 2, \ldots, r.$

12. In discussing the problem of calculating frequencies for different relatives, of pairs of genotypes at a single two-allele locus, Li (1955, page 34) uses two matrices of conditional probabilities. They are

$$P = \begin{bmatrix} p^2 & 2pq & q^2 \\ p^2 & 2pq & q^2 \\ p^2 & 2pq & q^2 \end{bmatrix} \quad \text{and} \quad T = \begin{bmatrix} p & q & 0 \\ \frac{1}{2}p & \frac{1}{2} & \frac{1}{2}q \\ 0 & p & q \end{bmatrix},$$

P when the relatives have no genes identical by descent and T is when they have one gene identical by descent. Show that, with $p + q = 1$,

(*a*) $PT = P = TP$, (*c*) $P^2 = P$,
(*b*) $T^2 = \frac{1}{2}(P + T)$, (*d*) $T^n = P + (\frac{1}{2})^{n-1}(T - P)$,

and if $S = \frac{1}{4}I + \frac{1}{2}T + \frac{1}{4}P$, then

(*e*) $ST = TS = T^2$ and (*f*) $S^2 = \frac{1}{16}(I + 6T + 9P)$.

13. The generation matrix for selfing is given by Kempthorne (1957, page 120) as $A = \begin{bmatrix} 1 & \frac{1}{2} \\ 0 & \frac{1}{2} \end{bmatrix}$. Given that $f^{(i)} = Af^{(i-1)}$, show that

$$f^{(2)} = \begin{bmatrix} 1 & \frac{3}{4} \\ 0 & \frac{1}{4} \end{bmatrix} f^{(0)}, \qquad f^{(3)} = \begin{bmatrix} 1 & \frac{7}{8} \\ 0 & \frac{1}{8} \end{bmatrix} f^{(0)}$$

and

$$f^{(n)} = A^n f^{(0)} = \begin{bmatrix} 1 & 1 - 1/2^n \\ 0 & 1/2^n \end{bmatrix} f^{(0)}.$$

14. (*a*) Under what conditions does $(A + B)(A - B) = A^2 - B^2$?
(*b*) If $A = A^2$ (e.g., Exc. 6*b*), prove that $(I - A)^2 = I - A$.

(*c*) Given that $AA' = I$ and $BB' = I$ (e.g., Exc. 6*c*) show that $AB(AB)' = I$.
(*d*) Describe why $X'X = 0$ implies $X = 0$.
(*e*) Show that the only real symmetric matrix whose square is null is the null matrix itself.
(*f*) Show that if $X'X = X$, $X = X' = X^2$.
(*g*) Show that trace (AB') equals the sum of the elements in the term-by-term product of A and B.
(*h*) Prove that $\text{tr}(AA') = \text{tr}(A'A) = \sum_i \sum_j a_{ij}^2$

REFERENCES

Anderson, V. L. and O. Kempthorne (1954). A model for the study of quantitative inheritance. *Genetics*, **39**, 883–898.
Bishop, S. E., J. K. Loosli, G. W. Trimberger and K. L. Turk (1963). Effects of pelleting and varying grain intakes on milk yield and composition. *J. Dairy Sci.*, **46**, 22–26.
Dorn, Gordon L. and Allan B. Burdick (1962). On the recombinatorial structure of complementation relationships in the *m-dy* complex of *Drosophila melanogaster*. *Genetics*, **47**, 503–518.
Feller, William (1957). *An Introduction to Probability Theory and Its Applications*. Vol. I, Second Edition, Wiley, New York.
Fisher, R. A. (1949). *The Theory of Inbreeding*. Oliver and Boyd, Edinburgh.
Gard, Richard (1961). Effects of beaver on trout in Sagehen Creek, California. *J. Wildlife Management*, **25**, 221–242.
Hayman, B. I. and K. Mather (1955). The description of gene interaction in continuous variation. *Biometrics*, **11**, 69–82.
Kamar, Gamal A. R. (1962). Growth of various breeds of ducks under Egyptian conditions. *Poultry Sci.*, **41**, 1344–1346.
Kemeny, John G., Laurie J. Snell and Gerald L. Thompson (1956). *Introduction to Finite Mathematics*. Prentice-Hall, Englewood Cliffs, New Jersey.
Kempthorne, Oscar (1957). *An Introduction to Genetic Statistics*. Wiley, New York.
Li, C. C. (1955). *Population Genetics*. The University of Chicago Press, Chicago.
MacLeod, John and Joseph Donnelly (1962). Microgeographic aggregations in blowfly populations. *J. Animal Ecology*, **31**, 525–544.
Peek, James M. (1962). Studies of moose in the Gravelly and Snowcrest Mountains, Montana. *J. Wildlife Management*, **26**, 360–365.
Shoffner, R. N., C. R. Polley, R. E. Burger and E. L. Johnson (1962). Light regulation in turkey management. I. Effect on body weight (growth). *Poultry Sci.*, **41**, 1560–1562.
Tallis, G. M. (1962). The maximum likelihood estimation of correlation from contingency tables. *Biometrics*, **18**, 342–353.

CHAPTER 3

DETERMINANTS

Matrices and many of the operations associated with them have been introduced in Chapters 1 and 2. A procedure not yet considered, division, is dealt with in Chapter 4, but before proceeding to it we discuss an operation applicable to the elements of a square matrix that leads to a scalar value known as the determinant of the matrix. Knowledge of this operation is necessary for our discussion of the counterpart of division in matrix algebra, and it is also useful in succeeding chapters when dealing with aspects of matrix theory proper that have direct application to problems in biology. Moreover, since the reader will undoubtedly encounter determinants elsewhere in his reading on matrices, he should find it appropriate to pursue a general discourse on them here.

The literature of determinants is extensive and forms part of many texts on matrices. The development in this book will be relatively brief, will deal with elementary methods of evaluation and will discuss selected topics arising therefrom. Extensive use is made of small numerical examples and there is minimal emphasis on rigorous, mathematical proof. Several portions of the chapter can well be omitted at a first reading.

We begin with some general descriptions and then develop a widely accepted method for calculating a determinant from its associated matrix, a method that is primarily useful for deriving the determinant of a small-sized matrix, but one that is more helpful in divulging the details of a determinant than is a formal definition. A formal definition is given in Section 2.

1. SIMPLE EVALUATION

A determinant is a polynomial of the elements of a square matrix. It is a scalar. It is the sum of certain products of the elements of the matrix from which it is derived, each product being multiplied by $+1$ or -1 according to certain rules.

Determinants are defined only for square matrices—the determinant of a non-square matrix is undefined and does not exist. The determinant of a (square) matrix of order n is referred to as an *n-order* determinant and the customary notation for the determinant of the matrix A is $|A|$, it being assumed that A is square. Some texts use the notation $\|A\|$ or $[A]$, but $|A|$ is more common and is used throughout this book. Obtaining the value of $|A|$ by adding the appropriate products of the elements of A (with the correct $+1$ or -1 factor included in the product) is variously referred to as *evaluating* the determinant, *expanding* the determinant or *reducing* the determinant. The procedure for evaluating a determinant is illustrated by a series of simple numerical examples.

To begin with we have the elementary result that the determinant of a 1×1 matrix is the value of its sole element. The value of a second-order determinant is scarcely more difficult. For example, the determinant of

$$A = \begin{bmatrix} 7 & 3 \\ 4 & 6 \end{bmatrix}$$

is written

$$|A| = \begin{vmatrix} 7 & 3 \\ 4 & 6 \end{vmatrix}$$

and is calculated as

$$|A| = 7(6) - 3(4) = 30.$$

This illustrates the general result for expanding a second-order determinant: from the product of the diagonal terms subtract the product of the off-diagonal terms. In other words, the determinant of a 2×2 matrix consists of the product (multiplied by $+1$) of the diagonal terms plus the product (multiplied by -1) of the off-diagonal terms. Hence in general

$$\begin{aligned} |A| &= \begin{vmatrix} a_{11} & a_{12} \\ a_{21} & a_{22} \end{vmatrix} \\ &= a_{11}a_{22} + (-1)a_{12}a_{21} \\ &= a_{11}a_{22} - a_{12}a_{21}. \end{aligned}$$

Further examples are

$$\begin{vmatrix} 6 & 8 \\ 17 & 21 \end{vmatrix} = 6(21) - 8(17) = -10,$$

and

$$\begin{vmatrix} 9 & -3.81 \\ 7 & -1.05 \end{vmatrix} = 9(-1.05) - (-3.81)(7) = 17.22.$$

The evaluation of a second-order determinant is patently simple.

A third-order determinant can be expanded as a linear function of three second order determinants derived from it. Their coefficients are elements of a row (or column) of the main determinant, each product being multiplied by $+1$ or -1. For example, the expansion of

$$|A| = \begin{vmatrix} 1 & 2 & 3 \\ 4 & 5 & 6 \\ 7 & 8 & 10 \end{vmatrix}$$

based on the elements of the first row, 1, 2 and 3, is

$$|A| = 1(+1)\begin{vmatrix} 5 & 6 \\ 8 & 10 \end{vmatrix} + 2(-1)\begin{vmatrix} 4 & 6 \\ 7 & 10 \end{vmatrix} + 3(+1)\begin{vmatrix} 4 & 5 \\ 7 & 8 \end{vmatrix}$$

$$= 1(50 - 48) - 2(40 - 42) + 3(32 - 35) = -3.$$

The determinant that multiplies each element of the chosen row (in this case the first row) is the determinant derived from $|A|$ by crossing out the row and column containing the element concerned. For example, the first element, 1, is multiplied by the determinant $\begin{vmatrix} 5 & 6 \\ 8 & 10 \end{vmatrix}$ which is obtained from $|A|$ through crossing out the first row and column, and the element 2 is multiplied (apart from the factor of -1) by the determinant derived from $|A|$ by deleting the row and column containing that element—namely, the first row and second column, so leaving $\begin{vmatrix} 4 & 6 \\ 7 & 10 \end{vmatrix}$. Determinants obtained in this way are called *minors* of $|A|$, that is to say, $\begin{vmatrix} 5 & 6 \\ 8 & 10 \end{vmatrix}$ is the minor of the element 1 in $|A|$, and $\begin{vmatrix} 4 & 6 \\ 7 & 10 \end{vmatrix}$ is the minor of the element 2. The $(+1)$ and (-1) factors are decided on according to the following rule: if A is written in the form $A = \{a_{ij}\}$, the product of a_{ij} and its minor in the expansion of the determinant $|A|$ is multiplied by $(-1)^{i+j}$. Thus, because the element 1 in the example is the element a_{11}, its product with its minor is multiplied by $(-1)^{1+1} = +1$; and the product with its minor of the element 2, a_{12}, is multiplied by $(-1)^{1+2} = -1$.

This method of expanding a determinant is known as *expansion by the elements of a row* (*or column*) or sometimes as *expansion by minors*. It has been illustrated using elements of the first row, but can also be applied to the elements of any row (or column). For example, the expansion of $|A|$ just considered, using elements of the second row gives

$$|A| = 4(-1)\begin{vmatrix} 2 & 3 \\ 8 & 10 \end{vmatrix} + 5(+1)\begin{vmatrix} 1 & 3 \\ 7 & 10 \end{vmatrix} + 6(-1)\begin{vmatrix} 1 & 2 \\ 7 & 8 \end{vmatrix}$$

$$= -4(-4) + 5(-11) - 6(-6) = -3$$

as before, and using elements of the first column the expansion is the same:

$$|A| = 1(+1)\begin{vmatrix} 5 & 6 \\ 8 & 10 \end{vmatrix} + 4(-1)\begin{vmatrix} 2 & 3 \\ 8 & 10 \end{vmatrix} + 7(+1)\begin{vmatrix} 2 & 3 \\ 5 & 6 \end{vmatrix}$$

$$= 1(2) - 4(-4) + 7(-3) = -3.$$

The minors in these expansions are derived in exactly the same manner as in the first, by crossing out rows and columns of $|A|$; and so are the $(+1)$ and (-1) factors; for example, the minor of the element 4 is $|A|$ with second row and first column deleted, and since 4 is a_{21} its product with its minor is multiplied by $(-1)^{2+1} = -1$. Other terms are obtained in a similar manner.

The foregoing example illustrates the expansion of the general third-order determinant

$$|A| = \begin{vmatrix} a_{11} & a_{12} & a_{13} \\ a_{21} & a_{22} & a_{23} \\ a_{31} & a_{32} & a_{33} \end{vmatrix}.$$

Expanding this by elements of the first row gives

$$|A| = a_{11}(+1)\begin{vmatrix} a_{22} & a_{23} \\ a_{32} & a_{33} \end{vmatrix} + a_{12}(-1)\begin{vmatrix} a_{21} & a_{23} \\ a_{31} & a_{33} \end{vmatrix} + a_{13}(+1)\begin{vmatrix} a_{21} & a_{22} \\ a_{31} & a_{32} \end{vmatrix}$$

$$= a_{11}a_{22}a_{33} - a_{11}a_{23}a_{32} - a_{12}a_{21}a_{33} + a_{12}a_{23}a_{31} + a_{13}a_{21}a_{32} - a_{13}a_{22}a_{31}.$$

The reader should satisfy himself that expansion by the elements of any other row or column leads to the same result.

Example.

For
$$|A| = \begin{vmatrix} 1 & 0 & 4 \\ 3 & 1 & 2 \\ 17 & -3 & 5 \end{vmatrix}$$

expansion by elements of the first row is

$$|A| = 1(+1)\begin{vmatrix} 1 & 2 \\ -3 & 5 \end{vmatrix} + 0(-1)\begin{vmatrix} 3 & 2 \\ 17 & 5 \end{vmatrix} + 4(+1)\begin{vmatrix} 3 & 1 \\ 17 & -3 \end{vmatrix}$$

$$= 1(11) - 0 + 4(-26) = -93$$

and expansion by elements of the second row is

$$|A| = 3(-1)\begin{vmatrix} 0 & 4 \\ -3 & 5 \end{vmatrix} + 1(+1)\begin{vmatrix} 1 & 4 \\ 17 & 5 \end{vmatrix} + 2(-1)\begin{vmatrix} 1 & 0 \\ 17 & -3 \end{vmatrix}$$

$$= -3(12) + (-63) - 2(-3) = -93.$$

The second column might also be used, in which case

$$|A| = 0(-1)\begin{vmatrix} 3 & 2 \\ 17 & 5 \end{vmatrix} + 1(+1)\begin{vmatrix} 1 & 4 \\ 17 & 5 \end{vmatrix} + (-3)(-1)\begin{vmatrix} 1 & 4 \\ 3 & 2 \end{vmatrix}$$

$$= 0 + (-63) + 3(-10) = -93.$$

No matter by what row or column the expansion is made, the value of the determinant is the same. Note that once a row or column is decided on and the sign calculated for the product of the first element therein with its minor, the signs for the following products alternate from plus to minus and minus to plus.

The expansion of an n-order determinant by this method is an extension of the expansion of a third-order determinant as just given. Thus the determinant of the $n \times n$ matrix $A = \{a_{ij}\}$ for $i, j = 1, 2, \ldots, n$, is obtained as follows. Consider the elements of any one row (or column): multiply each element, a_{ij}, of this row (or column) by its minor, $|M_{ij}|$, the determinant derived from $|A|$ by crossing out the row and column containing a_{ij}; multiply the product by $(-1)^{i+j}$; add the signed products and their sum is the determinant $|A|$; that is, when expanding by elements of a row

$$|A| = \sum_{j=1}^{n} a_{ij}(-1)^{i+j}|M_{ij}| \qquad \text{for any } i, \tag{1}$$

and when expanding by elements of a column

$$|A| = \sum_{i=1}^{n} a_{ij}(-1)^{i+j}|M_{ij}| \qquad \text{for any } j. \tag{2}$$

This expansion is used recurrently when n is large, i.e. each $|M_{ij}|$ is expanded by the same procedure. Thus a fourth-order determinant is first expanded as four signed products each involving a third-order minor, and each of these is expanded as a sum of three signed products involving a second-order determinant. Consequently a fourth-order determinant ultimately involves $4 \times 3 \times 2 = 24$ products of its elements, each product containing four elements. This leads us to the general statement that the determinant of a square matrix of order n is a sum of

$(n!)^1$ signed products, each product involving n elements of the matrix. The determinant is referred to as an *n-order* determinant. Utilizing methods given in Aitken (1948) it can be shown that each product has one and only one element from each row and column and that all such products are included and none occur more than once.

This method of evaluating a determinant involves tedious calculations for determinants of order greater than three. Fortunately, easier methods exist, but because the method already discussed forms the basis of these easier methods it has been considered in detail. Furthermore, it is found useful in developing additional properties of determinants.

*2. FORMAL DEFINITION

As indicated before, the determinant of an n-order square matrix is the sum of $n!$ signed products of the elements of the matrix, each product containing one and only one element from every row and column of the matrix. Evaluation through expansion by elements of a row or column, as represented in equations (1) and (2), yields the requisite products with their correct signs when the minors are successively expanded at each stage by the same procedure. We have already referred to this as the method of expansion by minors.

There are many formal definitions of a determinant, all, of course, consistent with each other, and all leading to the same results as expansion by minors. The following is adapted from Ferrar (1941). The determinant of a square matrix A, of order n, $A = \{a_{ij}\}\quad i, j = 1, 2, \ldots, n$, is (i) the sum of all possible products of n elements of A such that (ii) each product has one and only one element from every row and column of A, (iii) the sign of a product being $(-1)^p$ for $p = \sum_{i=1}^{n} n_i$, where (iv) by writing the product with its i subscripts in natural order $a_{1j_1}a_{2j_2}\cdots a_{ij_i}\cdots a_{nj_n}$, the j subscripts j_i, $i = 1, 2, \ldots, n$, being the first n integers in some order, n_i is defined as the number of j's less than j_i that follow j_i in this order.

Example. In the expansion of

$$|A| = \begin{vmatrix} a_{11} & a_{12} & a_{13} \\ a_{21} & a_{22} & a_{23} \\ a_{31} & a_{32} & a_{33} \end{vmatrix}$$

the j_1, j_2, j_3 array in the product $a_{12}a_{23}a_{31}$ is $j_1 = 2$, $j_2 = 3$ and $j_3 = 1$. Now n_1 is the number of j's in this array that are less than j_1 but follow it; i.e. that are less than 2 but follow it. There is only one, namely $j_3 = 1$; therefore $n_1 = 1$. Likewise $n_2 = 1$ because $j_2 = 3$ is followed by only one j less than 3, and finally $n_3 = 0$. Thus

$$n_1 + n_2 + n_3 = 1 + 1 + 0 = 2$$

and the sign of the product $a_{12}a_{23}a_{31}$ in $|A|$ is therefore $(-1)^2 = +1$. That the sign is

[1] $n!$, read as "factorial n", is the product of all integers 1 through n inclusive; e.g. $4! = 1(2)(3)(4) = 24$.

also positive in the expansion of $|A|$ by minors is easily shown, for in expanding by elements of the first row the term involving a_{12} is

$$-a_{12}\begin{vmatrix} a_{21} & a_{23} \\ a_{31} & a_{33} \end{vmatrix} = -a_{12}(a_{21}a_{33} - a_{23}a_{31})$$

which includes the term $+a_{12}a_{23}a_{31}$.

We shall now indicate the generality of the result implicit in the above example, that the formal definition of $|A|$ agrees with the procedure of expansion by minors. The definition states that each product of n elements in $|A|$ contains one element from every row and column. Suppose we set out to form one such product by selecting elements for it one at a time from each of the rows of A. Starting with the first row there are n possible choices of an element as the first element of the product. Having made a choice there are then only $n - 1$ possible choices in the second row because the column containing the element chosen from the first row must be excluded from the choices in subsequent rows (in order to have one and only one element from each column as well as from each row). Similarly the column containing the element chosen from the second row has to be excluded from the possible choices in the third and subsequent rows. Thus there are $(n - 2)$ possible choices in the third row, $(n - 3)$ in the fourth row and so on. Hence, based on the definition, the total number of products in $|A|$ is

$$n(n - 1)(n - 2) \ldots (3)(2)(1) = (n!).$$

On the other hand, expansion of $|A|$ by minors initially gives $|A|$ as a sum of n elements each multiplied (apart from sign) by its minor, a determinant of order $n - 1$. Each minor can be expanded in similar fashion as the sum of $(n - 1)$ elements multiplied (apart from sign) by minors that are determinants of order $n - 2$. In this way we see that the complete determinant consists of $n(n - 1)(n - 2) \ldots (3)(2)(1) = (n!)$ signed products each containing n elements. And at each stage of the expansion the method of deleting a row and a column from a determinant to obtain the minor of an element ensures that in each product the n elements consist of one from every row and column of A. Hence (apart from sign) both the definition and the procedure of expansion by minors lead to the same set of products. We will now show that the sign of each product is the same in both cases.

The sign of any product in a determinant expanded by minors is the product of the signs applied to each minor involved in the derivation of the product. Consider the product as

$$a_{1j_1}a_{2j_2}a_{3j_3} \cdots a_{nj_n}.$$

Supposing the first expansion of $|A|$ is by elements of the first row, the sign attached to the minor of a_{1j_1} is $(-1)^{1+j_1}$. If the expansion of this minor is by elements of the second row of $|A|$ which is now the first row of the minor, the minor therein of a_{2j_2} will have attached to it the sign $(-1)^{1+j_2-1}$ if j_1 is less than j_2 and $(-1)^{1+j_2}$ if j_1 exceeds j_2. Likewise the sign attached to the minor of a_{3j_3} in the minor of a_{2j_2} in the minor of a_{1j_1} in the expansion of $|A|$ will be $(-1)^{1+j_3-2}$ if both j_1 and j_2 are less than j_3. It will be $(-1)^{1+j_3-1}$ if either of j_1 and j_2 is less than j_3 and the other exceeds it, and it will be $(-1)^{1+j_3}$ if both j_1 and j_2 exceed j_3. Hence the sign will be $(-1)^{1+j_3-m_3}$ where m_3 is the number of j's less than j_3 which precede it in the array $j_1j_2j_3 \cdots j_n$. In general the sign of the ith minor involved will be $(-1)^{1+j_i-m_i}$ where m_i is the number of j's less than j_i which precede it in the array. Therefore the combined sign is

$$\prod_{i=1}^{n} (-1)^{1+j_i-m_i} = (-1)^q \text{ for } q = \sum_{i=1}^{n} (1+j_i - m_i).$$

Now, since the j_i's are the first n integers in some order, the number of them that are less than some particular j_i is $j_i - 1$. Therefore, by the definitions of n_i and m_i their sum is $j_i - 1$, i.e. $n_i + m_i = j_i - 1$, which gives

$$q = \sum_{i=1}^{n} (1 + j_i - m_i) = \sum_{i=1}^{n} (2 + n_i) = 2n + \sum_{i=1}^{n} n_i.$$

Therefore $(-1)^q = (-1)^p$ for $p = \sum_{i=1}^{n} n_i$, so that when the determinant is expanded by minors the sign of a product is $(-1)^p$ as specified by the definition. Hence the definition and the method of expansion by minors are equivalent.

3. ELEMENTARY EXPANSIONS

a. Determinant of a transpose

Since each product in the expansion of a determinant contains one and only one element from every row and column, every row can be interchanged with its corresponding column without affecting the value of the determinant, that is, for $i = 1, 2, \ldots, n$, row i can be interchanged with column i. This means the determinant of the transpose of a matrix is the same as the determinant of the matrix: $|A'| = |A|$.

Example.

$$\begin{vmatrix} 1 & -1 & 0 \\ 2 & 1 & 2 \\ 4 & 4 & 9 \end{vmatrix} = \begin{vmatrix} 1 & 2 \\ 4 & 9 \end{vmatrix} + \begin{vmatrix} 2 & 2 \\ 4 & 9 \end{vmatrix} = 1 + 10 = 11,$$

and

$$\begin{vmatrix} 1 & 2 & 4 \\ -1 & 1 & 4 \\ 0 & 2 & 9 \end{vmatrix} = \begin{vmatrix} 1 & 4 \\ 2 & 9 \end{vmatrix} - 2\begin{vmatrix} -1 & 4 \\ 0 & 9 \end{vmatrix} + 4\begin{vmatrix} -1 & 1 \\ 0 & 2 \end{vmatrix} = 1 + 18 - 8 = 11.$$

Three properties relating to the rows of a determinant are now considered. Because of the interchangeability of rows with columns just discussed the three properties hold true for columns also, but for simplicity's sake they are presented only in terms of rows.

b. Interchanging two rows

Consider the two determinants

$$|A| = \begin{vmatrix} a_{11} & a_{12} & a_{13} \\ a_{21} & a_{22} & a_{23} \\ a_{31} & a_{32} & a_{33} \end{vmatrix} \quad \text{and} \quad |B| = \begin{vmatrix} a_{21} & a_{22} & a_{23} \\ a_{11} & a_{12} & a_{13} \\ a_{31} & a_{32} & a_{33} \end{vmatrix}.$$

The first and second rows of $|B|$ are the second and first rows of $|A|$ respectively, i.e. $|B|$ is $|A|$ with the first and second rows interchanged.

Now in expanding $|A|$ by the formal definition just given, the term in $a_{13}a_{21}a_{32}$ has $n_1 = 2$, $n_2 = 0$ and $n_3 = 0$, so that its sign is

$$(-1)^{2+0+0} = +1.$$

If we write

$$|B| = \begin{vmatrix} b_{11} & b_{12} & b_{13} \\ b_{21} & b_{22} & b_{23} \\ a_{31} & a_{32} & a_{33} \end{vmatrix}$$

with

$$\begin{bmatrix} b_{11} & b_{12} & b_{13} \\ b_{21} & b_{22} & b_{23} \end{bmatrix} = \begin{bmatrix} a_{21} & a_{22} & a_{23} \\ a_{11} & a_{12} & a_{13} \end{bmatrix} \tag{3}$$

then in the expansion of $|B|$ in this form the term $b_{11}b_{23}a_{32}$ has $n_1 = 0$, $n_2 = 1$ and $n_3 = 0$, so that the sign of the term is $(-1)^{0+1+0} = -1$. Hence, because of the equalities represented by (3), the term is

$$-b_{11}b_{23}a_{32} = -a_{21}a_{13}a_{32} = -a_{13}a_{21}a_{32},$$

the same as in $|A|$ except for a factor of (-1). This result will be found true of all terms in the expansion of $|B|$. Thus $|B| = -|A|$, i.e., interchanging two rows of a determinant changes its sign.[1]

Example.

$$\begin{vmatrix} 6 & 1 & 0 \\ 3 & -1 & 2 \\ 4 & 0 & -1 \end{vmatrix} = 6\begin{vmatrix} -1 & 2 \\ 0 & -1 \end{vmatrix} - 1\begin{vmatrix} 3 & 2 \\ 4 & -1 \end{vmatrix} = 6 + 11 = 17;$$

$$\begin{vmatrix} 4 & 0 & -1 \\ 3 & -1 & 2 \\ 6 & 1 & 0 \end{vmatrix} = 4\begin{vmatrix} -1 & 2 \\ 1 & 0 \end{vmatrix} - 1\begin{vmatrix} 3 & -1 \\ 6 & 1 \end{vmatrix} = -8 - 9 = -17.$$

Corollary. If two rows of a determinant are the same the determinant is zero. For then interchanging the rows changes the sign as just discussed, but since the rows are the same the value of the determinant must otherwise be unaltered. Hence $|A| = -|A|$ so that $|A| = 0$.

Example.

$$\begin{vmatrix} 4 & 17 & 3 \\ 4 & 17 & 3 \\ 1 & 4 & 1 \end{vmatrix} = 4\begin{vmatrix} 17 & 3 \\ 4 & 1 \end{vmatrix} - 17\begin{vmatrix} 4 & 3 \\ 1 & 1 \end{vmatrix} + 3\begin{vmatrix} 4 & 17 \\ 1 & 4 \end{vmatrix} = 20 - 17 - 3 = 0.$$

c. Factorization

If λ (a scalar) is a factor of a row it is also a factor of the determinant. Suppose λ is a factor of the kth row. Then, because every term in the expansion of the determinant contains an element from every row and

[1] And this is so for the interchanging of any two rows, be they adjacent or otherwise.

therefore one from the kth row, λ can be factored out from every term in the expansion; i.e. λ is a factor of the determinant. A consequence of this is that the determinant can be written with the factor λ removed from the kth row and expressed as a factor.

Example.

$$|A| = \begin{vmatrix} 2 & 10 & 8 \\ 1 & 7 & 3 \\ 6 & 1 & 4 \end{vmatrix} = (2)\begin{vmatrix} 1 & 5 & 4 \\ 1 & 7 & 3 \\ 6 & 1 & 4 \end{vmatrix}.$$

Expanding the first form gives

$$|A| = 2(28 - 3) - 10(4 - 18) + 8(1 - 42) = -138$$

and expanding the second gives

$$|A| = 2[(28 - 3) - 5(4 - 18) + 4(1 - 42)] = -138.$$

When appropriate, this process can be carried out for several rows simultaneously. For example, the matrix

$$\begin{bmatrix} p^3 & p^2q & 0 \\ p^2q & pq & pq^2 \\ 0 & pq^2 & q^3 \end{bmatrix}$$

is given by Li (1955, page 17) as the array of frequencies of different mother-child combinations relative to one pair of genes in a random mating population. If the determinant of this matrix were to be evaluated p^2 could be factorized out from the first row, pq from the second and q^2 from the third, to give

$$p^3q^3\begin{vmatrix} p & q & 0 \\ p & 1 & q \\ 0 & p & q \end{vmatrix}.$$

The factor p can now be removed from the first column and q can be removed from the last column giving

$$p^4q^4\begin{vmatrix} 1 & q & 0 \\ 1 & 1 & 1 \\ 0 & p & 1 \end{vmatrix} = p^4q^4(1 - p - q).$$

Three useful corollaries can be derived from this property of factorization.

(i) **Corollary 1.** If one row of a determinant is a multiple of another row, the determinant is zero. Factoring out the multiple reduces the determinant to having two rows which are the same. Hence it is zero.

Example.

$$\begin{vmatrix} -3 & 6 & 12 \\ 2 & -4 & -8 \\ 7 & 5 & 9 \end{vmatrix} = -(1.5)\begin{vmatrix} 2 & -4 & -8 \\ 2 & -4 & -8 \\ 7 & 5 & 9 \end{vmatrix} = 0.$$

(ii) **Corollary 2.** If a determinant has a row of zeros the value of the determinant is zero. Zero is here a factor of one row and hence a factor of the determinant, which is therefore zero.

Example.

$$\begin{vmatrix} 0 & 0 \\ 3 & 7 \end{vmatrix} = 0.$$

(iii) **Corollary 3.** If A is an $n \times n$ matrix and λ is a scalar, the determinant of the matrix λA is $\lambda^n|A|$; i.e. $|\lambda A| = \lambda^n|A|$. In this case λ is a factor of each of the n rows of λA and when the λ factor is removed from each row the determinant $|A|$ remains.

Example.

$$\begin{vmatrix} 3 & 0 & 27 \\ -9 & 3 & 0 \\ 15 & 6 & -3 \end{vmatrix} = 3^3\begin{vmatrix} 1 & 0 & 9 \\ -3 & 1 & 0 \\ 5 & 2 & -1 \end{vmatrix} = -2700.$$

d. Adding multiples of a row

Adding to one row of a determinant any multiple of another row does not affect the value of the determinant. For example,

$$|A| = \begin{vmatrix} 1 & 3 & 2 \\ 8 & 17 & 21 \\ 2 & 7 & 1 \end{vmatrix}$$

$$= 1(17 - -147) - 3(8 - 42) + 2(56 - 34) = 16;$$

and adding four times row 1 to row 2 does not affect the value of $|A|$:

$$|A| = \begin{vmatrix} 1 & 3 & 2 \\ 8+4 & 17+12 & 21+8 \\ 2 & 7 & 1 \end{vmatrix} = \begin{vmatrix} 1 & 3 & 2 \\ 12 & 29 & 29 \\ 2 & 7 & 1 \end{vmatrix}$$

$$= 1(29 - 203) - 3(12 - 58) + 2(84 - 58) = -174 + 138 + 52 = 16.$$

The generality of this result can be readily demonstrated by constructing from

$$|B| = \begin{vmatrix} a_1 & b_1 & c_1 \\ a_2 & b_2 & c_2 \\ a_3 & b_3 & c_3 \end{vmatrix}$$

the determinant which is $|B|$ with λ times row 2 added to the first row, namely,

$$\begin{vmatrix} a_1 + \lambda a_2 & b_1 + \lambda b_2 & c_1 + \lambda c_2 \\ a_2 & b_2 & c_2 \\ a_3 & b_3 & c_3 \end{vmatrix}.$$

Expansion by elements of the first row gives

$$(a_1 + \lambda a_2)\begin{vmatrix} b_2 & c_2 \\ b_3 & c_3 \end{vmatrix} - (b_1 + \lambda b_2)\begin{vmatrix} a_2 & c_2 \\ a_3 & c_3 \end{vmatrix} + (c_1 + \lambda c_2)\begin{vmatrix} a_2 & b_2 \\ a_3 & b_3 \end{vmatrix}$$

$$= a_1\begin{vmatrix} b_2 & c_2 \\ b_3 & c_3 \end{vmatrix} - b_1\begin{vmatrix} a_2 & c_2 \\ a_3 & c_3 \end{vmatrix} + c_1\begin{vmatrix} a_2 & b_2 \\ a_3 & b_3 \end{vmatrix}$$

$$+ \lambda\left\{a_2\begin{vmatrix} b_2 & c_2 \\ b_3 & c_3 \end{vmatrix} - b_2\begin{vmatrix} a_2 & c_2 \\ a_3 & c_3 \end{vmatrix} + c_2\begin{vmatrix} a_2 & b_2 \\ a_3 & b_3 \end{vmatrix}\right\}$$

$$= \begin{vmatrix} a_1 & b_1 & c_1 \\ a_2 & b_2 & c_2 \\ a_3 & b_3 & c_3 \end{vmatrix} + \lambda\begin{vmatrix} a_2 & b_2 & c_2 \\ a_2 & b_2 & c_2 \\ a_3 & b_3 & c_3 \end{vmatrix} = |B|,$$

because the determinant multiplying λ is zero since it has two rows which are the same.

This property is used repeatedly in expanding determinants, adding both positive and negative multiples of rows to other rows. For example, in

$$|A| = \begin{vmatrix} 1 & 3 & 2 \\ 8 & 17 & 21 \\ 2 & 7 & 1 \end{vmatrix} \tag{4}$$

adding (-8) times the first row to the second does not affect $|A|$ and gives

$$|A| = \begin{vmatrix} 1 & 3 & 2 \\ 0 & -7 & 5 \\ 2 & 7 & 1 \end{vmatrix}. \tag{5}$$

If we now add (-2) times the first row to the third row we get

$$|A| = \begin{vmatrix} 1 & 3 & 2 \\ 0 & -7 & 5 \\ 0 & 1 & -3 \end{vmatrix}. \tag{6}$$

Expansion by elements of the first column is now straightforward because two of its elements are zero:

$$|A| = 1\begin{vmatrix} -7 & 5 \\ 1 & -3 \end{vmatrix} = 16,$$

as before. Thus, by repeated use of the property of adding a multiple of one row to another we reduce one column to having all elements except one equal to zero; expansion by elements of that column then involves only one minor, which is of course a determinant of order one less than the original one. This in turn can be reduced also, and so on until the original determinant is reduced to a single 2×2 determinant. At each stage the process of adding multiples of one row to the other rows can be combined into one operation; e.g. in the previous example the steps represented by (5) and (6) can be made simultaneously and the form (6) written directly from (4).

The process can be started with any row; for example, consider basing the reduction of the foregoing example on its third row. If -21 times row 3 is added to row 2 and -2 times row 3 is added to row 1 we get

$$|A| = \begin{vmatrix} -3 & -11 & 0 \\ -34 & -130 & 0 \\ 2 & 7 & 1 \end{vmatrix}$$

so that expansion by elements of the third column gives

$$|A| = \begin{vmatrix} -3 & -11 \\ -34 & -130 \end{vmatrix} = 390 - 374 = 16$$

as before. The choice of which row to base the reduction on in any particular instance is dependent on observing which one might lead to the simplest arithmetic. For example, in expanding

$$|D| = \begin{vmatrix} 1 & 13 & 2 \\ 7 & 23 & 11 \\ -9 & 6 & 5 \end{vmatrix}$$

the arithmetic will be easier if multiples of the first row are added to the other rows than if multiples of either the second or third rows are added to the others.

The foregoing properties can be applied in endless variation in expanding determinants. Efficiency in perceiving a procedure that leads to a minimal amount of effort in any particular case is largely a matter of practice, and beyond describing the possible steps available, as has been done, there is little more that can be said. The underlying method might be summarized as follows. Through adding multiples of one row to other rows of the determinant reduce a column to having only one non-zero element. Expansion by elements of that column then involves only one non-zero minor, which is a determinant of order one less than the original determinant. Successive applications of this method reduce the determinant to one of order 2×2 whose expansion is "obvious". If at any stage these reductions lead to the elements of a row containing a common factor, this can be factorized out as a factor of the determinant, and if they lead to a row of zeros or to two rows being identical, the determinant is zero.

As a final example consider

$$|F| = \begin{vmatrix} 1 & 1 & 1 \\ a & b & c \\ a^2 & b^2 & c^2 \end{vmatrix}, \quad \text{which equals} \quad \begin{vmatrix} 1 & 0 & 0 \\ a & b - a & c - a \\ a^2 & b^2 - a^2 & c^2 - a^2 \end{vmatrix}$$

after subtracting the first column from each of the others. Expanding by elements of the first row and factoring out $(b - a)$ and $(c - a)$ from the resultant minor gives

$$|F| = (b - a)(c - a)\begin{vmatrix} 1 & 1 \\ b + a & c + a \end{vmatrix} = (b - a)(c - a)(c - b).$$

This particular form of determinant is called an *alternant*. It is referred to again in Chapter 7.

e. Adding a row to a multiple of a row

We have seen that adding a multiple of a row to another row does not affect the value of a determinant. Thus for

$$|A| = \begin{vmatrix} a_1 & a_2 & a_3 \\ b_1 & b_2 & b_3 \\ c_1 & c_2 & c_3 \end{vmatrix} \quad \text{and} \quad B = \begin{vmatrix} a_1 + \lambda a_2 & b_1 + \lambda b_2 & c_1 + \lambda c_2 \\ a_2 & b_2 & c_2 \\ a_3 & b_3 & c_3 \end{vmatrix}$$

we have shown that $|B| = |A|$. Notice, however, that adding a row to a multiple of another row is not the same thing and leads to a different result. For example, adding row 2 to λ times row 1 gives the determinant

$$|C| = \begin{vmatrix} \lambda a_1 + a_2 & \lambda b_1 + b_2 & \lambda c_1 + c_2 \\ a_2 & b_2 & c_2 \\ a_3 & b_3 & c_3 \end{vmatrix} = \lambda|A|.$$

This is because $|C|$ is derived from $|A|$ by multiplying the first row of $|A|$ by λ, and adding the second row to it. The first of these two steps changes $|A|$ to $\lambda|A|$, and the second does not alter this value. Hence $|C| = \lambda|A|$. Thus, adding a row to the multiple of another row has the effect of multiplying the determinant by the factor involved.

Example.

$$\begin{vmatrix} -2 & 1 \\ 4 & 3 \end{vmatrix} = \begin{vmatrix} -2 & 1 \\ 0 & 5 \end{vmatrix} = -10,$$

but

$$\begin{vmatrix} 2(-2) + 4 & 2(1) + 3 \\ 4 & 3 \end{vmatrix} = \begin{vmatrix} 0 & 5 \\ 4 & 3 \end{vmatrix} = 2(-10).$$

4. ADDITION AND SUBTRACTION OF DETERMINANTS

In general, the sum of two determinants (or difference between them) cannot be written as a determinant. The simplest demonstration of this is in the case of second-order determinants:

$$|A| + |B| = \begin{vmatrix} a_{11} & a_{12} \\ a_{21} & a_{22} \end{vmatrix} + \begin{vmatrix} b_{11} & b_{12} \\ b_{21} & b_{22} \end{vmatrix}$$

$$= a_{11}a_{22} - a_{12}a_{21} + b_{11}b_{22} - b_{12}b_{21}$$

cannot be written in determinantal form other than as $|A| + |B|$, and certainly not as $|A + B|$. The same applies to the difference $|A| - |B|$.

We may note in passing that both $|A| + |B|$ and $|A| - |B|$ have meaning even when A and B are square matrices of different orders, because the value of a determinant is a scalar. This is in contrast to the matrix expressions $A + B$ and $A - B$ which have meaning only when the matrices are conformable for addition (have the same order).

Another point of interest is that although $|A| + |B|$ does not generally equal $|A + B|$ the latter determinant can be written as the sum of certain other determinants. For example,

$$|A + B| = \left| \begin{bmatrix} a_{11} & a_{12} \\ a_{21} & a_{22} \end{bmatrix} + \begin{bmatrix} b_{11} & b_{12} \\ b_{21} & b_{22} \end{bmatrix} \right|$$

$$= \begin{vmatrix} a_{11} + b_{11} & a_{12} + b_{12} \\ a_{21} + b_{21} & a_{22} + b_{22} \end{vmatrix}$$

$$= \begin{vmatrix} a_{11} & a_{12} \\ a_{21} + b_{21} & a_{22} + b_{22} \end{vmatrix} + \begin{vmatrix} b_{11} & b_{12} \\ a_{21} + b_{21} & a_{22} + b_{22} \end{vmatrix}$$

$$= \begin{vmatrix} a_{11} a_{12} \\ a_{21} a_{22} \end{vmatrix} + \begin{vmatrix} a_{11} a_{12} \\ b_{21} b_{22} \end{vmatrix} + \begin{vmatrix} b_{11} b_{12} \\ a_{21} a_{22} \end{vmatrix} + \begin{vmatrix} b_{11} b_{12} \\ b_{21} b_{22} \end{vmatrix}.$$

In general if A and B are $n \times n$, $|A + B|$ can be expanded as the sum of 2^n n-order determinants.

Example.

The reader should verify that

$$\begin{vmatrix} a + b & a & a \\ a & a + b & a \\ a & a & a + b \end{vmatrix} = (3a + b)b^2,$$

by expanding it as $|A + B|$, where A is the matrix of the a's and $B = bI$.

5. DIAGONAL EXPANSION

Often a matrix can be expressed as the sum of two matrices one of which is a diagonal matrix, that is, as $(A + D)$ where $A = \{a_{ij}\}$ for $i, j = 1, 2, \ldots, n$ and D is a diagonal matrix of order n. (Diagonal matrices are defined in Section 1.6.) The determinant of such a matrix can then be obtained as a polynomial of the elements of D.

First we introduce an abbreviated notation for minors of the determinant of A,

$$|A| = \begin{vmatrix} a_{11} & a_{12} & a_{13} & \cdots & a_{1n} \\ a_{21} & a_{22} & a_{23} & \cdots & a_{2n} \\ a_{31} & a_{32} & a_{33} & \cdots & a_{3n} \\ \cdot & \cdot & \cdot & & \cdot \\ \cdot & \cdot & \cdot & & \cdot \\ \cdot & \cdot & \cdot & & \cdot \\ a_{n1} & a_{n2} & a_{n3} & \cdots & a_{nn} \end{vmatrix}.$$

They will be denoted by just their diagonal elements; for example, $\begin{vmatrix} a_{11} & a_{12} \\ a_{21} & a_{22} \end{vmatrix}$ is written as $|a_{11} \quad a_{22}|$ and in similar fashion $\begin{vmatrix} a_{12} & a_{13} \\ a_{22} & a_{23} \end{vmatrix}$ is written $|a_{12} \quad a_{23}|$. Combined with the notation $A = \{a_{ij}\}$ no confusion can arise. For example, $|a_{21} \quad a_{32}|$ denotes the 2×2 minor having a_{21} and a_{32} as diagonal elements, and from $|A|$ we see that the elements in the same rows and columns as these are a_{22} and a_{31}, so that

$$|a_{21} \quad a_{32}| = \begin{vmatrix} a_{21} & a_{22} \\ a_{31} & a_{32} \end{vmatrix}.$$

Similarly

$$|a_{21} \quad a_{33} \quad a_{44}| = \begin{vmatrix} a_{21} & a_{23} & a_{24} \\ a_{31} & a_{33} & a_{34} \\ a_{41} & a_{43} & a_{44} \end{vmatrix}.$$

We will now consider the determinant $|A + D|$, initially for a 2×2 case, denoting the diagonal elements of D by x_1 and x_2; for example,

$$|A + D| = \begin{vmatrix} a_{11} + x_1 & a_{12} \\ a_{21} & a_{22} + x_2 \end{vmatrix}.$$

By direct expansion

$$|A + D| = (a_{11} + x_1)(a_{22} + x_2) - a_{12}a_{21}.$$

Written as a function of x_1 and x_2 this is

$$x_1x_2 + x_1a_{22} + x_2a_{11} + \begin{vmatrix} a_{11} & a_{12} \\ a_{21} & a_{22} \end{vmatrix}.$$

In similar fashion it can be shown that

$$\begin{vmatrix} a_{11}+x_1 & a_{12} & a_{13} \\ a_{21} & a_{22}+x_2 & a_{23} \\ a_{31} & a_{32} & a_{33}+x_3 \end{vmatrix}$$

$$= x_1x_2x_3 + x_1x_2a_{33} + x_1x_3a_{22} + x_2x_3a_{11}$$

$$+ x_1\begin{vmatrix} a_{22} & a_{23} \\ a_{32} & a_{33} \end{vmatrix} + x_2\begin{vmatrix} a_{11} & a_{13} \\ a_{31} & a_{33} \end{vmatrix} + x_3\begin{vmatrix} a_{11} & a_{12} \\ a_{21} & a_{22} \end{vmatrix} + \begin{vmatrix} a_{11} & a_{12} & a_{13} \\ a_{21} & a_{22} & a_{23} \\ a_{31} & a_{32} & a_{33} \end{vmatrix}$$

which, using the abbreviated notation, can be written as

$$x_1x_2x_3 + x_1x_2a_{33} + x_1x_3a_{22} + x_2x_3a_{11} + x_1|a_{22}a_{33}| + x_2|a_{11}a_{33}| + x_3|a_{11}a_{22}| + |a_{11}a_{22}a_{33}|. \quad (7)$$

Considered as a polynomial in the x's we see that the coefficient of the product of all the x's is unity; the coefficients of the second-degree terms in the x's are the diagonal elements of A; the coefficients of the first-degree terms in the x's are the second-order minors of $|A|$ having their diagonals coincident with the diagonal of $|A|$; and the term independent of the x's is $|A|$ itself. The minors of $|A|$ in these coefficients, namely those whose diagonals are coincident with the diagonal of $|A|$, are called the *principal minors* of $|A|$.

This method of expansion is useful on many occasions because the determinantal form $|A + D|$ occurs quite often, and when $|A|$ is such that many of its principal minors are zero the expansion of $|A + D|$ by this method is greatly simplified.

Example.

If $$|X| = \begin{vmatrix} 7 & 2 & 2 \\ 2 & 8 & 2 \\ 2 & 2 & 9 \end{vmatrix},$$

we have

$$|X| = |A + D| = \left|\begin{bmatrix} 2 & 2 & 2 \\ 2 & 2 & 2 \\ 2 & 2 & 2 \end{bmatrix} + \begin{bmatrix} 5 & 0 & 0 \\ 0 & 6 & 0 \\ 0 & 0 & 7 \end{bmatrix}\right|.$$

Every element of A is a 2 so that $|A|$ and all its 2×2 minors are zero. Consequently $|X|$ evaluated by (7) consists of only the first four terms:

$$|X| = 5(6)7 + 5(6)2 + 5(7)2 + 6(7)2 = 424.$$

Evaluating a determinant in this manner is also useful when all elements of the diagonal matrix D are the same, i.e. when the x_i's are equal. The expansion (7) then becomes

$$x^3 + x^2(a_{11} + a_{22} + a_{33}) + x(|a_{11}a_{22}| + |a_{11}a_{33}| + |a_{22}a_{33}|) + |A|,$$

which is generally written as

$$x^3 + x^2 tr_1(A) + x tr_2(A) + |A|$$

where $tr_1(A)$ is the trace of A (sum of diagonal elements—see Section 1.6) and where $tr_2(A)$ is the sum of the principal minors of order 2 of $|A|$. This method of expansion is known as expansion by *diagonal elements* or simply as *diagonal expansion.*

The general diagonal expansion of a determinant of order n,

$$|A + D| = \begin{vmatrix} a_{11} + x_1 & a_{12} & a_{13} & \cdots & a_{1n} \\ a_{21} & a_{22} + x_2 & a_{23} & \cdots & a_{2n} \\ a_{31} & a_{32} & a_{33} + x_3 & \cdots & a_{3n} \\ \cdot & \cdot & \cdot & \cdot & \cdot \\ \cdot & \cdot & \cdot & \cdot & \cdot \\ \cdot & \cdot & \cdot & \cdot & \cdot \\ a_{n1} & a_{n2} & a_{n3} & \cdots & a_{nn} + x_n \end{vmatrix},$$

consists of the sum of all possible products of the x_i taken r at a time for $r = n, n - 1, \ldots, 2, 1, 0$, each product being multiplied by its complementary principal minor of order $n - r$ in $|A|$. By *complementary principal minor* in $|A|$ is meant the principal minor whose diagonal elements are other than those associated in $|A + D|$ with the x's of the particular product concerned; for example, the complementary principal minor associated with $x_1x_3x_6$ is $|a_{22}a_{44}a_{55}a_{77}a_{88} \ldots a_{nn}|$. When the x's are all equal the expansion becomes $\sum_{i=0}^{n} x^{n-i} tr_i(A)$ where $tr_i(A)$ is the sum of the principal minors of order i of $|A|$ and, by definition, $tr_0(A) = 1$. Note in passing that $tr_n(A) = |A|$.

Example.

$$|A + D| = \begin{vmatrix} a + b & a & a & a \\ a & a + b & a & a \\ a & a & a + b & a \\ a & a & a & a + b \end{vmatrix}$$

By diagonal expansion we get

$$|A + D| = b^4 + b^3 tr_1(A) + b^2 tr_2(A) + b tr_3(A) + |A|$$

where A is the 4×4 matrix whose every element is a. Thus $|A|$ and all minors of order 2 or more are zero. Hence

$$|A + D| = b^4 + b^3(4a) = (4a + b)b^3.$$

*6. THE LAPLACE EXPANSION

In the expansion of

$$|A| = \begin{vmatrix} a_{11} & a_{12} & a_{13} & a_{14} \\ a_{21} & a_{22} & a_{23} & a_{24} \\ a_{31} & a_{32} & a_{33} & a_{34} \\ a_{41} & a_{42} & a_{43} & a_{44} \end{vmatrix}$$

the minor of a_{11} is $|a_{22}a_{33}a_{44}|$. An extension of this, easily verified, is that the coefficient of $|a_{11}\ \ a_{22}|$ is $|a_{33}\ \ a_{44}|$; namely, the coefficient of

$$\begin{vmatrix} a_{11} & a_{12} \\ a_{21} & a_{22} \end{vmatrix} = a_{11}a_{22} - a_{12}a_{21}$$

in $|A|$ is

$$\begin{vmatrix} a_{33} & a_{34} \\ a_{43} & a_{44} \end{vmatrix} = a_{33}a_{44} - a_{34}a_{43}.$$

Likewise the coefficient of $|a_{11}\ \ a_{24}|$ is $|a_{32}\ \ a_{43}|$: the coefficient of

$$|a_{11}\ \ a_{24}| = \begin{vmatrix} a_{11} & a_{14} \\ a_{21} & a_{24} \end{vmatrix} = a_{11}a_{24} - a_{21}a_{14}$$

in the expansion of $|A|$ is

$$|a_{32}\ \ a_{43}| = \begin{vmatrix} a_{32} & a_{33} \\ a_{42} & a_{43} \end{vmatrix} = a_{32}a_{43} - a_{33}a_{42}.$$

Each determinant just described as the coefficient of a particular minor of $|A|$ is the complementary minor in $|A|$ of that particular minor: it is the determinant obtained from $|A|$ by deleting from it all the rows and columns containing the particular minor. This is simply an extension of the procedure for finding the coefficient of an individual element in $|A|$ as derived in the expansion by elements of a row or column discussed earlier. In that case the particular minor is a single element and its coefficient in $|A|$ is $|A|$ amended by deletion of the row and column containing the element concerned. A sign factor is also involved, namely $(-1)^{i+j}$ for the coefficient of a_{ij} in $|A|$. In the extension to coefficients of minors the sign factor is minus one raised to the power of the sum of the subscripts of the diagonal elements of the chosen minor: for example, the sign factor for the coefficient of $|a_{32}\ \ a_{43}|$ is $(-1)^{3+2+4+3} = +1$, as just given. The complementary minor multiplied by this sign factor can be appropriately referred to as the coefficient of the particular minor concerned. Furthermore, just as the expansion of a determinant is the sum of products of elements of a row (or column) with their coefficients, so also is it the sum of products of all minors of order m that can be derived from any set of m rows, each multiplied by its coefficient as just defined. This generalization of the method of expanding a determinant by elements of a row to expanding it by minors of a set of rows was first established by the eighteenth century mathematician Laplace and so bears his name. Aitken (1948) and Ferrar (1941) are two places where proof of the procedure is given; we shall be satisfied here with a general statement of the method and an example illustrating its use.

The Laplace expansion of a determinant $|A|$ of order n can be obtained as follows. (i) Consider any m rows of $|A|$. They contain $n!/[m!(n-m)!]$ minors of order m (where $n!$ is the expression defined in the footnote of Section 3.1). (ii) Multiply each of these minors, M say, by its complementary minor and by a sign factor, where (*a*) the complementary minor of M is the $n - m$ order minor derived from $|A|$ by deleting the m rows and columns containing M, and (*b*) the sign factor is $(-1)^{\mu}$ where μ is the sum of the subscripts of the diagonal elements of M, A being defined as $A = \{a_{ij}\}$, $i, j = 1, 2, \ldots, n$. (iii) The sum of all such products is $|A|$.

Example.

For

$$|A| = \begin{vmatrix} 1 & 2 & 3 & 0 & 0 \\ 1 & 0 & 4 & 2 & 3 \\ 2 & 0 & 1 & 4 & 5 \\ -1 & 2 & -1 & 0 & 0 \\ 0 & 2 & 1 & 2 & 3 \end{vmatrix}$$

interchanging the second and fourth rows gives

$$|A| = -\begin{vmatrix} 1 & 2 & 3 & 0 & 0 \\ -1 & 2 & -1 & 0 & 0 \\ 2 & 0 & 1 & 4 & 5 \\ 1 & 0 & 4 & 2 & 3 \\ 0 & 2 & 1 & 2 & 3 \end{vmatrix}.$$

In this form we will expand $|A|$ using the Laplace expansion based on the first 2 rows, $(m = 2)$. There are ten minors of order 2 in these two rows; seven of them are zero because they involve a column of zeros. Hence $|A|$ can be expanded as the sum of three products involving the three 2×2 non-zero minors in the first two rows, namely as

$$-|A| = (-1)^{1+1+2+2}\begin{vmatrix} 1 & 2 \\ -1 & 2 \end{vmatrix}\begin{vmatrix} 1 & 4 & 5 \\ 4 & 2 & 3 \\ 1 & 2 & 3 \end{vmatrix} + (-1)^{1+1+2+3}\begin{vmatrix} 1 & 3 \\ -1 & -1 \end{vmatrix}\begin{vmatrix} 0 & 4 & 5 \\ 0 & 2 & 3 \\ 2 & 2 & 3 \end{vmatrix}$$
$$+ (-1)^{1+2+2+3}\begin{vmatrix} 2 & 3 \\ 2 & -1 \end{vmatrix}\begin{vmatrix} 2 & 4 & 5 \\ 1 & 2 & 3 \\ 0 & 2 & 3 \end{vmatrix}.$$

The sign factors in these terms have been derived by envisaging A as $\{a_{ij}\}$. Consequently the first 2×2 minor, $\begin{vmatrix} 1 & 2 \\ -1 & 2 \end{vmatrix}$, is $\begin{vmatrix} a_{11} & a_{12} \\ a_{21} & a_{22} \end{vmatrix}$, leading to $(-1)^{1+1+2+2}$ as its sign factor; likewise for the other terms. Simplification of the whole expression gives

$$-|A| = 4\begin{vmatrix} 1 & 4 & 5 \\ 3 & 0 & 0 \\ 1 & 2 & 3 \end{vmatrix} - 2(2)\begin{vmatrix} 4 & 5 \\ 2 & 3 \end{vmatrix} + (-8)\begin{vmatrix} 2 & 4 & 5 \\ 1 & 0 & 0 \\ 0 & 2 & 3 \end{vmatrix} = -24 - 8 + 16$$

and hence $|A| = 16$. It will be found that expansion by a more direct method leads to the same result.

7. MULTIPLICATION OF DETERMINANTS

Two applications of the Laplace expansion lead to the multiplicative property of determinants. Suppose we have two square matrices A and B that are both of order n. Then, by the use of the Laplace expansion, it can be shown that

$$\begin{vmatrix} A & O \\ X & B \end{vmatrix} = |A||B| \tag{8}$$

where X is any square matrix of order n, and if

$$D = \begin{bmatrix} O & A \\ -I & B \end{bmatrix}$$

then

$$|D| = \begin{vmatrix} O & A \\ -I & B \end{vmatrix} = |A|. \tag{9}$$

***Proofs.**

Equation (8) follows immediately from using the Laplace expansion on the first n rows of the left-hand side. The only non-zero minor of order n among them is $|A|$ and its coefficient in the remaining rows is $|B|$, so giving the value $|A||B|$. The reason for $|D|$ always being $+|A|$ in (9) and never $-|A|$ is as follows. The Laplace expansion by the first n rows gives $|D|$ as $|A||-I|(-1)^s$ where s is the sum of the subscripts of the diagonal elements of A and D is written as $D = \{d_{ij}\}\quad i,j = 1, 2, \ldots, 2n$. In this form the diagonal elements of A are $d_{1,1+n}, d_{2,2+n}, \ldots, d_{n,n+n}$, so that $s = \sum_{i=1}^{n} (i + i + n) = n(2n + 1)$. Hence $|D| = |A||-I|(-1)^{n(2n+1)}$. Furthermore, because A and B are both of order n, so is $(-I)$, and therefore $|-I| = (-1)^n$. Hence

$$|D| = |A|(-1)^{n(2n+2)} = |A|.$$

Now comes the important result that the determinant of the product of two square matrices of the same order is the product of the determinants of the individual matrices, that is,

$$|AB| = |A||B|. \tag{10}$$

***Proof.**

First, this result can be true only when A and B are square and of the same order, for their determinants exist only if they are square and their product exists only if they are conformable for multiplication. Now consider the product of two partitioned matrices involving such matrices A and B:

$$\begin{bmatrix} I & A \\ 0 & I \end{bmatrix}\begin{bmatrix} A & 0 \\ -I & B \end{bmatrix} = \begin{bmatrix} 0 & AB \\ -I & B \end{bmatrix}. \tag{11}$$

Recalling the procedure of matrix multiplication (Section 2.5) we observe that on the left-hand side of (11) the effect of the pre-multiplication by $\begin{bmatrix} I & A \\ 0 & I \end{bmatrix}$ on the matrix $\begin{bmatrix} A & 0 \\ -I & B \end{bmatrix}$ is to add multiples of the rows of $[-I \quad B]$ to the rows of $[A \quad 0]$. This has no effect on the determinant of $\begin{bmatrix} A & 0 \\ -I & B \end{bmatrix}$. Therefore the determinant of the left-hand side of (11) is $\begin{vmatrix} A & 0 \\ -I & B \end{vmatrix}$, and by (8) this equals $|A||B|$. But by (9) the determinant of the right-hand side of (11) is $|AB|$. Hence (10) is true.

The matrices in (10) are square and of the same order. Therefore both AB and BA exist. That the determinants of these products are equal can be shown in many ways. One approach is to use (10) in conjunction with the commutative law of multiplication in general (scalar) algebra. Thus

$$|AB| = |A||B| = |B||A| = |BA|.$$

8. CONCLUSION

Some of the more useful techniques for expanding determinants have been discussed. Others could be considered, but it is unlikely that they will be needed by a biologist and they are therefore omitted. Several of these other methods are based on extending the Laplace expansion, using it recurrently to expand a determinant not only by minors and their complementary minors but also to expand these minors themselves. Many of these expansions are identified by the names of their originators, for example Cauchy, Binet-Cauchy, and Jacoby. A good account of some of them is to be found in Aitken (1948) and Ferrar (1941).

9. EXERCISES

1. Show that both

(*a*) $\begin{vmatrix} 1 & 5 & -5 \\ 3 & 2 & -5 \\ 6 & -2 & -5 \end{vmatrix}$ and $\begin{vmatrix} -3 & 2 & -6 \\ -3 & 5 & -7 \\ -2 & 3 & -4 \end{vmatrix}$ equal -5;

(*b*) $\begin{vmatrix} 2 & 6 & 5 \\ -2 & 7 & -5 \\ 2 & -7 & 9 \end{vmatrix}$ and $\begin{vmatrix} 2 & -1 & 9 \\ -1 & 7 & 2 \\ 3 & -21 & 2 \end{vmatrix}$ equal 104;

(*c*) $\begin{vmatrix} 1 & 6 & 4 & 3 \\ 2 & 8 & 5 & 4 \\ 3 & 8 & 7 & 5 \\ 4 & 9 & 7 & 7 \end{vmatrix}$ and $\begin{vmatrix} 1 & -1 & -1 & -1 \\ -1 & 1 & -1 & -1 \\ -1 & -1 & 1 & -1 \\ -1 & -1 & -1 & 1 \end{vmatrix}$ equal -16;

(*d*) $\begin{vmatrix} 21 & 6 & 3 & 9 \\ 12 & 16 & 36 & 4 \\ 13 & 10 & 19 & 5 \\ 1 & 93 & 81 & 6 \end{vmatrix}$ and $\begin{vmatrix} 4 & 6 & 8 & 1 & 2 \\ -1 & -7 & 2 & 3 & 1 \\ 2 & -8 & 12 & 7 & 4 \\ 7 & 9 & 17 & 27 & -5 \\ 8 & 3 & 6 & 2 & 37 \end{vmatrix}$ are zero.

2. Show that $\begin{vmatrix} 1 & 0 & 0 \\ 2 & 3 & 0 \\ 4 & 5 & 6 \end{vmatrix} = 18$ but that $\begin{vmatrix} 0 & 0 & 1 \\ 0 & 3 & 2 \\ 6 & 5 & 4 \end{vmatrix} = -18,$

whereas
$$\begin{vmatrix} 1 & 0 & 0 & 0 \\ 2 & 3 & 0 & 0 \\ 4 & 5 & 6 & 0 \\ 7 & 8 & 9 & 10 \end{vmatrix} = \begin{vmatrix} 0 & 0 & 0 & 1 \\ 0 & 0 & 3 & 2 \\ 0 & 6 & 5 & 4 \\ 10 & 9 & 8 & 7 \end{vmatrix} = 180.$$

3. Expand
$$\begin{vmatrix} 1 & 1 & 1 & 1 \\ 2 & 3 & -1 & 5 \\ 4 & 9 & 1 & 25 \\ 8 & 27 & -1 & 125 \end{vmatrix} \text{ as } (-1)(3)(-3)(4)(-2)(-6) = 432.$$

4. Expand:

(*a*)
$$\begin{vmatrix} 6-\lambda & 3 & 1 \\ 2 & 4-\lambda & 2 \\ 1 & 5 & 7-\lambda \end{vmatrix} \text{ as } -\lambda^3 + 17\lambda^2 - 77\lambda + 78;$$

(*b*)
$$\begin{vmatrix} 2-\lambda & -2 & 3 \\ 10 & -4-\lambda & 5 \\ 5 & -4 & 6-\lambda \end{vmatrix} \text{ as } -(\lambda - 1)^2(\lambda - 2);$$

(*c*)
$$\begin{vmatrix} -1-\lambda & -2 & 1 \\ -2 & 2-\lambda & -2 \\ -1 & -2 & -1-\lambda \end{vmatrix} \text{ as } -\lambda^3 + 10\lambda + 12.$$

5. Without expanding the determinants explain why
$$\begin{vmatrix} 1 & 1 & 1 \\ x & y & z \\ yz & xz & xy \end{vmatrix} = \begin{vmatrix} 1 & 1 & 1 \\ x & y & z \\ x^2 & y^2 & z^2 \end{vmatrix}.$$

6. Show that
$$\begin{vmatrix} a+b+c & a+b & a & a \\ a+b & a+b+c & a & a \\ a & a & a+b+c & a+b \\ a & a & a+b & a+b+c \end{vmatrix}$$
$$= c^2(4a + 2b + c)(2b + c).$$

7. Li (1955, page 97) shows that in a random mating population the zygotic proportions of genotypes determined by only two pairs of genes can be expressed as
$$2\begin{bmatrix} g_{11}^2 & 2g_{11}g_{13} & g_{13}^2 \\ g_{11}g_{31} & g_{11}g_{33} + g_{13}g_{31} & g_{13}g_{33} \\ g_{31}^2 & 2g_{31}g_{33} & g_{33}^2 \end{bmatrix}$$
where g_{11}, g_{13}, g_{31} and g_{33} are the initial gametic outputs. He states that the determinant of this matrix is
$$8\begin{vmatrix} g_{11} & g_{13} \\ g_{31} & g_{33} \end{vmatrix}^3.$$
Prove this.

8. Explain why the determinant of a diagonal matrix is the product of its diagonal elements. Is the same true for a triangular matrix? Give examples.

9. Without expanding the determinants, suggest values of x that satisfy the following equations.

(*a*) $\begin{vmatrix} x & x & x \\ 2 & -1 & 0 \\ 7 & 4 & 5 \end{vmatrix} = 0$ (*b*) $\begin{vmatrix} 1 & x & x^2 \\ 1 & 2 & 4 \\ 1 & -1 & 1 \end{vmatrix} = 0$

(*c*) $\begin{vmatrix} 4 & x & x \\ x & 4 & x \\ x & x & 4 \end{vmatrix} = 0$ (*d*) $\begin{vmatrix} x & 4 & 4 \\ 4 & x & 4 \\ 4 & 4 & x \end{vmatrix} = 0$

10. Explain why

$$\left| \begin{bmatrix} 1 & 2 & 3 \\ 4 & 5 & 6 \\ 5 & 7 & 9 \end{bmatrix} + \begin{bmatrix} x & x & x \\ x & x & x \\ x & x & x \end{bmatrix} \right|$$

$$= \begin{vmatrix} 1 & 2 & 3 \\ 4 & 5 & 6 \\ x & x & x \end{vmatrix} + \begin{vmatrix} 1 & 2 & 3 \\ x & x & x \\ 5 & 7 & 9 \end{vmatrix} + \begin{vmatrix} x & x & x \\ 4 & 5 & 6 \\ 5 & 7 & 9 \end{vmatrix}$$

$$= 0.$$

REFERENCES

Aitken, A. C. (1948). *Determinants and Matrices.* Fifth Edition; Oliver and Boyd, Edinburgh.

Ferrar, W. L. (1941). *Algebra, a Text-Book of Determinants, Matrices and Algebraic Forms.* First Edition; Oxford University Press.

Li, C. C. (1955). *Population Genetics.* The University of Chicago Press, Chicago.

CHAPTER 4

THE INVERSE OF A MATRIX

The arithmetic operations of addition, subtraction and multiplication in matrix algebra have already been dealt with, but division has not. As shown in Chapter 2 matrix multiplication is somewhat more involved than is multiplication in regular arithmetic, and consequently the definition of its complementary operation, division, is not immediately "obvious". This chapter is therefore devoted to discussing it.

1. INTRODUCTION

Division in its usual sense does not exist in matrix algebra, and the concept of "dividing" by a matrix A is replaced by the concept of multiplying by a matrix called the inverse of A. This is a matrix whose product with A is the identity matrix. It is usually denoted by the symbol A^{-1}, often read as "A to the (power of) minus one", and referred to as the inverse of A. We begin discussing its derivation by considering the problem of solving equations.

Illustration. Suppose an entomologist counted 186 dead larvae in a laboratory experiment on the efficacy of a pesticide which had begun with 300 active larvae. He very easily calculates the fraction dead as

$$186/300 = 0.62.$$

In doing so he is effectively solving a simple equation $300x = 186$, where x is a scalar. One way of doing this is to multiply both sides of the equation by $1/300$, which gives $(1/300)300x = (1/300)186$, i.e. $x = 186/300 = 0.62$. And in general, for scalars a, b and x where a and b are known values and a is not zero, the equation

$$ax = b$$

can be solved by multiplying both sides of it by $1/a$ to give $(1/a)ax = (1/a)b$ or

$$x = (1/a)b = a^{-1}b. \tag{1}$$

A more frequent problem that arises is that of solving a set of simultaneous linear equations.

Illustration. In Section 2.6 we had the equations (all symbols again being scalars)

$$\begin{aligned} m + a &= GG \\ m + d &= Gg \\ m - a &= gg, \end{aligned} \tag{2}$$

where GG, Gg and gg represent the three genotypes for a single locus at which only two possible alleles can occur. The symbols, m, a and d represent gene effects, namely a general mean and measures of gene substitution and dominance respectively. Falconer (1960, page 113) uses equations such as these in referring to experimental work done with mice in which the effect on body weight of a dwarfing gene known as "pygmy" was to be estimated. The mean three-week weights of mice having the three different genotypes GG, Gg and gg were 14, 12 and 6 grams respectively, and using these values in equations (2) gives

$$\begin{aligned} m + a &= 14 \\ m + d &= 12 \\ m - a &= 6. \end{aligned} \tag{3}$$

Adding the first and last of these gives $2m = 20$ or $m = 10$, so that from the first equation $a = 14 - 10 = 4$ and from the second $d = 12 - 10 = 2$. Neither this particular method of solving the equations nor any like it can be put in a form similar to (1). With the aid of matrices, however, the solution *can* be put in this form; x would then be a column vector of solutions, a^{-1} would be what is known as the inverse of a matrix, for which the notation A^{-1} would be used, and b would be a column vector. The solution would read $x = A^{-1}b$, and this form would apply whether we were solving two simultaneous linear equations in two unknowns or a hundred equations in a hundred unknowns. No matter how many such equations there were the symbolic form of their solution would always be the same. The advantages so bestowed by matrix algebra are clearly manifold.

Observe that equations (3) can be written as a matrix equation

$$\begin{bmatrix} 1 & 1 & 0 \\ 1 & 0 & 1 \\ 1 & -1 & 0 \end{bmatrix} \begin{bmatrix} m \\ a \\ d \end{bmatrix} = \begin{bmatrix} 14 \\ 12 \\ 6 \end{bmatrix}$$

embodying the principles of multiplication and equality given in Chapter 2. Let us multiply both sides of this equation by the matrix

$$\begin{bmatrix} \frac{1}{2} & 0 & \frac{1}{2} \\ \frac{1}{2} & 0 & -\frac{1}{2} \\ -\frac{1}{2} & 1 & -\frac{1}{2} \end{bmatrix}.$$

Since such an operation involves matrix multiplication we must specify whether it is to be pre-multiplication or post-multiplication; we shall use pre-multiplication. Thus

$$\begin{bmatrix} \frac{1}{2} & 0 & \frac{1}{2} \\ \frac{1}{2} & 0 & -\frac{1}{2} \\ -\frac{1}{2} & 1 & -\frac{1}{2} \end{bmatrix}\begin{bmatrix} 1 & 1 & 0 \\ 1 & 0 & 1 \\ 1 & -1 & 0 \end{bmatrix}\begin{bmatrix} m \\ a \\ d \end{bmatrix} = \begin{bmatrix} \frac{1}{2} & 0 & \frac{1}{2} \\ \frac{1}{2} & 0 & -\frac{1}{2} \\ -\frac{1}{2} & 1 & -\frac{1}{2} \end{bmatrix}\begin{bmatrix} 14 \\ 12 \\ 6 \end{bmatrix},$$

and after obtaining the two products we get

$$\begin{bmatrix} 1 & 0 & 0 \\ 0 & 1 & 0 \\ 0 & 0 & 1 \end{bmatrix}\begin{bmatrix} m \\ a \\ d \end{bmatrix} = \begin{bmatrix} 10 \\ 4 \\ 2 \end{bmatrix}.$$

This is equivalent to $I_3 \begin{bmatrix} m \\ a \\ d \end{bmatrix} = \begin{bmatrix} 10 \\ 4 \\ 2 \end{bmatrix}$, or simply $\begin{bmatrix} m \\ a \\ d \end{bmatrix} = \begin{bmatrix} 10 \\ 4 \\ 2 \end{bmatrix}$

which is the solution $m = 10$, $a = 4$ and $d = 2$ obtained previously. Hence we have found that the equations

$$\begin{bmatrix} 1 & 1 & 0 \\ 1 & 0 & 1 \\ 1 & -1 & 0 \end{bmatrix}\begin{bmatrix} m \\ a \\ d \end{bmatrix} = \begin{bmatrix} 14 \\ 12 \\ 6 \end{bmatrix}$$

have the solution

$$\begin{bmatrix} m \\ a \\ d \end{bmatrix} = \begin{bmatrix} \frac{1}{2} & 0 & \frac{1}{2} \\ \frac{1}{2} & 0 & -\frac{1}{2} \\ -\frac{1}{2} & 1 & -\frac{1}{2} \end{bmatrix}\begin{bmatrix} 14 \\ 12 \\ 6 \end{bmatrix} = \begin{bmatrix} 10 \\ 4 \\ 2 \end{bmatrix}$$

where the 3×3 matrix introduced on the right-hand side is so related to the 3×3 matrix of the original equations that the product of the two matrices is an identity matrix; i.e.

$$\begin{bmatrix} \frac{1}{2} & 0 & \frac{1}{2} \\ \frac{1}{2} & 0 & -\frac{1}{2} \\ -\frac{1}{2} & 1 & -\frac{1}{2} \end{bmatrix}\begin{bmatrix} 1 & 1 & 0 \\ 1 & 0 & 1 \\ 1 & -1 & 0 \end{bmatrix} = \begin{bmatrix} 1 & 0 & 0 \\ 0 & 1 & 0 \\ 0 & 0 & 1 \end{bmatrix}.$$

Let us now write the previous matrices and vectors in symbols:

$$A = \begin{bmatrix} 1 & 1 & 0 \\ 1 & 0 & 1 \\ 1 & -1 & 0 \end{bmatrix}, \quad x = \begin{bmatrix} m \\ a \\ d \end{bmatrix}, \quad b = \begin{bmatrix} 14 \\ 12 \\ 10 \end{bmatrix}$$

and

$$A^{-1} = \begin{bmatrix} \frac{1}{2} & 0 & \frac{1}{2} \\ \frac{1}{2} & 0 & -\frac{1}{2} \\ -\frac{1}{2} & 1 & -\frac{1}{2} \end{bmatrix}.$$

We have then solved the equation

$$Ax = b \tag{4}$$

as

$$x = A^{-1}b \tag{5}$$

where

$$A^{-1}A = I. \tag{6}$$

Since I is the "one" of matrix algebra the matrix A^{-1} is called the *reciprocal of A*, or *inverse of A*. We define it for the moment by equation (6), namely as a matrix which, when post-multiplied by A gives the identity matrix I. With A^{-1} so defined the solution for x is given by (5). Notice this solution has not been derived from equation (4) by diving both sides of (4) by A, but by pre-multiplying both sides by the matrix A^{-1} so that

$$A^{-1}Ax = A^{-1}b;$$

and by using the definition $A^{-1}A = I$ this becomes $Ix = A^{-1}b$ and hence $x = A^{-1}b$. At no stage has there been division by A. The solution is obtained through pre-multiplication by a matrix A^{-1} defined such that $A^{-1}A = I$. As stated at the outset, division in any ordinary sense is undefined in matrix algebra and has no meaning. Instead, we deal with inverse matrices and use them as multipliers, the product of a matrix and its inverse being an identity matrix.

Before considering the wider implications of the definition of A^{-1} as so far derived we give another example of its use in solving linear equations.

Example. It is suggested that the reader verify that when the equations

$$\begin{aligned} 2u - 2v - w &= 5 \\ u + v - 2w &= 1 \\ u \phantom{{}+2v} - w &= 4 \end{aligned}$$

in scalars u, v and w are written in matrix form, the solution obtained by pre-multiplying them by

$$\begin{bmatrix} -1 & -2 & 5 \\ -1 & -1 & 3 \\ -1 & -2 & 4 \end{bmatrix} \quad \text{is} \quad \begin{bmatrix} u \\ v \\ w \end{bmatrix} = \begin{bmatrix} 13 \\ 6 \\ 9 \end{bmatrix}.$$

If the matrix of coefficients of u, v and w in the equations is denoted by A, the matrix used as the pre-multiplier is the inverse A^{-1}:

$$A = \begin{bmatrix} 2 & -2 & -1 \\ 1 & 1 & -2 \\ 1 & 0 & -1 \end{bmatrix} \quad \text{and} \quad A^{-1} = \begin{bmatrix} -1 & -2 & 5 \\ -1 & -1 & 3 \\ -1 & -2 & 4 \end{bmatrix}.$$

It will be found that $A^{-1}A = I$ and if the equations are written as $Ax = b$ the solution is $x = A^{-1}b$.

The remainder of this chapter is devoted first to narrowing the definition of A^{-1} as given by the equation $A^{-1}A = I$, and then to specifying how A^{-1} is obtained from A. When this is known, any set of simultaneous linear equations that have a unique solution can be put in the form $Ax = b$ and solved as $x = A^{-1}b$, no matter how many such equations there are. Computational difficulties may sometimes arise, but the form of the solution and the procedure for obtaining it remain the same.

2. PRODUCTS EQUAL TO I

The concept of a matrix inverse has been established by considering the solution of a set of simultaneous linear equations. This is a situation in which the inverse of a matrix is widely used and therefore seems highly suitable for introducing the idea of a matrix inverse. It is, of course, not essential, for from the wider algebraic viewpoint all that is involved is the question "is there a matrix R whose product with the matrix A is I?" There are three answers to this question, depending on the characteristics of A: (i) in some cases R exists and is unique for a given A; (ii) sometimes numerous matrices R exist for a particular A, i.e. R exists but is not unique; and (iii) in some instances R does not exist at all. Examples of these three situations are now given.

i. If

$$A_1 = \begin{bmatrix} 2 & 5 \\ 3 & 8 \end{bmatrix} \quad \text{and} \quad R_1 = \begin{bmatrix} 8 & -5 \\ -3 & 2 \end{bmatrix}$$

$R_1A_1 = I$, and R_1 is the unique matrix for this particular A such that $RA = I$.

ii. For

$$A_2 = \begin{bmatrix} 1 & 1 \\ -1 & 0 \\ 3 & -1 \end{bmatrix}$$

there is an endless number of matrices R such that $RA_2 = I$, for example,

$$R_2 = \begin{bmatrix} 1 & 3 & 1 \\ 2 & 5 & 1 \end{bmatrix} \quad \text{and} \quad R_2^* = \begin{bmatrix} 4 & 15 & 4 \\ 7 & 25 & 6 \end{bmatrix}.$$

iii. When

$$A_3 = \begin{bmatrix} 0 & 3 & 7 \\ 0 & 2 & 5 \end{bmatrix}$$

there is no matrix R such that $RA_3 = I$, for the leading element of RA_3 will always be zero and so the product matrix can never equal I.

One might notice a further characteristic of these three situations:

i. $A_1R_1 = I$ as well as $R_1A_1 = I$;
ii. $A_2R_2 \neq I$ even though $R_2A_2 = I$;
iii. For $S = \begin{bmatrix} 4 & 8 \\ 5 & -7 \\ -2 & 3 \end{bmatrix}$, $A_3S = I$ even though there is no matrix R such that $RA_3 = I$.

The A^{-1} established in considering examples of solving linear equations was a matrix such that $A^{-1}A = I$. But we have just illustrated how $RA = I$ does not necessarily yield a matrix R that is unique for a given A, nor one for which AR necessarily equals I even if RA does. Only in the case of A_1 did R have both these properties; and they seem reasonable properties to ask of an inverse—that it be unique for a given A, and that both pre- and post-multiplication by A lead to the identity matrix. Analogous to the familiar reciprocal of an ordinary number in scalar algebra, these properties seem appropriate for the inverse (reciprocal) of a matrix. Accordingly we will derive a matrix from A which has these properties. The derived matrix will be the inverse of A and instead of satisfying only $A^{-1}A = I$ it will have the two properties (*a*) $A^{-1}A = AA^{-1} = I$ and (*b*) A^{-1} unique for given A.

The first of these implies that further discussion of the inverse matrix A^{-1} must be confined to instances in which the two products $A^{-1}A$ and AA^{-1} both exist. As explained in Section 2.5*d*, this occurs only when A and A^{-1} are both square matrices of the same order. This means that an A^{-1} such that (*a*) is true exists only when A is square. What we are about to describe as the inverse of a matrix can therefore exist only if that matrix is square (and its inverse will also be square, of the same order). In contrast, rectangular matrices do not have inverses, although some of them do have a restricted kind of inverse (see Section 4.7).

Having established that the inverse A^{-1} can exist only when A is square, it is also clear that it can exist only when A has a determinant. In consequence of this the inverse is now derived in a manner that utilizes the determinant. This form of derivation has been chosen because it is particularly suitable for describing how the elements of the inverse of a matrix can be obtained directly from those of the matrix itself, although it is tedious in operation in practically all instances except those of small-sized matrices. It is therefore seldom used for calculating the inverse of a matrix in practical situations. Nevertheless, it is a method that presents with clarity the underlying form of the elements of an inverse, and furthermore, it readily leads to demonstrating properties (*a*) and (*b*) just discussed. These and others are dealt with following the derivation. Brief discussion is also given to some arithmetic problems in calculating inverses.

3. THE COFACTORS OF A DETERMINANT

We saw in equation (1) of Chapter 3 how the determinant of a matrix $A = \{a_{ij}\}$ for $i, j = 1, 2, \ldots, n$ could be expanded as

$$|A| = \sum_{j=1}^{n} a_{ij}(-1)^{i+j}|M_{ij}|, \qquad \text{for any } i,$$

where $|M_{ij}|$ is the minor of a_{ij} and is the determinant obtained from $|A|$ by crossing out the ith row and jth column. The product $(-1)^{i+j}|M_{ij}|$ is known as the *cofactor* of a_{ij} in $|A|$. It is simply the minor of a_{ij} with sign attached. We might say that the cofactor is the *signed minor*. If we write μ_{ij} for the coefficient of a_{ij} in $|A|$ we have

$$\mu_{ij} = (-1)^{i+j}|M_{ij}|$$

and hence

$$|A| = \sum_{j=1}^{n} a_{ij}\mu_{ij} = \sum_{i=1}^{n} a_{ij}\mu_{ij}, \tag{7}$$

and the determinant $|A|$ is the sum of products of the elements of a row (or column) with their cofactors.

A property of cofactors important for developing the inverse of a matrix is that the sum of products of the elements of one row (or column) with the cofactors of the elements of another row (or column) is zero:

$$\sum_{j=1}^{n} a_{ij}\mu_{hj} = 0, \quad \text{for} \quad i \neq h \qquad \text{and} \qquad \sum_{i=1}^{n} a_{ij}\mu_{ik} = 0, \quad \text{for} \quad j \neq k. \tag{8}$$

* The truth of this result is made apparent by observing that $\sum_{j=1}^{n} a_{ij}\mu_{hj}$ for $i \neq h$ can be expanded as a determinant having $a_{i1}, a_{i2}, \ldots, a_{in}$ as a row, with the other $n - 1$ rows being those of $|A|$ from which the cofactors $\mu_{h1}, \mu_{h2}, \ldots, \mu_{hn}$ have come. But since these are the cofactors in $|A|$ of the elements $a_{h1}, a_{h2}, \ldots, a_{hn}$, these rows consist of all

rows of A save $(a_{h1}, a_{h2}, \ldots, a_{hn})$. Therefore they contain $(a_{i1}, a_{i2}, \ldots, a_{in})$ as a row; that is, the determinant form of $\sum_{j=1}^{n} a_{ij}\mu_{hj}$ has two rows the same and hence is zero.

Example. In the determinant

$$|A| = \begin{vmatrix} 1 & 2 & 3 \\ 4 & 5 & 6 \\ 7 & 8 & 10 \end{vmatrix}$$

the cofactors of the elements of the first row are as follows:

that of the 1 is $\quad (-1)^{1+1}\begin{vmatrix} 5 & 6 \\ 8 & 10 \end{vmatrix} = 50 - 48 = 2;$

that of the 2 is $\quad (-1)^{1+2}\begin{vmatrix} 4 & 6 \\ 7 & 10 \end{vmatrix} = -40 + 42 = 2;$

that of the 3 is $\quad (-1)^{1+3}\begin{vmatrix} 4 & 5 \\ 7 & 8 \end{vmatrix} = 32 - 35 = -3.$

Multiplying these elements of the first row by their cofactors gives $|A|$:

$$|A| = 1(2) + 2(2) + 3(-3) = -3.$$

But multiplying the elements of another row, the second say, by these cofactors gives zero:

$$4(2) + 5(2) + 6(-3) = 0.$$

The determinantal form of this last expression is

$$4(-1)^{1+1}\begin{vmatrix} 5 & 6 \\ 8 & 10 \end{vmatrix} + 5(-1)^{1+2}\begin{vmatrix} 4 & 6 \\ 7 & 10 \end{vmatrix} + 6(-1)^{1+3}\begin{vmatrix} 4 & 5 \\ 7 & 8 \end{vmatrix} = \begin{vmatrix} 4 & 5 & 6 \\ 4 & 5 & 6 \\ 7 & 8 & 10 \end{vmatrix},$$

which is clearly zero because two rows are the same. The same is true if the cofactors of the first row are multiplied by the elements of the last row:

$$7(2) + 8(2) + 10(-3) = \begin{vmatrix} 7 & 8 & 10 \\ 4 & 5 & 6 \\ 7 & 8 & 10 \end{vmatrix} = 0.$$

As further demonstration of these properties the reader might satisfy himself that the cofactors of the elements of the second row are 4, -11 and 6, and that

$$\begin{aligned} 1(4) + 2(-11) + 3(6) &= 0 \\ 4(4) + 5(-11) + 6(6) &= -3 = |A| \end{aligned}$$

and

$$7(4) + 8(-11) + 10(6) = 0.$$

4. DERIVATION OF THE INVERSE

The inverse of a square matrix is now derived from properties of the cofactors of elements of its determinant. The derivation is in terms of the numerical example and involves the cofactors of all the elements of the matrix—column by column, rather than row by row.

Consider the matrix

$$A = \begin{bmatrix} 1 & 2 & 3 \\ 4 & 5 & 6 \\ 7 & 8 & 10 \end{bmatrix}.$$

First we derive the cofactors of each column in turn and demonstrate the properties just discussed. The cofactors of the elements of the first column are respectively

$$(-1)^{1+1}\begin{vmatrix} 5 & 6 \\ 8 & 10 \end{vmatrix}, \quad (-1)^{2+1}\begin{vmatrix} 2 & 3 \\ 8 & 10 \end{vmatrix} \quad \text{and} \quad (-1)^{3+1}\begin{vmatrix} 2 & 3 \\ 5 & 6 \end{vmatrix},$$
$$= 2, 4 \text{ and } -3. \tag{9}$$

Adding the products of these cofactors with the elements of each column of A gives, in turn,

$$2(1) + 4(4) - 3(7) = -3 = |A|, \tag{10}$$
$$2(2) + 4(5) - 3(8) = 0, \tag{11}$$

and

$$2(3) + 4(6) - 3(10) = 0. \tag{12}$$

Similarly the cofactors of elements of the second column are

$$(-1)\begin{vmatrix} 4 & 6 \\ 7 & 10 \end{vmatrix}, \quad (+1)\begin{vmatrix} 1 & 3 \\ 7 & 10 \end{vmatrix} \quad \text{and} \quad (-1)\begin{vmatrix} 1 & 3 \\ 4 & 6 \end{vmatrix},$$
$$= 2, -11 \text{ and } 6. \tag{13}$$

and the corresponding sums of products with the elements of the three columns are

$$2(1) - 11(4) + 6(7) = 0, \tag{14}$$
$$2(2) - 11(5) + 6(8) = -3 = |A| \tag{15}$$

and

$$2(2) - 11(6) + 6(10) = 0. \tag{16}$$

And finally, the cofactors of the elements of the third column are

$$(+1)\begin{vmatrix} 4 & 5 \\ 7 & 8 \end{vmatrix}, \quad (-1)\begin{vmatrix} 1 & 2 \\ 7 & 8 \end{vmatrix} \quad \text{and} \quad (+1)\begin{vmatrix} 1 & 2 \\ 4 & 5 \end{vmatrix},$$
$$= -3, 6 \text{ and } -3. \tag{17}$$

Sums of products of these with the elements of the columns of A are

$$-3(1) + 6(4) - 3(7) = 0, \qquad (18)$$
$$-3(2) + 6(5) - 3(8) = 0 \qquad (19)$$

and

$$-3(3) + 6(6) - 3(10) = -3 = |A|. \qquad (20)$$

Now consider the matrix obtained by replacing the terms of A by their cofactors—that is, from

$$\begin{bmatrix} 1 & 2 & 3 \\ 4 & 5 & 6 \\ 7 & 8 & 10 \end{bmatrix} \quad \text{obtaining} \quad \begin{bmatrix} 2 & 2 & -3 \\ 4 & -11 & 6 \\ -3 & 6 & -3 \end{bmatrix}.$$

The columns of the new matrix are the values in (9), (13) and (17). Let us transpose this newly formed matrix and multiply it by the scalar

$$1/(-3) = (1/|A|).$$

This gives

$$\frac{1}{-3}\begin{bmatrix} 2 & 4 & -3 \\ 2 & -11 & 6 \\ -3 & 6 & -3 \end{bmatrix}. \qquad (21)$$

Now pre-multiply A by the matrix we have just derived:

$$\frac{1}{-3}\begin{bmatrix} 2 & 4 & -3 \\ 2 & -11 & 6 \\ -3 & 6 & -3 \end{bmatrix}\begin{bmatrix} 1 & 2 & 3 \\ 4 & 5 & 6 \\ 7 & 8 & 10 \end{bmatrix}.$$

The elements in the rows of this product are the sums of products in equations (10), (11), (12); (14), (15), (16); and (18), (19), (20) respectively. Therefore the product is

$$\frac{1}{-3}\begin{bmatrix} -3 & 0 & 0 \\ 0 & -3 & 0 \\ 0 & 0 & -3 \end{bmatrix} = \begin{bmatrix} 1 & 0 & 0 \\ 0 & 1 & 0 \\ 0 & 0 & 1 \end{bmatrix}.$$

In this instance, when A is pre-multiplied by the matrix derived in (21) the product is the identity matrix I. Thus (21) is the kind of matrix just discussed, namely A^{-1} such that $A^{-1}A = I$. So far this is the sole property of A^{-1}. We have not shown that for A^{-1} so obtained AA^{-1} also equals I, nor have we shown that it is the unique matrix for which $A^{-1}A = AA^{-1} = I$. These are properties (*a*) and (*b*) mentioned earlier. We shall discuss them shortly, but first let us reconstruct the derivation of A^{-1} as obtained in (21).

Initially we had A as

$$A = \begin{bmatrix} a_{11} & a_{12} & a_{13} \\ a_{21} & a_{22} & a_{23} \\ a_{31} & a_{32} & a_{33} \end{bmatrix}$$

and then formed a new matrix by replacing each element of A by its cofactor:

$$\begin{bmatrix} \mu_{11} & \mu_{12} & \mu_{13} \\ \mu_{21} & \mu_{22} & \mu_{23} \\ \mu_{31} & \mu_{32} & \mu_{33} \end{bmatrix}.$$

This was transposed, giving

$$\begin{bmatrix} \mu_{11} & \mu_{21} & \mu_{31} \\ \mu_{12} & \mu_{22} & \mu_{32} \\ \mu_{13} & \mu_{23} & \mu_{33} \end{bmatrix} \tag{22}$$

and multiplied by the scalar $1/|A|$ to produce

$$\frac{1}{|A|}\begin{bmatrix} \mu_{11} & \mu_{21} & \mu_{31} \\ \mu_{12} & \mu_{22} & \mu_{32} \\ \mu_{13} & \mu_{23} & \mu_{33} \end{bmatrix} = A^{-1},$$

the inverse of A. Pre-multiplying A by this matrix gives the identity matrix because in the product each diagonal element is a sum of products of elements of a column of A with their cofactors in $|A|$, and therefore equal to $|A|$ by equation (7); the scalar $1/|A|$ reduces them to 1. And the off-diagonal elements of the product matrix are sums of products of elements of columns of A with cofactors in $|A|$ of elements of other columns, and therefore equal to zero by equation (8). Hence the product is I and so $A^{-1}A = I$. (The product AA^{-1} is considered later.)

The matrix (22), namely A with its elements replaced by their cofactors and then transposed, is called the *adjugate*, or sometimes the *adjoint*, matrix of A. Thus the inverse, A^{-1}, can be described as the adjugate of A multiplied by the scalar $1/|A|$.

Example. The determinant of

$$A = \begin{bmatrix} 2 & 5 \\ 3 & 9 \end{bmatrix} \quad \text{is} \quad |A| = \begin{vmatrix} 2 & 5 \\ 3 & 9 \end{vmatrix} = 18 - 15 = 3.$$

The adjugate matrix is $\begin{bmatrix} 9 & -5 \\ -3 & 2 \end{bmatrix}$, and so the inverse is

$$\mathrm{A}^{-1} = \tfrac{1}{3}\begin{bmatrix} 9 & -5 \\ -3 & 2 \end{bmatrix}.$$

5. CONDITIONS FOR EXISTENCE OF INVERSE

The inverse A^{-1} of a matrix A, such that $A^{-1}A = I$, has been developed in the preceding paragraph as the adjugate matrix of A multiplied by the scalar $1/|A|$. The conditions imposed on A in order for this matrix to exist are two.

i. A^{-1} *can* exist only when A is square—as discussed in Section 4.2.
ii. A^{-1} *does* exist only if $|A|$ is non-zero. (If $|A|$ is zero the scalar factor $1/|A|$ in A^{-1} is infinite and A^{-1} does not exist. $|A|$ must therefore be non-zero for A^{-1} to exist.)

A square matrix is said to be *singular* when its determinant is zero and *nonsingular* when its determinant is non-zero. Singularity is therefore a property of square matrices only, not of rectangular matrices, and it is only nonsingular matrices that have inverses. Just as conformability is necessary for the existence of a matrix product, so is nonsingularity necessary for the existence of a matrix inverse. In both instances the necessary condition is sometimes not mentioned explicitly but it must always be satisfied.

6. PROPERTIES OF THE INVERSE

If A is a square, nonsingular matrix its inverse, A^{-1}, has the following properties.

i. The inverse commutes with A, both products being the identity matrix: $A^{-1}A = AA^{-1} = I$.
ii. The inverse of A is unique.
iii. The determinant of the inverse of A is the reciprocal of the determinant of A: $|A^{-1}| = 1/|A|$.
vi. The inverse matrix is nonsingular.
v. The inverse of A^{-1} is A: $(A^{-1})^{-1} = A$.
vi. The inverse of a transpose is the transpose of the inverse:

$$(A')^{-1} = (A^{-1})'.$$

vii. If A is symmetric so is its inverse: if $A' = A$, then $(A^{-1})' = A^{-1}$.
viii. The inverse of a product is the product of the inverses taken in reverse order, provided the inverses exist: if A^{-1} and B^{-1} exist,

$$(AB)^{-1} = B^{-1}A^{-1}.$$

ix. If A is such that its inverse equals its transpose, A is said to be an *orthogonal* matrix and $AA' = I$.

The first two of the foregoing have been referred to earlier as (*a*) and (*b*). Proofs of these properties are now indicated; examples are left to the reader.

i. It has been shown that for A^{-1} as developed $A^{-1}A = I$. This is because the elements of the product matrix $A^{-1}A$ are sums of products of elements of columns of A multiplied by cofactors of elements of the same and different columns of A. In like fashion the product AA^{-1} has elements that are sums of products of elements of rows of A multiplied by cofactors of elements of the same and different rows of A. Hence $AA^{-1} = I$.

ii. Suppose that A^{-1} is not a unique inverse of A and that S is another inverse different from A^{-1}, such that $SA = I$. Then

$$\begin{aligned} S - A^{-1} &= (S - A^{-1})I = (S - A^{-1})AA^{-1} \\ &= SAA^{-1} - A^{-1}AA^{-1} = IA^{-1} - IA^{-1} \end{aligned}$$

because we are assuming $SA = I$ and $A^{-1}A$ does equal I. Therefore $S - A^{-1} = 0$, S is the same as A^{-1} and so A^{-1} is the unique inverse of A.

iii and iv. It was shown in Section 3.7 that if A and B are two square matrices of the same order, $|A|\,|B| = |AB|$. Therefore

$$|A|\,|A^{-1}| = |AA^{-1}| = |I| = 1, \qquad \text{and so} \quad |A^{-1}| = 1/|A|.$$

The nonsingularity of A^{-1} follows from this.

v, vi and vii. Consider the identity $I = A^{-1}A$. Pre-multiplying it by $(A^{-1})^{-1}$ leads to the result $(A^{-1})^{-1} = A$, and transposing it leads to $(A')^{-1} = (A^{-1})'$. For $A' = A$ the latter becomes $A^{-1} = (A^{-1})'$.

viii. Provided A and B are both square, of the same order and nonsingular, it is clear that

$$B^{-1}A^{-1}AB = B^{-1}(A^{-1}A)B = B^{-1}IB = B^{-1}B = I,$$

and so, by post-multiplying by $(AB)^{-1}$, we have $(AB)^{-1} = B^{-1}A^{-1}$. The reversal rule that applies to transposing a product therefore applies also to inverting it.

ix. It is obvious that if $A^{-1} = A'$, $AA' = A'A = I$. The reader might satisfy himself that this is true for $A = \frac{1}{15}\begin{bmatrix} 5 & -14 & 2 \\ -10 & -5 & -10 \\ 10 & 2 & -11 \end{bmatrix}$. As already stated, such a matrix is said to be *orthogonal.*

*7. LEFT AND RIGHT INVERSES

In introducing the idea of an inverse matrix in Section 2 it was pointed out that A^{-1} could exist only for A being square, for this is the only condition under which both of the products $A^{-1}A$ and AA^{-1} exist. Thereafter inverse matrices have been considered only in terms of square matrices. However, pertaining to rectangular matrices the following question can be asked: does there exist, for any rectangular matrix $A_{r\times c}$ $(r \neq c)$, a matrix $B_{c\times r}$ for which $AB = I_r$ and $BA = I_c$? The answer is "No", there is no such matrix B.

Proof. There is no loss of generality in assuming $r < c$ and in writing A and B in partitioned form as

$$A = [X_{r\times r} \quad Y_{r\times(c-r)}] \qquad \text{and} \qquad B = \begin{bmatrix} Z_{r\times r} \\ W_{(r-c)\times r} \end{bmatrix}.$$

Then $BA = \begin{bmatrix} ZX & ZY \\ WX & WY \end{bmatrix}$ and in order for this to equal I_c it is necessary that $ZX = I_r$, which occurs only if X is nonsingular and $Z = X^{-1}$. It is also necessary that $WX = 0$, which implies $W = 0$ because X is nonsingular. As a result $BA = \begin{bmatrix} I & ZY \\ 0 & 0 \end{bmatrix}$ which can never be an identity matrix and therefore no B of the required form exists. Thus for a rectangular matrix $A_{r\times c}$ there is no matrix B such that $AB = I_r$ and $BA = I_c$.

However, for certain rectangular matrices, $A_{r\times c}$ for $r \neq c$, there are matrices $B_{c\times r}$ for which $AB = I_r$ (but $BA \neq I_c$); B in such cases is called the *right inverse* of A. And for other matrices $A_{r\times c}$ there are matrices $D_{c\times r}$ for which $DA = I_c$ (but $AD \neq I_r$); in these cases D is called the *left inverse* of A. As has just been shown, no matrix can be both the left and right inverse of the same rectangular matrix. As is shown in Chapter 5, no rectangular matrix has both a left and a right inverse. It will have neither of them or one or the other, but not both. Only square (nonsingular) matrices have both a left and a right inverse, and they are the same, namely the (unique) inverse that has already been described.

Example.

$$L = \begin{bmatrix} 1 & 3 & 1 \\ 2 & 5 & 1 \end{bmatrix} \text{ is the left inverse of } A = \begin{bmatrix} 1 & 1 \\ -1 & 0 \\ 3 & -1 \end{bmatrix}$$

because $LA = I_2$. Likewise A is the right inverse of L. But there is no right inverse of A (no matrix B for which $AB = I_3$) nor is there a left inverse of L (no matrix D for which $DL = I_3$).

8. SOME USES OF INVERSES

Uses of the inverse matrix are so many and varied that only a selection can be given. Four have been chosen for illustration.

a. Equations

Repeated reference has already been made to the problem of solving equations. A further example comes from ecology.

Illustration. A problem that occurs in studying the dynamics of a living population is that of assessing its age distribution. Suppose for some animal population being studied, deer say, that there are n clearly defined age groupings and the proportion of the population in each age group is $r_1, r_2, \ldots, r_n$. For a population of wild deer these values can never be known exactly because not all deer can be located, nor can their ages be known with exactitude. But the ages of individuals that can be located may be estimated, from which estimates $\hat{r}_1, \hat{r}_2, \ldots, \hat{r}_n$ of the population proportions can be calculated. (The notation $\hat{r}$—read as "r hat"—is commonly used to indicate an estimate of any quantity r whose true value is unknown.) Deer whose exact ages are known might also be available, either in the form of a colony born and reared in captivity, or of deer born in captivity, tagged with their birth date, released and subsequently recaptured. From such animals one can assess the efficiency of personnel whose normal task it is to estimate the age of deer captured from the wild population, achieving this by having such personnel estimate the ages of a number of deer whose true ages are available. In this way it is possible to estimate p_{ij}, the probability that an animal estimated as of age i is truly of age j. Then, for example, the estimated fraction in age 1 in the population, $\hat{r}_1$, is the sum of the true fractions $r_1, r_2, \ldots, r_n$ each multiplied by the probability that such an animal is estimated as of age 1. Thus

$$\hat{r}_1 = p_{11}r_1 + p_{12}r_2 + \ldots + p_{1n}r_n.$$

And for all age groups together

$$\begin{bmatrix} \hat{r}_1 \\ \hat{r}_2 \\ \cdot \\ \cdot \\ \cdot \\ \hat{r}_n \end{bmatrix} = \begin{bmatrix} p_{11} & p_{12} & \cdots & p_{1n} \\ p_{21} & p_{22} & \cdots & p_{2n} \\ \cdot & \cdot & & \cdot \\ \cdot & \cdot & & \cdot \\ \cdot & \cdot & & \cdot \\ p_{n1} & p_{n2} & \cdots & p_{nn} \end{bmatrix} \begin{bmatrix} r_1 \\ r_2 \\ \cdot \\ \cdot \\ \cdot \\ r_n \end{bmatrix},$$

so that if $\hat{r}$ denotes the vector of estimated proportions, r the vector of true proportions and P the matrix of probabilities, $\hat{r} = Pr$. Improved estimates of the proportions can be obtained by solving this equation for r using the inverse of P, namely $r = P^{-1}\hat{r}$.

b. Algebraic simplifications

There are numerous ways in which inverse matrices occur in matrix algebra in some sense analogous to division in scalar algebra. One is as follows. For x being a scalar other than unity the result

$$1 + x + x^2 + \cdots + x^{n-1} = \frac{x^n - 1}{x - 1}$$

has as its matrix counterpart

$$I + X + X^2 + \ldots + X^{n-1} = (X - I)^{-1}(X^n - I)$$

where $(X-I)$ is a nonsingular matrix.

Illustration. Alling (1958) defines six states of pulmonary tuberculosis diagnosis relative to the disease being active or arrested. The matrix of transition probabilities (Section 1.6), namely the probabilities that an individual diagnosed in one state of the disease will have changed to another state at the next diagnosis, he writes as a partitioned matrix (Section 2.11)

$$P = \begin{bmatrix} I & 0 \\ C & B \end{bmatrix},$$

where C and B take particular forms based on description of the six states of the disease. The matrix of probabilities after two diagnoses (made at equal intervals of time) then depends on

$$P^2 = \begin{bmatrix} I & 0 \\ C & B \end{bmatrix}^2 = \begin{bmatrix} I & 0 \\ C + BC & B^2 \end{bmatrix},$$

and that after t diagnoses depends on

$$P^t = \begin{bmatrix} I & 0 \\ C & B \end{bmatrix}^t,$$

which, by use of the result in X just given, is

$$P^t = \begin{bmatrix} I & 0 \\ (I - B)^{-1}(I - B^t)C & B^t \end{bmatrix}.$$

c. Determinants of partitioned matrices

Evaluating the determinant of a partitioned matrix is sometimes aided by the availability of inverse matrices. Suppose

$$M = \begin{bmatrix} A & B \\ C & D \end{bmatrix}$$

is a partitioned matrix such that D is square and nonsingular; i.e., its inverse, D^{-1}, exists. We seek to evaluate the determinant $|M|$. Recalling equation (8) of Section 3.7 we see that for any matrix X

$$\begin{vmatrix} I & 0 \\ X & I \end{vmatrix} = |I|\,|I| = 1.$$

In particular this result is true for $X = -D^{-1}C$, so that

$$\begin{vmatrix} I & 0 \\ -D^{-1}C & I \end{vmatrix} = 1.$$

Therefore

$$|M| = \begin{vmatrix} A & B \\ C & D \end{vmatrix} = \begin{vmatrix} A & B \\ C & D \end{vmatrix} \begin{vmatrix} I & 0 \\ -D^{-1}C & I \end{vmatrix}$$

$$= \left| \begin{bmatrix} A & B \\ C & D \end{bmatrix} \begin{bmatrix} I & 0 \\ -D^{-1}C & I \end{bmatrix} \right| = \begin{vmatrix} A - BD^{-1}C & B \\ 0 & D \end{vmatrix},$$

and on using a Laplace expansion of the right-hand side this gives

$$|M| = |D|\,|A - BD^{-1}C|;$$

that is,

$$\begin{vmatrix} A & B \\ C & D \end{vmatrix} = |D|\,|A - BD^{-1}C|. \tag{23}$$

And with A^{-1} existing it can be shown in similar fashion that

$$\begin{vmatrix} A & B \\ C & D \end{vmatrix} = |A|\,|D - CA^{-1}B|. \tag{24}$$

Example. Partitioning

$$Q = \left[\begin{array}{rr|rrr} 21 & 37 & 1 & -1 & 2 \\ 18 & 12 & 3 & 5 & 7 \\ \hline -4 & 6 & 2 & 0 & 0 \\ 6 & 0 & 0 & 3 & 0 \\ 8 & 4 & 0 & 0 & 4 \end{array}\right] = \begin{bmatrix} A & B \\ C & D \end{bmatrix}$$

$$= 24 \left| \begin{bmatrix} 21 & 37 \\ 18 & 12 \end{bmatrix} - \begin{bmatrix} 1 & -1 & 2 \\ 3 & 5 & 7 \end{bmatrix} \begin{bmatrix} -2 & 3 \\ 2 & 0 \\ 2 & 1 \end{bmatrix} \right| = 24 \begin{vmatrix} 21 & 32 \\ 0 & -4 \end{vmatrix} = -2016.$$

d. Quadratic forms

Section 2.9 contains a brief discussion of quadratic forms. Although $x'Ax$ is the usual expression for a quadratic form situations sometimes arise where for the inverse of a matrix, P^{-1} say, we want to compute the quadratic form $x'P^{-1}x$, x being a known vector; for example, one such situation arises in regression analysis (see Chapter 9). We proceed to show how $x'P^{-1}x$ can be computed without inverting P, requiring instead the evaluation of two determinants resulting from the determinant of a partitioned matrix.

Direct application of equation (23) gives

$$\begin{vmatrix} P & x \\ x' & 1 \end{vmatrix} = |1|\,|P - xx'| = |P - xx'|.$$

And from (24) we also have the result

$$\begin{vmatrix} P & x \\ x' & 1 \end{vmatrix} = |P|\,|1 - x'P^{-1}x| = (1 - x'P^{-1}x)|P|.$$

Equating these two gives

$$x'P^{-1}x = 1 - \frac{|P - xx'|}{|P|}. \tag{25}$$

An alternative result obtainable from the determinant $\begin{vmatrix} P & x \\ x' & -1 \end{vmatrix}$ is

$$x'P^{-1}x = \frac{|P + xx'|}{|P|} - 1, \tag{26}$$

so giving

$$|P| = \tfrac{1}{2}(|P + xx'| + |P - xx'|) \tag{27}$$

Example. By direct expansion

$$|P| = \begin{vmatrix} 2 & 5 \\ 3 & 8 \end{vmatrix} = 1, \qquad \text{and for} \quad x' = [4 \quad 6]$$

$$|P + xx'| = \left|\begin{bmatrix} 2 & 5 \\ 3 & 8 \end{bmatrix} + \begin{bmatrix} 16 & 24 \\ 24 & 36 \end{bmatrix}\right| = \begin{vmatrix} 18 & 29 \\ 27 & 44 \end{vmatrix} = 9,$$

and

$$|P - xx'| = \left|\begin{bmatrix} 2 & 5 \\ 3 & 8 \end{bmatrix} - \begin{bmatrix} 16 & 24 \\ 24 & 36 \end{bmatrix}\right| = \begin{vmatrix} -14 & -19 \\ -21 & -28 \end{vmatrix} = -7.$$

Equation (27) then gives $|P| = \frac{1}{2}(9 - 7) = 1$.

The three equations (25), (26) and (27) are, of course, true only when P is nonsingular, but they are true for any vector x. Many additional results can therefore be derived by considering special values of x. Furthermore, these equations also hold for any vector x and for any other vector y' used in place of x'; for example, for the bilinear form $y'P^{-1}x$ we have, analogous to (26),

$$y'P^{-1}x = \frac{|P + xy'|}{|P|} - 1.$$

The reader should investigate these results for special values of the vectors involved (for example, Exercise 10).

Illustration. An interesting use of the quadratic $x'P^{-1}x$ is afforded by Searle (1963) in discussing the efficiency of ancestor records in animal selection. It is shown there that the correlation between true additive genetic merit and an estimate thereof can be expressed as $r = \sqrt{w'V^{-1}w/h}$, where $V\sigma^2$ is the variance-covariance matrix appropriate to the vector of records used in the estimation process, $w\sigma^2$ is the vector of covariances of these records with true additive genetic merit and h is the genetic parameter known as heritability. When records on n generations of ancestors are available the correlation takes the form $r_n^2 = t_n' A_n^{-1} t_n$ where

$$t_n' = [p \quad p^2 \quad p^3 \quad \ldots \quad p^n]$$

and

$$A_n = \begin{bmatrix} kp & p^2 & p^3 & \ldots & p^n \\ p^2 & kp^2 & p^3 & \ldots & p^n \\ p^3 & p^3 & kp^3 & \ldots & p^n \\ \cdot & \cdot & \cdot & & \cdot \\ \cdot & \cdot & \cdot & & \cdot \\ \cdot & \cdot & \cdot & & \cdot \\ p^n & p^n & p^n & \ldots & kp^n \end{bmatrix},$$

with $p = \frac{1}{2}$ and $k = 1/h$. By utilizing many of the properties of inverse matrices discussed in this chapter it is shown that

$$r_n^2 = \frac{2h - 1}{2h} + \frac{(1 - h)^2}{2h(1 - hr_{n-1}^2)}.$$

The correlation r_n represents the efficiency of n generations of ancestor records, and from this result the relative efficiency of successive generations may be derived.

9. INVERTING MATRICES ON HIGH-SPEED COMPUTERS

Our development of the matrix inverse has been entirely in terms of determinants. This procedure has been described in detail because it demonstrates so clearly the derivation of each element of the inverse. But it is not an efficient procedure for inverting matrices of large order, since it results in excessive computing. Alternative methods are available, however, and even though they involve no mean effort (especially if carried out on a desk calculator) they are preferable to calculating determinants. These alternatives have been described by many writers. All of them involve irksome and lengthy arithmetic work, which today is made light of by high-speed computers. The decision of which method to use and how has therefore largely become the concern of computer programmers, numerical analysts and other personnel associated with computer installations. As a result, the presence of computers has almost entirely removed this decision from the concern of those having a specific matrix to invert. And since computers are rapidly becoming more available, even to high school pupils in many instances, anyone having a matrix to invert is likely to find in his own institution or close at hand a computing facility equipped for doing the necessary arithmetic. Possibly the time element might be less using one's own desk calculator for small matrices rather than waiting for computer time, but the effort would usually be appreciably greater, even for a matrix as small as a 6×6. These conditions render it unnecessary to discuss the mechanics of the different methods available for inverting a matrix. The reader who wishes to pursue the subject in detail might turn to Dwyer (1951), Hildebrand (1956), Bodewig (1959), Fadeeva (1958) or Wilkinson (1963), to mention only a few of the many sources available.

Two matters only will be discussed in connection with inverting matrices by computers. The first is the general procedure used in solving a set of linear equations with the aid of a desk calculator—a procedure which is essentially the foundation of any method of inverting a matrix. The second is the subject of rounding error—since both desk calculators and computers have limited capacity in the number of decimal digits that they carry for any particular value entered or generated therein, there is always a problem of rounding errors, the accumulative effect of which can sometimes lead to erroneous results. We now briefly discuss these two topics.

a. Solving equations

Suppose we have a matrix

$$A = \begin{bmatrix} 2 & 5 & 7 \\ 3 & 9 & 15 \\ 5 & 16 & 20 \end{bmatrix}$$

whose inverse we wish to find by a method other than evaluating $|A|$ and the cofactor of every element. One method would be to imagine having the equations

$$A \begin{bmatrix} x \\ y \\ z \end{bmatrix} = \begin{bmatrix} a \\ b \\ c \end{bmatrix} \tag{28}$$

which could be solved explicitly for x, y and z by customary procedures of successive elimination. (Both here and for the remainder of the chapter the symbols x, y, z and a, b, c are scalar quantities.) The result, we know, would be

$$\begin{bmatrix} x \\ y \\ z \end{bmatrix} = A^{-1} \begin{bmatrix} a \\ b \\ c \end{bmatrix}; \tag{29}$$

for if (28) is true and A^{-1} exists, then pre-multiplying (28) by A^{-1} leads to (29). Hence if the solutions of (28) for x, y and z obtained explicitly in terms of a, b and c were written down in matrix form, the resulting matrix would be A^{-1}. Let us carry through this process.

Equations (28) are

$$2x + 5y + 7z = a, \tag{30}$$

$$3x + 9y + 15z = b \tag{31}$$

and

$$5x + 16y + 20z = c. \tag{32}$$

Dividing (30) by 2 gives

$$x + 2.5y + 3.5z = 0.5a, \tag{33}$$

and subtracting 3 times this from (31) and 5 times it from (32) gives

$$1.5y + 4.5z = b - 1.5a \tag{34}$$

and

$$3.5y + 2.5z = c - 2.5a. \tag{35}$$

Dividing (34) by 1.5 and multiplying (35) by 2 yields

$$y + 3z = b/1.5 - a \tag{36}$$

and

$$7y + 5z = 2c \quad -5a. \tag{37}$$

Subtract 7 times (36) from (37) and we obtain

$$-16z = 2c - 5a - 7(b/1.5 - a)$$

so that

$$\begin{aligned} z &= (2a - 7b/1.5 + 2c)/(-16) \\ &= (-3a + 7b - 3c)/24. \end{aligned} \tag{38}$$

Substituting z into (36) leads to

$$y = (-15a - 5b + 9c)/24 \tag{39}$$

and putting both y and z in (33) gives

$$x = (60a - 12b - 12c)/24. \tag{40}$$

The last three results can be assembled in matrix form as

$$\begin{bmatrix} x \\ y \\ z \end{bmatrix} = \tfrac{1}{24} \begin{bmatrix} 60 & -12 & -12 \\ -15 & -5 & 9 \\ -3 & 7 & -3 \end{bmatrix} \begin{bmatrix} a \\ b \\ c \end{bmatrix} \tag{41}$$

and from comparison with (29) we conclude that

$$A^{-1} = \tfrac{1}{24}\begin{bmatrix} 60 & -12 & -12 \\ -15 & -5 & 9 \\ -3 & 7 & -3 \end{bmatrix}.$$

Multiplication with A will show that this is so. Now this whole procedure is, of course, very tedious and would be even more so were A to be of order larger than three. Though tedious, it is not difficult in concept, and sustained effort combined with arithmetic accuracy will always yield correct results.

The arithmetic processes just outlined are the foundation of most computer methods of inverting a matrix, although no dummy variables such as x, y, z, a, b and c are used, only the arithmetic operations which determine their coefficients at each stage. Several names are familiarly associated with procedures for organizing (and hence programming) these calculations, for example, Gauss, Doolittle, Fox, Crout and others. The methods which bear these names are largely, though not entirely, variants one of another directed towards improving the efficiency of organizing the calculations rather than being distinctly different methods of calculation. The reader who follows the arithmetic just described will therefore have a basic understanding of the calculations involved in any computer method, even if he has no insight into better ways of arranging them than that just shown. Calculating additional examples of his own will assuredly establish a feeling for the procedures and the effort required. The advantages of having computers carry out the work should then be clear.

Other benefits also accrue from using computers to invert matrices. Since individual steps of the calculations are essentially the same for both large and small matrices, although more numerous for large, computers can readily be made to cope with matrices of any order (limited solely by the size of the computer). Further, the accuracy of the computer far exceeds that of a desk calculator operated by *homo sapiens*, who needs to repeat specific steps of a calculation in order to check on accuracy. Generally speaking, a computer makes no errors (although it is a high-speed idiot and will carry out erroneous instructions just as quickly as it will correct ones) and requires no repetitive operations for the sake of accuracy. The greatest advantage of computers, though, is their speed, for it makes feasible tasks that might otherwise never be undertaken. For example, modern computers can invert a matrix of order 100 in a matter of minutes, whereas in contrast, instances are known of desk calculator work requiring six weeks for the inversion of a 40 × 40 matrix.

In general, then, for reasons of size, accuracy and speed, computers are a most useful tool for inverting matrices, and their existence has eliminated the chore of the bothersome arithmetic involved.

b. Rounding error

Although the accuracy of high-speed computers has just been put forward as one of their advantages, there is nevertheless one problem concerning accuracy which must be appreciated—the problem of rounding error. It does, of course, arise with desk calculators also, since in both computers and calculators there is a limited number of decimal digits that can be effectively carried in each value handled by the machine. Even though modern computers can handle numbers containing many many digits, the rounding error problem still exists. No matter how large the computer, nor how great its capacity, it can never contain the value $4\frac{2}{3}$ as a decimal number with complete, 100 per cent accuracy. Some computers will simply truncate their numbers, taking $4\frac{2}{3}$ for example, as 4.666 . . . to as many sixes as capacity will allow; others round off their numbers taking $4\frac{2}{3}$ as 4.6667, again with as many sixes as capacity allows. Understandably this can lead to error when large numbers of multiplication and division operations take place as in inverting a large matrix. This creates a problem, especially since the situations in which error will occur cannot be specified with certainty, and a mathematical discussion of the cumulative effects of rounding errors becomes very involved in anything but simple situations (see Wilkinson, 1963). At best one can usually say that trouble does not occur very often. At worst one must be cognizant of the possibility of trouble of this sort on any occasion that a computer is used.

Computer methods for inverting a matrix are not based on computing the determinant of the matrix and the cofactors of each element. But suppose they were. And suppose the determinant of a matrix was truly zero: there would then be no inverse. However, rounding errors incurred while evaluating the determinant might lead to its being given a non-zero value.

Example. The determinant

$$|M| = \begin{vmatrix} 0.12 & -0.21 & 0.12 \\ 0.23 & 0.17 & 0.19 \\ 0.15 & 0.31 & 0.11 \end{vmatrix}$$

expanded by elements of its first row is

$$0.12\begin{vmatrix} 0.17 & 0.19 \\ 0.31 & 0.11 \end{vmatrix} + 0.21\begin{vmatrix} 0.23 & 0.19 \\ 0.15 & 0.11 \end{vmatrix} + 0.12\begin{vmatrix} 0.23 & 0.17 \\ 0.15 & 0.31 \end{vmatrix}$$

$$= 0.12(0.0187 - 0.0589) + 0.21(0.0253 - 0.0285) + 0.12(0.0713 - 0.0255) \qquad (42)$$

$$= 0.12(-0.0402) + 0.21(-0.0032) + 0.12(0.0458)$$

$$= -0.004824 - 0.000672 + 0.005496 = 0.$$

Suppose this determinant were evaluated on a computer of very limited capacity, one capable of carrying only three decimal digits, meaning by this that at every stage numbers would be rounded off to the nearest integer in the third decimal place, a 5 in the fourth place being rounded up (a common procedure in computers). Were exactly the same method of calculation used as just shown expression (42) would become, as a result of the rounding-off procedure,

$$\begin{aligned} &0.12(0.019 - 0.059) + 0.21(0.025 - 0.029) + 0.12(0.071 - 0.026) \\ &= 0.12(-0.040) + 0.21(-0.004) + 0.12(0.045). \end{aligned}$$

The exact value of this is

$$-0.00480 - 0.00084 + 0.00540,$$

but because of rounding off its computed value would be derived as

$$\begin{aligned} &-0.005 - 0.001 + 0.005 \\ &= -0.001, \end{aligned}$$

which is not the correct value of the determinant, namely zero.

This simple example demonstrates how rounding error can lead to computers producing erroneous results. It is, of course, an oversimplification of the general problem, which is one that becomes quite difficult to follow when more numerous arithmetic operations are involved. Furthermore, the example might suggest that rounding error problems can readily become acute and perhaps frequently do so. For several reasons this is not the case at all. First of all, any respectable computer can efficiently handle many more than three decimal digits, and rounding errors always originate in the least significant one. By accumulation through numerous arithmetic operations they may, however, affect the last two, three or even four digits and thus a computer capable of handling eleven significant digits might reliably give results correct to seven or eight digits, which, of course, is sufficient for many problems. A second reason that rounding error is not a frequent problem is that computational methods for inverting matrices have been devised to minimize its effect (see Dwyer, 1951, and Wilkinson, 1963, for example); and third, computer programming techniques have been developed for handling numbers with very many decimals (for example, floating decimal arithmetic and multiple-word length operations). The result is that computing establishments seldom return an inverted matrix that is "bugged" by rounding error. But such an event *can* occur and those who, perhaps rightly, rely on computers for obtaining their inverse matrices should be aware of the possibility and the nature of the consequences.

Consider what would have happened in the foregoing example if our "three-decimal" computer, having found $|M| = -0.001$, had proceeded to derive M^{-1} using the method of cofactors. With rounding all arithmetic operations off to three decimal places the result would have been

$$\frac{1}{-0.001}\begin{bmatrix} -0.040 & 0.060 & -0.060 \\ 0.004 & -0.005 & 0.005 \\ 0.045 & -0.069 & 0.068 \end{bmatrix} = \begin{bmatrix} 40 & -60 & 60 \\ -4 & 5 & -5 \\ -45 & 69 & -68 \end{bmatrix}.$$

This is the kind of result that may sometimes occur: rounding errors can cause true zero values to be computed as very small, non-zero quantities, and division by these small quantities leads to an alleged inverse which may have some unduly large elements. This might be just the clue that something is in error, and those who rely on computers to invert matrices should be aware of this possibility. Only experience improves one's ability to suspect errors of this nature, but until such experience is gained we need not be overawed by the problem, as it is more than counterbalanced by the speed and accuracy with which inverses are generally obtained with satisfaction from a computing facility. Furthermore, most facilities have experienced the problem of rounding error and are familiar with methods of investigating it.

Safeguards against erroneous inverses resulting from rounding error are few. The important thing is to be aware of the possible consequences, to scrutinize results carefully and even to check them by using a desk calculator to derive an element or two of the product of the matrix with its inverse. Total prevention of rounding errors is impossible, but one useful method of reducing them is to see that the elements of a matrix that has to be inverted are approximately all of the same order of magnitude. This can often be achieved by pre- or post-multiplying the matrix by a diagonal matrix before inverting it in order to amend the elements of a row or column to bring them into line with the others. It is necessary, of course, to remember to appropriately multiply the inverse matrix obtained from the computer in order to derive the inverse of the original matrix.

Example.

$$\begin{bmatrix} 2 & 1365 \\ 3 & 2050 \end{bmatrix} = \begin{bmatrix} 2 & 2.73 \\ 3 & 4.10 \end{bmatrix}\begin{bmatrix} 1 & 0 \\ 0 & 500 \end{bmatrix}$$

and

$$\begin{bmatrix} 2 & 2.73 \\ 3 & 4.10 \end{bmatrix}^{-1} = \frac{1}{0.01}\begin{bmatrix} 4.10 & -2.73 \\ -3 & 2 \end{bmatrix}.$$

Therefore

$$\begin{bmatrix} 2 & 1365 \\ 3 & 2050 \end{bmatrix}^{-1} = \begin{bmatrix} 1 & 0 \\ 0 & 500 \end{bmatrix}^{-1}\begin{bmatrix} 2 & 2.73 \\ 3 & 4.10 \end{bmatrix}^{-1}$$

$$= \begin{bmatrix} 1 & 0 \\ 0 & 0.002 \end{bmatrix}\frac{1}{0.01}\begin{bmatrix} 4.10 & -2.73 \\ -3 & 2 \end{bmatrix} = \begin{bmatrix} 410 & -273 \\ -0.6 & 0.4 \end{bmatrix}.$$

10. EXERCISES

1. Show that if

(*a*) $A = \begin{bmatrix} 6 & 13 \\ 5 & 12 \end{bmatrix}$, $|A| = 7$, $A^{-1} = \frac{1}{7}\begin{bmatrix} 12 & -13 \\ -5 & 6 \end{bmatrix}$,

and $AA^{-1} = A^{-1}A = I$;

(*b*) $B = \begin{bmatrix} 3 & -4 \\ 7 & 14 \end{bmatrix}$, $|B| = 70$, $B^{-1} = \frac{1}{70}\begin{bmatrix} 14 & 4 \\ -7 & 3 \end{bmatrix}$,

and $BB^{-1} = B^{-1}B = I$;

2. Demonstrate the reversal rule for the inverse of a product of two matrices, using A and B given in Exercise 1.

3. Show that if

(*a*) $A = \begin{bmatrix} -1 & 3 & 0 \\ 0 & 2 & 1 \\ 1 & 0 & 4 \end{bmatrix}$, $|A| = -5$, $A^{-1} = \frac{1}{5}\begin{bmatrix} -8 & 12 & -3 \\ -1 & 4 & -1 \\ 2 & -3 & 2 \end{bmatrix}$,

and $A^{-1}A = AA^{-1} = I$;

(*b*) $B = \begin{bmatrix} 10 & 6 & -1 \\ 6 & 5 & 4 \\ -1 & 4 & 17 \end{bmatrix}$, $|B| = 25$, $B^{-1} = \begin{bmatrix} 2.76 & -4.24 & 1.16 \\ -4.24 & 6.76 & -1.84 \\ 1.16 & -1.84 & 0.56 \end{bmatrix}$,

and $B^{-1}B = BB^{-1} = I$;

(*c*) $C = \frac{1}{10}\begin{bmatrix} 0 & -6 & 8 \\ -10 & 0 & 0 \\ 0 & -8 & -6 \end{bmatrix}$, $|C| = 1$, $C^{-1} = C'$,

and $CC' = C'C = I$.

4. For each of the following matrices derive the determinant and inverse. Check each inverse by multiplication.

(*a*) $\begin{bmatrix} 7 & 3 \\ 8 & 9 \end{bmatrix}$ (*b*) $\begin{bmatrix} 6 & 31 \\ 8 & 29 \end{bmatrix}$ (*c*) $\begin{bmatrix} -7 & -4 \\ 3 & 1 \end{bmatrix}$

(*d*) $\begin{bmatrix} 1 & 5 & -5 \\ 3 & 2 & -5 \\ 6 & -2 & -5 \end{bmatrix}$ (*e*) $\begin{bmatrix} -3 & 2 & -6 \\ -3 & 5 & -7 \\ -2 & 3 & -4 \end{bmatrix}$ (*f*) $\begin{bmatrix} 2 & 1 & 3 \\ -5 & 1 & 0 \\ 1 & 4 & -2 \end{bmatrix}$

(*g*) $\begin{bmatrix} -1 & -2 & 1 \\ -2 & 2 & -2 \\ 1 & -2 & -1 \end{bmatrix}$ (*h*) $\begin{bmatrix} 1 & -2 & 1 \\ -2 & 4 & -2 \\ 1 & -2 & 1 \end{bmatrix}$ (*k*) $\begin{bmatrix} 7 & 4 & -1 \\ 4 & 7 & -1 \\ -4 & -4 & 4 \end{bmatrix}$

5. Derive the inverse of

$$P = \begin{bmatrix} 1 & 0 & 2 & 0 \\ 0 & 4 & 0 & 1 \\ 2 & 0 & 3 & 2 \\ 0 & 1 & 2 & 1 \end{bmatrix} \quad \text{as} \quad P^{-1} = \tfrac{1}{19}\begin{bmatrix} 7 & 4 & 6 & -16 \\ 4 & 5 & -2 & -1 \\ 6 & -2 & -3 & 8 \\ -16 & -1 & 8 & 4 \end{bmatrix}.$$

Show that P^{-1} can be obtained very easily after calculating $|P|$ and the cofactors of three appropriately chosen elements.

6. Calculate the inverse of $\begin{bmatrix} 1 & 0 & 6 & 8 \\ 0 & 1 & 5 & 4 \\ 0 & 0 & -1 & 0 \\ 0 & 0 & 0 & -1 \end{bmatrix}$.

7. For $A = \begin{bmatrix} 6 & -1 & 4 \\ 2 & 5 & -3 \\ 1 & 1 & 2 \end{bmatrix}$

i. Calculate the transpose of A^{-1} and the inverse of A'.
ii. Calculate the inverse of A^{-1}.

8. Find the inverse of $\begin{bmatrix} 3 & 0 & 0 & 0 \\ 0 & 7 & 0 & 0 \\ 0 & 0 & 1 & 0 \\ 0 & 0 & 0 & 5 \end{bmatrix}$.

9. Prove the following statements:

(*a*) If A and B are symmetric, $[(AB)']^{-1} = A^{-1}B^{-1}$.
(*b*) If $C = X(X'X)^{-1}X'$, $C^2 = C = C'$.
(*c*) $(ABCD)^{-1} = D^{-1}C^{-1}B^{-1}A^{-1}$.
(*d*) If $A^{-1} = A'$ and $B^{-1} = B'$, $(AB)(AB)' = I$.
(*e*) If $A = \begin{bmatrix} I & B \\ 0 & -I \end{bmatrix}$, $A^2 = I$.
(*f*) If B is an $n \times n$ matrix and λ is a scalar, and if $A = \lambda B$, then

i. A is an $n \times n$ matrix;
ii. $|A| = \lambda^n|B|$;
iii. $A^{-1} = \dfrac{1}{\lambda}B^{-1}$. When is iii not true?

(*g*) If A is a nonsingular square matrix of order n, $|\text{adjugate}\,(A)| = |A|^{n-1}$.
(*h*) The product of two matrices being a null matrix does not necessarily imply that one of them is null.

10. Prove the following results:

(*a*) The leading element of P^{-1} is $(1 - |Q|/|P|)$ where Q is P with its leading element reduced by unity.

(*b*) The sum of the elements of P^{-1} is $(1 - |R|/|P|)$ where R is P with every element reduced by unity.

(*c*) If x and y are vectors of order n

$$\sum_{i=1}^{n} x_i y_i = |I + xy'| - 1.$$

11. Show that

(*a*) if A is an orthogonal matrix its determinant has the value $+1$ or -1,

(*b*) if λ_1, λ_2, λ_3 and λ_4 are scalars, and if A is a square matrix of order n, then

$$\begin{vmatrix} \lambda_1 A & \lambda_2 A \\ \lambda_3 A & \lambda_4 A \end{vmatrix} = \left\{ \begin{vmatrix} \lambda_1 & \lambda_2 \\ \lambda_3 & \lambda_4 \end{vmatrix} \right\}^n (|A|)^2;$$

(*c*) for an appropriate X,

i. $$I + X + X^2 + \ldots + X^{t-1} = (I - X)^{-1}(I - X^t) = (X^t - I)(X - I)^{-1},$$

ii. $$(I - X)^{-1}(I - X^t)(I - X) = I - X^t,$$

iii. $$\begin{bmatrix} I & 0 \\ Y & X \end{bmatrix}^t = \begin{bmatrix} I & 0 \\ (I - X)^{-1}(I - X^t)Y & X^t \end{bmatrix}.$$

REFERENCES

Alling, David W. (1958). The after-history of pulmonary tuberculosis: a stochastic model. *Biometrics*, **14**, 527–547.

Bodewig, E. (1959). *Matrix Calculus*. North Holland Publishing Company, Amsterdam.

Dwyer, Paul. (1951). *Linear Computations*. Wiley, New York.

Fadeeva, V. N. (1958). *Computational Methods of Linear Algebra*. Dover, New York.

Falconer, D. S. (1960). *Introduction to Quantitative Genetics*. Oliver and Boyd, Edinburgh.

Hildebrande, F. B. (1956). *Introduction to Numerical Analysis*. McGraw-Hill, New York.

Searle, S. R. (1963). The efficiency of ancestor records in animal selection. *Heredity*, **18**, 351–360.

Wilkinson, J. H. (1963). *Rounding Errors in Algebraic Processes*. Prentice-Hall, Englewood Cliffs, New Jersey.

CHAPTER 5

RANK AND LINEAR INDEPENDENCE

Previous chapters deal with determinants and the fundamental operations of matrix algebra. We now consider certain aspects of matrix theory proper. To some readers the discussion may appear somewhat removed from practical uses of matrix algebra, for opportunities become less frequent for demonstrating procedures with illustrations taken from real life. Nevertheless, these procedures do lead to important applications of matrix algebra to quantitative problems in biology even though the intermediary stages of establishing them may appear unduly mathematical. Any extent to which this is so could, of course, have been overcome by a "cookbook" presentation which would have omitted both the discussion of underlying theory and the proofs of the "recipes" involved. The sterility of such an approach is, however, not attractive, and so the requisite theory has been included. Several of the necessary results are stated formally as theorems, with their proofs marked as optional for a first reading. Numerical illustrations are given in nearly all instances.

1. SOLVING LINEAR EQUATIONS

One of the greatest uses of matrices is in the solution of systems of linear equations, as demonstrated at the beginning of Chapter 4. Let us take another look at the illustration used there concerning the estimation of gene effects on the body weight of mice. We had the equations

$$\begin{bmatrix} 1 & 1 & 0 \\ 1 & 0 & 1 \\ 1 & -1 & 0 \end{bmatrix}\begin{bmatrix} m \\ a \\ d \end{bmatrix} = \begin{bmatrix} 14 \\ 12 \\ 6 \end{bmatrix}, \tag{1}$$

which can be written more generally as

$$Ax = y \tag{2}$$

where x is the vector of unknowns, A is the matrix of the coefficients of the elements of x in the equations and y is the vector of values on the right-hand sides. In first discussing equation (1) we found that pre-multiplying both sides by the matrix $\begin{bmatrix} \frac{1}{2} & 0 & \frac{1}{2} \\ \frac{1}{2} & 0 & -\frac{1}{2} \\ -\frac{1}{2} & 1 & -\frac{1}{2} \end{bmatrix}$ led to solutions for m, a and d because the product of this matrix with $\begin{bmatrix} 1 & 1 & 0 \\ 1 & 0 & 1 \\ 1 & -1 & 0 \end{bmatrix}$ is the identity matrix. This is so because the first of these two matrices is the inverse of the second, as may be readily verified. Given equations (1) as they stand, or in more general terms equations (2), we would now, after reading Chapter 4, utilize the concept of an inverse matrix directly and derive the solutions as $x = A^{-1}y$ where A^{-1} is the inverse of A. The crux of these solutions is, of course, the existence of the inverse A^{-1}. Only if it exists can the solution be $x = A^{-1}y$; and A^{-1} does exist only if the determinant of A is non-zero. Therefore, in any situation in which we wish to have equations of the form $Ax = y$ solved as $x = A^{-1}y$, we must first investigate the value of the determinant of A to see if it is non-zero. Now if A is a matrix of many rows and columns, finding its determinant by the methods discussed in Chapter 3 may be almost as lengthy a procedure as solving the equations themselves, and although the determinant is an integral part of the description of A^{-1} given in Chapter 4, its explicit value is not required for the more useful (computer) methods of calculating A^{-1}. But we must know that $|A|$ is non-zero before attempting to calculate A^{-1} in order to know that A^{-1} does exist. Methods for doing this are now discussed, methods which provide for an investigation of the matrix A to see whether its determinant is zero or not without having to expand the determinant in full. But first an illustration that follows from the one just used.

Illustration. If in a breeding experiment using back-crossing, the observed averages of the two back-crosses and the F_2 generation resulting from homozygous parents GG and gg are y_1, y_2 and y_3, the following equations in the gene effects m, a and d can be written down:

$$\begin{aligned} m - \tfrac{1}{2}a + \tfrac{1}{2}d &= y_1 \\ m + \tfrac{1}{2}a + \tfrac{1}{2}d &= y_2 \\ m \qquad\quad + \tfrac{1}{2}d &= y_3. \end{aligned}$$

Attempting to solve these uniquely for m, a and d is fruitless. This is easily verified by writing the equations in matrix form

$$\begin{bmatrix} 1 & -\frac{1}{2} & \frac{1}{2} \\ 1 & \frac{1}{2} & \frac{1}{2} \\ 1 & 0 & \frac{1}{2} \end{bmatrix} \begin{bmatrix} m \\ a \\ d \end{bmatrix} = \begin{bmatrix} y_1 \\ y_2 \\ y_3 \end{bmatrix} \tag{3}$$

for it is immediately apparent that the determinant of the 3×3 matrix on the left is zero (because its last column is a multiple of its first). Hence the inverse of this matrix does not exist and there is no solution of the form $x = A^{-1}y$.

Investigating whether or not a determinant is zero has been simple in this and previous examples because the determinants concerned are of small order. This simplicity is not necessarily present when large-sized matrices are involved, and other properties of matrices must then be utilized.

2. LINEAR INDEPENDENCE

The sum of two scalar variables x_1 and x_2 say, each multiplied by a non-zero constant, k_1 and k_2 say, is $k_1x_1 + k_2x_2$. This is called a *linear combination* of the variables x_1 and x_2. It is called linear because in the simple case of just one variable the plot of k_1x_1 against values of x_1 is a straight line. If the linear combination of the x's is zero only when all the k's are zero the x's are said to be *linearly independent*, that is to say, if the equation

$$k_1x_1 + k_2x_2 = 0 \tag{4}$$

is true only when $k_1 = k_2 = 0$ then x_1 and x_2 are linearly independent. Conversely, if for some non-zero values of the k's the linear combination of the x's is zero, i.e. if (4) is true for some non-zero k's, then x_1 and x_2 are said to be *linearly dependent*.

Illustration. If x_1 represents the weight of a piece of lead and x_2 its volume, the two measures are related by the equation $x_1 = 11.34x_2$ because 11.34 is the specific gravity of lead. Written in the same form as (4), namely,

$$x_1 - 11.34x_2 = 0, \tag{5}$$

we say that x_1 and x_2 are linearly dependent, with $k_1 = 1$ and $k_2 = -11.34$.

Notice that when the variables are linearly dependent the constants k_1 and k_2 need not be considered unique, for if $k_1x_1 + k_2x_2 = 0$ so does $ck_1x_1 + ck_2x_2 = 0$, where c is any constant. Thus if (5) is true so is, for example, $10x_1 - 113.4x_2 = 0$ and $25x_1 - 283.5x_2 = 0$. The values of the constants can be altered so long as their relative values stay the same.

Although linear independence and linear dependence have been defined in terms of only two variables, the definitions apply equally as well to more than two. Thus x_1, x_2, x_3 and x_4 are linearly independent if

$$k_1x_1 + k_2x_2 + k_3x_3 + k_4x_4 = 0 \tag{6}$$

is true only when $k_1 = k_2 = k_3 = k_4 = 0$. Likewise they are linearly dependent if (6) is true for some set of k's other than all of them equal to zero. Thus if λ_1 and λ_2 are two non-zero constants for which

$$\lambda_1x_1 + \lambda_2x_2 = 0,$$

the x's are said to be linearly dependent because (6) is then true for the set of k's $k_1 = \lambda_1$, $k_2 = \lambda_2$, $k_3 = 0$ and $k_4 = 0$, not all of which are zero. Conversely, the x's are linearly independent only if (6) is true for all the k's being zero, and on no other occasion.

Note that when a set of x's is a linearly dependent set there must be at least two k's that are non-zero. If only one were, k_2 say, equation (6) would reduce to $k_2x_2 = 0$, so giving $k_2 = 0$ also, and all the k's would then be zero. Therefore, if (6) is satisfied for any non-zero k's, at least two of them are non-zero.

An important consequence of linear dependence is that it implies that at least one of the x's concerned can be expressed as a linear combination of the others. This is easily proved. Let us assume four variables x_1, x_2, x_3 and x_4 are linearly dependent. Then (6) is true for some set of k's not all zero. Suppose $k_1 \neq 0$. Then, since at least one of the other k's is also non-zero, equation (6) is equivalent to

$$x_1 = (-k_2/k_1)x_2 + (-k_3/k_1)x_3 + (-k_4/k_1)x_4,$$

so demonstrating that x_1 is a linear combination of x_2, x_3 and x_4.

Linear independence has so far been discussed in terms of no more than four variables. Extension to more than four is quite straightforward: the n variables $x_1, x_2, \ldots, x_n$ are said to be linearly independent if the equation

$$k_1x_1 + k_2x_2 + \cdots + k_nx_n = 0$$

is satisfied only when all the k's are zero. If it is satisfied for some or all of the k's being non-zero, the x's are said to be linearly dependent.

3. LINEAR DEPENDENCE OF VECTORS

The foregoing discussion has been in terms of the x's as scalar variables, but the same concept of linear dependence also applies to vectors. For example, consider the vectors

$$v_1 = \begin{bmatrix} 2 \\ 6 \\ 4 \end{bmatrix} \quad \text{and} \quad v_2 = \begin{bmatrix} 3 \\ 9 \\ 6 \end{bmatrix}.$$

It is easily seen that

$$3v_1 - 2v_2 = 0 \tag{7}$$

and in the same way that equation (4) is evidence of the linear dependence of x_1 on x_2 and vice versa, so is equation (7) a statement of the linear dependence of the vectors v_1 and v_2 on each other. Indeed the definition of two vectors v_1 and v_2 being linearly dependent is that there exists two non-zero scalars k_1 and k_2 such that $k_1v_1 + k_2v_2 = 0$. This is, of course, a vector equation, representing as many scalar equations as there are elements in v_1, and it is meaningful only when v_1 and v_2 are of the same order.

Example.

$$v_1 = \begin{bmatrix} 84 \\ 91 \\ 119 \\ 161 \end{bmatrix} \quad \text{and} \quad v_2 = \begin{bmatrix} 3.6 \\ 3.9 \\ 5.1 \\ 6.9 \end{bmatrix}$$

are linearly dependent because $3v_1 = 70v_2$ and hence $3v_1 - 70v_2 = 0$.

Linear dependence has been defined and illustrated in terms of only two vectors, but the general definition is not confined to this situation and applies to any number of vectors. Thus in general, n vectors all of the same order, $u_1, u_2, \ldots, u_n$, are said to be linearly dependent if non-zero constants $k_1, k_2, \ldots, k_n$ exist such that

$$k_1u_1 + k_2u_2 + \cdots + k_nu_n = 0.$$

If zeros are the only values of the k's for which this equation is true, the vectors are said to be linearly independent. Otherwise they are linearly dependent.

Example. Consider

$$u_1 = \begin{bmatrix} 1 \\ 0 \\ -2 \end{bmatrix}, \quad u_2 = \begin{bmatrix} -4 \\ 3 \\ 5 \end{bmatrix}, \quad u_3 = \begin{bmatrix} 11 \\ -6 \\ -16 \end{bmatrix} \quad \text{and} \quad u_4 = \begin{bmatrix} 1 \\ 2 \\ 3 \end{bmatrix}.$$

Simple arithmetic shows that

$$3u_1 - 2u_2 - u_3 = 0,$$

and therefore u_1, u_2 and u_3 are linearly dependent. Similarly

$$3u_1 - 2u_2 - u_3 - 0(u_4) = 0$$

so that u_1, u_2, u_3 and u_4 are also linearly dependent. But no values k_1, k_2 and k_4 can be found, other than all of them zero, such that

$$k_1u_1 + k_2u_2 + k_4u_4 = 0.$$

Hence the three vectors u_1, u_2 and u_4 are linearly independent.

4. LINEAR DEPENDENCE AND DETERMINANTS

Definition. For the sake of verbal simplicity we will henceforth refer to "linear independence" as just "independence" and to "linear dependence" as "dependence", dropping the description "linear", but without changing the meaning in so doing.

At the end of Section 2 it was shown that a set of scalars being dependent implies that at least one of them can be expressed as a linear combination of the others. The same is true of vectors. Recall, now, a condition under which a determinant is zero: if multiples of columns can be subtracted from another column to give a column of zeros the determinant is zero. But this is just the condition under which the one column is a linear combination of the others. And this in turn is the condition that the columns are dependent. Hence we see that if the columns of a matrix are dependent the determinant of the matrix is zero and consequently the inverse of the matrix does not exist. The same is, of course, true for rows.

Example. We saw that the vectors u_1, u_2 and u_3 of the preceding example were linearly dependent because $3u_1 - 2u_2 - u_3 = 0$. Hence the determinant having these vectors as columns is zero:

$$\begin{vmatrix} 1 & -4 & 11 \\ 0 & 3 & -6 \\ -2 & 5 & -16 \end{vmatrix} = \begin{vmatrix} 1 & -4 & -3(1) + 2(-4) + 11 \\ 0 & 3 & -3(0) + 2(3) - 6 \\ -2 & 5 & -3(-2) + 2(5) - 16 \end{vmatrix}$$

$$= \begin{vmatrix} 1 & -4 & 0 \\ 0 & 3 & 0 \\ -2 & 5 & 0 \end{vmatrix} = 0.$$

And conversely, because u_1, u_2 and u_4 are independent, the determinant formed from them is non-zero:

$$\begin{vmatrix} 1 & -4 & 1 \\ 0 & 3 & 2 \\ -2 & 5 & 3 \end{vmatrix} = \begin{vmatrix} 1 & -4 & 1 \\ 0 & 3 & 2 \\ 0 & -3 & 5 \end{vmatrix} = 21.$$

We have here developed a most important result: that if the rows (or columns) of a matrix are linearly dependent its determinant is zero and its inverse does not exist; conversely, if they are independent its determinant is non-zero and the inverse does exist.

The minimum condition for the linear dependence of rows (or columns) of a matrix is that one of them be a linear combination of the others. Investigating a matrix to see whether or not this is explicitly the case is not necessarily an easy task, especially when the matrix is of large order. Fortunately, alternative procedures are available.

5. SETS OF LINEARLY INDEPENDENT VECTORS

A question that might be asked about independent vectors is "How many independent vectors are there?" The answer is contained in the following theorem.

Theorem. A set of linearly independent non-null vectors of order n cannot contain more than n such vectors.

Corollary. A set of r non-null vectors of order n can be linearly independent only if $r \leqq n$.

Before proving this theorem let us note that it excludes null vectors, namely those whose every element is a zero. This is done because the dependence or independence of a group of vectors is unaffected by adding one or more null vectors to the group. Thus, where u's are vectors and k's are scalars, the equation $k_1u_1 + k_2u_2 = 0$ is unaltered by adding to its left-hand side k_3u_3 if u_3 is a null vector.

Note also the generality of the theorem. It does not state that there is only one, unique set of n independent vectors of order n, but that in a set of vectors of order n, that is to say, in any such set, there cannot be more than n (non-null) vectors if they are to be independent. In fact there is an infinite number of such sets, but none of them can consist of more than n vectors and still be a set of independent vectors.

Examples. The vectors $\begin{bmatrix} 1 \\ 3 \end{bmatrix}$ and $\begin{bmatrix} 2 \\ -1 \end{bmatrix}$ form a set of independent vectors of order 2; but form a set of vectors consisting of these two and any other, $\begin{bmatrix} 8 \\ 3 \end{bmatrix}$ say, and the set will *not* be independent, for it will be found that the third vector can be expressed as a linear combination of the other two; for example,

$$\begin{bmatrix} 8 \\ 3 \end{bmatrix} = 2\begin{bmatrix} 1 \\ 3 \end{bmatrix} + 3\begin{bmatrix} 2 \\ -1 \end{bmatrix}.$$

The vectors $\begin{bmatrix} 0 \\ 1 \end{bmatrix}$ and $\begin{bmatrix} -1 \\ 2 \end{bmatrix}$ also form an independent set, and again any other vector (of order 2) can be expressed as a linear combination of them:

$$\begin{bmatrix} 8 \\ 3 \end{bmatrix} = 19\begin{bmatrix} 0 \\ 1 \end{bmatrix} - 8\begin{bmatrix} -1 \\ 2 \end{bmatrix}.$$

It is clear, of course, that not *any* two vectors of order 2 form an independent set, nor do *any* three vectors of order 3. In all cases the vectors must satisfy the definition of independence in order to form an independent set. But if, as a further example, three vectors of order 3 are independent, then any other third-order vector can be expressed as a linear combination of them. For example, the vectors

$$\begin{bmatrix} 1 \\ 3 \\ 4 \end{bmatrix}, \quad \begin{bmatrix} 2 \\ -1 \\ 2 \end{bmatrix} \quad \text{and} \quad \begin{bmatrix} 0 \\ 1 \\ 3 \end{bmatrix}$$

are independent, and any other vector of order 3, $\begin{bmatrix} a \\ b \\ c \end{bmatrix}$, can be expressed as a linear combination of them in the form

$$\begin{bmatrix} a \\ b \\ c \end{bmatrix} = \tfrac{1}{15}(5a + 6b - 2c)\begin{bmatrix} 1 \\ 3 \\ 4 \end{bmatrix} + \tfrac{1}{15}(5a - 3b + c)\begin{bmatrix} 2 \\ -1 \\ 2 \end{bmatrix} + \tfrac{1}{15}(-10a - 6b + 7c)\begin{bmatrix} 0 \\ 1 \\ 3 \end{bmatrix}. \quad (8)$$

No matter what the values of a, b and c are, this expression holds true—that is, every third-order vector can be expressed as a linear combination of the set of three independent vectors. In general, every n-order vector can be expressed as a linear combination of any set of n independent vectors of order n. The maximum number of non-null vectors in a set of independent vectors is therefore n, the order of the vectors. This is simply a rewording of the theorem just given, for which a formal proof now follows.

***Proof.**

Let $u_1, u_2, \ldots, u_n$ be a set of n independent non-null vectors of order n. We show that any other non-null vector of order n, u_{n+1} say, can be expressed as a linear combination of $u_1, u_2, \ldots, u_n$, and is therefore not independent of them.

Because $u_1, u_2, \ldots, u_n$ are independent, the equation

$$k_1u_1 + k_2u_2 + \ldots + k_nu_n = 0 \tag{9}$$

is satisfied only when the scalars $k_1, k_2, \ldots, k_n$ are all zero. Consider the equation

$$\lambda_1u_1 + \lambda_2u_2 + \ldots + \lambda_nu_n + \lambda_{n+1}u_{n+1} = 0, \tag{10}$$

where the λ's are scalars. If $\lambda_{n+1} = 0$, equation (10) reduces to (9) and can therefore only be true if $\lambda_1, \lambda_2, \ldots, \lambda_n$ are also zero. But, if $\lambda_{n+1} \neq 0$, equation (10) can only be true if some of the $\lambda_1, \lambda_2, \ldots, \lambda_n$ are not zero, in which case it can be rewritten as

$$\frac{\lambda_1}{\lambda_{n+1}}u_1 + \frac{\lambda_2}{\lambda_{n+1}}u_2 + \ldots + \frac{\lambda_n}{\lambda_{n+1}}u_n = -u_{n+1}$$

This is equivalent to

$$[u_1\ u_2\ \ldots\ u_n]\begin{bmatrix} q_1 \\ q_2 \\ \cdot \\ \cdot \\ \cdot \\ q_n \end{bmatrix} = -u_{n+1} \tag{11}$$

where $q_i = \lambda_i/\lambda_{n+1}$. Now because the vectors $u_1, u_2, \ldots, u_n$ are independent, the determinant of the partitioned matrix $[u_1\ u_2\ \ldots\ u_n]$ is non-zero. Therefore (11) has a unique solution for the q's and consequently (10) is true for values of the λ's other than all of them zero. Thus u_{n+1} is a linear combination of the independent vectors $u_1, u_2, \ldots, u_n$ and is therefore not independent of them.

Example. The vector $\begin{bmatrix} a \\ b \end{bmatrix}$ is expressed as a linear combination of the independent vectors $\begin{bmatrix} 1 \\ 3 \end{bmatrix}$ and $\begin{bmatrix} 2 \\ -1 \end{bmatrix}$ by the equation

$$\begin{bmatrix} a \\ b \end{bmatrix} = \lambda_1\begin{bmatrix} 1 \\ 3 \end{bmatrix} + \lambda_2\begin{bmatrix} 2 \\ -1 \end{bmatrix},$$

where λ_1 and λ_2 are scalars. This can be written in the form

$$\begin{bmatrix} 1 & 2 \\ 3 & -1 \end{bmatrix}\begin{bmatrix} \lambda_1 \\ \lambda_2 \end{bmatrix} = \begin{bmatrix} a \\ b \end{bmatrix}, \tag{12}$$

and because the columns of the matrix $\begin{bmatrix} 1 & 2 \\ 3 & -1 \end{bmatrix}$ are independent vectors its determinant is nonsingular, the matrix has an inverse and therefore (12) has a solution for λ_1, and λ_2, namely $\lambda_1 = (a + 2b)/7$ and $\lambda_2 = (3a - b)/7$.

Hence

$$\begin{bmatrix} a \\ b \end{bmatrix} = \tfrac{1}{7}(a + 2b)\begin{bmatrix} 1 \\ 3 \end{bmatrix} + \tfrac{1}{7}(3a - b)\begin{bmatrix} 2 \\ -1 \end{bmatrix}.$$

The reader should verify equation (8) in this manner.

Definition. For convenience we shall use the abbreviation LINN to mean "linearly independent non-null". Thus a set of LINN vectors shall mean a set of linearly independent non-null vectors.

6. RANK

a. Independent rows and columns of a matrix

Explanation has been given of how a determinant is zero if any of its rows (or columns) are linear combinations of other rows (or columns). In other words, a determinant is zero if its rows (or columns) do not form a set of independent vectors. Evidently, therefore, a determinant cannot have its rows forming a dependent set and its columns an independent set, a statement which prompts the more general question of the relationship between the number of independent rows of a matrix and the number of independent columns. The relationship is simple.

Theorem. The number of independent rows in a matrix equals the number of independent columns and vice versa.

Example. In $A = \begin{bmatrix} 1 & 0 & -1 & 2 \\ 3 & 1 & 4 & 2 \\ 5 & 2 & 9 & 2 \end{bmatrix}$ the last two columns are linear combinations of the first two:

$$\begin{bmatrix} -1 \\ 4 \\ 9 \end{bmatrix} = -\begin{bmatrix} 1 \\ 3 \\ 5 \end{bmatrix} + 7\begin{bmatrix} 0 \\ 1 \\ 2 \end{bmatrix}$$

and

$$\begin{bmatrix} 2 \\ 2 \\ 2 \end{bmatrix} = 2\begin{bmatrix} 1 \\ 3 \\ 5 \end{bmatrix} - 4\begin{bmatrix} 0 \\ 1 \\ 2 \end{bmatrix}.$$

Therefore A has only two independent columns, and by the theorem it also has only two independent rows. This is indeed the case, since

$$[5 \quad 2 \quad 9 \quad 2] = 2[3 \quad 1 \quad 4 \quad 2] - [1 \quad 0 \quad -1 \quad 2].$$

**Proof.*

Consider A as a matrix of order $p \times q$ having k independent rows and m independent columns. The rows are then vectors of order q, so that by the corollary of the theorem in the previous Section, $k \leqq q$, and likewise $m \leqq p$. Note also that the property of independence of rows (or columns) of a matrix relates only to the rows (or columns) themselves, and not to their sequence in the matrix. Assuming the first k rows to be independent (and the first m columns) therefore has no effect on arguments based on consequences of linear independence. We find this convenient, and accordingly denote by X the leading $k \times m$ submatrix of A formed by the intersection of the k independent rows and m independent columns. A is then partitioned as

$$A = \begin{bmatrix} X_{k\times m} & Y_{k\times(q-m)} \\ Z_{(p-k)\times m} & W_{(p-k)\times(q-m)} \end{bmatrix}.$$

Further, let

$$X_{k\times m} = [u_1 \quad u_2 \quad \ldots \quad u_m]$$

and

$$Z_{(p-k)\times m} = [v_1 \quad v_2 \quad \ldots \quad v_m]$$

where the u's are vectors of order k and the v's are vectors of order $p - k$.

We seek to show that $k = m$. Let us assume that the columns of X are linearly dependent, i.e., for some non-zero scalars $\lambda_1, \lambda_2, \ldots, \lambda_m$,

$$\lambda_1 u_1 + \lambda_2 u_2 + \ldots + \lambda_m u_m = 0. \tag{13}$$

Now because the rows of A that contain X are linearly independent the rows of Z are linear combinations of those of X; i.e., the elements of the v's are linear combinations of those of the u's. Therefore, for the λ's for which (13) is true so is

$$\lambda_1 v_1 + \lambda_2 v_2 + \ldots + \lambda_m v_m = 0,$$

as is

$$\lambda_1 \begin{bmatrix} u_1 \\ v_1 \end{bmatrix} + \lambda_2 \begin{bmatrix} u_2 \\ v_2 \end{bmatrix} + \ldots + \lambda_m \begin{bmatrix} u_m \\ v_m \end{bmatrix} = 0.$$

But this means the m columns of A containing X and Z are dependent, which is contrary to what is given. Therefore the assumption that the columns of X are dependent is false; i.e., the m u's are independent. But the u's are of order k, and therefore by the previous theorem they can be independent only if $m \leqq k$. By the same argument we could also show that the k rows of X, having order m, are also independent, and hence $k \leqq m$. Therefore $k = m$ and the number of independent rows of A is the same as the number of independent columns.

b. Definition of rank

The definition of rank is now simple. The *rank* of any matrix is the number of linearly independent rows (or columns) therein.

Notation. We will use the notation $r(A)$ to mean the rank of A. Thus $r(A) = r$ means that A has rank r, that there are r linearly independent rows (columns) in A.

Certain properties and consequences of rank are worth stating.

i. Rank is zero only for a null matrix, otherwise it is always a positive integer.

ii. The rank of a rectangular matrix $A_{p\times q}$ is equal to or less than the smaller of p and q.

iii. The rank of a square matrix is equal to or is less than its order.

iv. If $r(A) = r$ there is at least one nonsingular minor of order r in A, and all minors of order greater than r are zero.

v. If $r(A_{p\times q}) = r$, A can be factorized as $A_{p\times q} = X_{p\times r}Y_{r\times q}$.

vi. If $r(A_n) < n$, A^{-1} does not exist.

The first three of these statements are obvious from the definition of rank as the number of independent rows (columns) in a matrix. Result (iv) is true because $r(A) = r$ implies that A has r independent rows and r independent columns and therefore the determinant of the matrix formed by these rows and columns is non-zero. All minors of order greater than r are zero because at least one of their rows will be a linear combination of the other r. Results (v) and (vi) merit more detailed discussion.

c. Factorizing a matrix

An immediate consequence of the notion of rank is that a matrix of p rows and q columns having rank r can be partitioned into a group of r independent rows and a group of $p - r$ rows that are linear combinations of the first group. This leads to a factorization of A as $A_{p\times q} = X_{p\times r}Y_{r\times q}$, as in (v).

Suppose the first r rows of a matrix $M_{p\times q}$ are independent. M can then be partitioned as $M = \begin{bmatrix} F \\ KF \end{bmatrix}$ where $F_{r\times q}$ represents the r independent rows and KF is $K_{p-r\times r}F_{r\times q}$, the $p - r$ rows that are linear combinations of the first r rows. Now suppose the first r columns of M are independent and that M is further partitioned as $M = \begin{bmatrix} A & B \\ KA & KB \end{bmatrix}$ where A is $r \times r$ and B is $r \times (q - r)$, with the columns $\begin{bmatrix} B \\ KB \end{bmatrix}$ being linear combinations of the independent columns $\begin{bmatrix} A \\ KA \end{bmatrix}$, namely $\begin{bmatrix} B \\ KB \end{bmatrix} = \begin{bmatrix} AL \\ KAL \end{bmatrix}$. This gives

$$M = \begin{bmatrix} A & AL \\ KA & KAL \end{bmatrix} \tag{14}$$

$$= \begin{bmatrix} I \\ K \end{bmatrix} [A \quad AL]$$

$$= \begin{bmatrix} I_r \\ K_{(p-r)\times r} \end{bmatrix} [A_{r\times r} \quad A_{r\times r}L_{r\times(q-r)}], \tag{15}$$

the last of these expressions being of the form required, $M_{p\times q} = X_{p\times r}Y_{r\times q}$, a $p \times r$ matrix multiplying an $r \times q$ matrix. To find X and Y explicitly, M is partitioned appropriately and equated to (14). Thus

$$M = \begin{bmatrix} A & AL \\ KA & KAL \end{bmatrix} = \begin{bmatrix} M_{11} & M_{12} \\ M_{21} & M_{22} \end{bmatrix}$$

gives $A = M_{11}$, $L = M_{11}^{-1}M_{12}$ and $K = M_{21}M_{11}^{-1}$, so that

$$M = \begin{bmatrix} I \\ K \end{bmatrix} [A \quad AL] = \begin{bmatrix} I \\ M_{21}M_{11}^{-1} \end{bmatrix} [M_{11} \quad M_{12}].$$

Example.

$$M = \left[\begin{array}{cc|c} 1 & 2 & 9 \\ 2 & 0 & 2 \\ \hline 3 & -2 & -5 \end{array}\right] = \begin{bmatrix} I \\ [3 \quad -2]\begin{bmatrix} 1 & 2 \\ 2 & 0 \end{bmatrix}^{-1} \end{bmatrix} \begin{bmatrix} 1 & 2 & 9 \\ 2 & 0 & 2 \end{bmatrix}$$

$$= \begin{bmatrix} 1 & 0 \\ 0 & 1 \\ -1 & 2 \end{bmatrix} \begin{bmatrix} 1 & 2 & 9 \\ 2 & 0 & 2 \end{bmatrix}.$$

An obvious extension of these results is that if A has rank of unity it can be factorized in the form $A = xy'$ where x is a column vector and y' a row vector. For example,

$$\begin{bmatrix} 2 & -2 & 4 \\ 3 & -3 & 6 \end{bmatrix} = \begin{bmatrix} 2 \\ 3 \end{bmatrix} [1 \quad -1 \quad 2].$$

d. Rank and the inverse of a matrix

Ascertaining whether or not equations $Ax = y$ can be solved as $x = A^{-1}y$ is greatly facilitated by the notion of rank. The solution $x = A^{-1}y$ exists only if A^{-1} does; this occurs only if $|A| \neq 0$, and this is true only if the rows (and the columns) are linearly independent; i.e., only if the rank of A equals its order. By introducing the concept of rank we therefore alter the problem of finding whether the value of a determinant of a square matrix is zero or non-zero to one of finding if the rank of the matrix is less than its order. If it is, the determinant is zero, but if the rank equals its order the determinant is non-zero. Because investigating rank relative to order is easier than evaluating a determinant we use this relationship between rank and order for ascertaining the existence or non-existence of an inverse; for, if $r(A_n) = r$, then

$$r < n \text{ implies } |A| = 0 \text{ and the non-existence of } A^{-1};$$

and

$$r = n \text{ implies } |A| \neq 0 \text{ and the existence of } A^{-1}.$$

In the latter case A is nonsingular; it is also said to be of *full rank*. The rank of the inverse is also n; if it were less $|A^{-1}|$ would be zero, which we know is not true because if A^{-1} exists, $|A^{-1}|$ equals $1/|A|$, and $|A|$ is non-zero.

The problem of ascertaining the existence of an inverse is therefore equivalent to ascertaining if the rank of a square matrix is less than its order. More generally we may wish to derive the rank exactly, of both square and rectangular matrices. If the rank of a matrix is r, r of its rows are linearly independent, but locating a set of r such rows is not always an easy task. On the other hand, to actually derive the value of r in any particular case is not at all difficult. To do so, extensive use is made of matrices known as elementary operators, which we will now consider.

7. ELEMENTARY OPERATORS

a. Definitions

Three kinds of matrix known as elementary operators play an important role in matrix theory. They are all square matrices derived very simply from the identity matrix. When used in a product with another matrix the effect of each of them is to produce in that matrix the same kind of a manipulation of rows (or columns) as is used in the evaluation of determinants discussed in Chapter 3, that is, rows (or columns) are amended to become linear combinations of one another. Furthermore, the product matrix has the same rank as the original matrix. These two properties of elementary operators provide us, as we shall see, with methods for determining the rank of any matrix. The elementary operators are as follows.

i. E_{ij} is I with its ith and jth rows interchanged. For example,

$$E_{12} = \begin{bmatrix} 0 & 1 & 0 \\ 1 & 0 & 0 \\ 0 & 0 & 1 \end{bmatrix}.$$

Pre-multiplication of a matrix A by E_{ij} interchanges the ith and jth rows of A. Thus

$$E_{12}A = \begin{bmatrix} 0 & 1 & 0 \\ 1 & 0 & 0 \\ 0 & 0 & 1 \end{bmatrix}\begin{bmatrix} 1 & 1 & 1 \\ 2 & 2 & 2 \\ 3 & 3 & 3 \end{bmatrix} = \begin{bmatrix} 2 & 2 & 2 \\ 1 & 1 & 1 \\ 3 & 3 & 3 \end{bmatrix}.$$

ii. $R_{ii}(\lambda)$ is I with λ replacing the 1 in the ith diagonal term.

Pre-multiplication of A by $R_{ii}(\lambda)$ leads to the ith row of A being multiplied by λ:

$$[R_{22}(4)]A = \begin{bmatrix} 1 & 0 & 0 \\ 0 & 4 & 0 \\ 0 & 0 & 1 \end{bmatrix}\begin{bmatrix} 1 & 1 & 1 \\ 2 & 2 & 2 \\ 3 & 3 & 3 \end{bmatrix} = \begin{bmatrix} 1 & 1 & 1 \\ 8 & 8 & 8 \\ 3 & 3 & 3 \end{bmatrix}.$$

iii. $P_{ij}(\lambda)$ is I with the scalar λ replacing the zero in the ith row and jth column for $i \neq j$. Pre-multiplication of A by $P_{ij}(\lambda)$ results in adding λ times the jth row of A to its ith row:

$$[P_{12}(2)]A = \begin{bmatrix} 1 & 2 & 0 \\ 0 & 1 & 0 \\ 0 & 0 & 1 \end{bmatrix} \begin{bmatrix} 1 & 1 & 1 \\ 2 & 2 & 2 \\ 3 & 3 & 3 \end{bmatrix} = \begin{bmatrix} 5 & 5 & 5 \\ 2 & 2 & 2 \\ 3 & 3 & 3 \end{bmatrix}.$$

b. Post-multiplication

The effects of pre-multiplying a matrix by an elementary operator have just been stated and illustrated. The reader should satisfy himself that post-multiplication of A by an elementary operator performs the same manipulations on the columns of A as are performed on the rows of A by pre-multiplication; for example, post-multiplication by E_{ij} interchanges the ith and jth columns.

c. Determinants

From the general form of the elementary operators it is easily shown that

$$|E_{ij}| = -1, \qquad |R_{ii}(\lambda)| = \lambda \qquad \text{and} \quad |P_{ij}(\lambda)| = 1.$$

Applying the general result for the determinant of a product matrix we therefore have the following results when A is square:

$$|E_{ij}A| = -|A|, \qquad |R_{ii}(\lambda)A| = \lambda|A| \qquad \text{and} \quad |P_{ij}(\lambda)A| = |A|.$$

d. Transposes

Verification of the following results is straightforward:

$$E'_{ij} = E_{ij}, \qquad R'_{ii}(\lambda) = R_{ii}(\lambda) \qquad \text{and} \quad P'_{ij}(\lambda) = P_{ji}(\lambda).$$

Thus an elementary operator and its transpose are of the same form, and indeed transposing has no effect on the E- and R-type operators, i.e. these two are symmetric.

e. Inverses

The following will be found true:

$$E_{ij}^{-1} = E_{ij}, \qquad [R_{ii}(\lambda)]^{-1} = R_{ii}(1/\lambda) \qquad \text{and} \quad [P_{ij}(\lambda)]^{-1} = P_{ij}(-\lambda).$$

Thus an E-type operator is the same as its inverse, and the effect of inverting R- and P-type operators is simply that of changing the constants, replacing λ by $1/\lambda$ in the R-type, and λ by $-\lambda$ in the P-type. From this and the preceding paragraph we see that the underlying form of any one of these elementary operators is unchanged when the operator is either inverted or transposed.

8. RANK AND THE ELEMENTARY OPERATORS

The rank of a matrix is unaffected when it is multiplied by an elementary operator, that is to say, if the matrix A is multiplied by an E-, P- or R-type operator the rank of the product is the same as the rank of A. We discuss the validity of this statement for each operator in turn. First the E-type. If A is pre-multiplied by an E-type operator the product EA is simply A with two rows interchanged. Hence EA is the same as A except with the rows in a different sequence. The number of them that are linearly independent will be unaltered. Therefore $r(EA) = r(A)$. Similar reasoning holds for the products of A with the other elementary operators. In the product of A and an R-type operator every element of some row (or column) of A is multiplied by a constant, and the product of a P-type operator with A is the same as A except it has a multiple of one row (column) added to another. In both cases the independence of rows (columns) is unaffected and the same number will be linearly independent after making the product as before. Thus multiplication of any matrix by an elementary operator does not alter rank.

9. FINDING THE RANK OF A MATRIX

We have just seen that in multiplying a matrix by an elementary operator the rank of the product is the same as that of the original matrix: therefore the rank of a matrix is the same as the rank of its successive products with several elementary operators. And by the nature of the manipulations involved in such products (the same as in expanding determinants) they can be used to reduce various elements to zero. In practice the process is used to reduce to zero the elements below the diagonal; the number of non-zero elements remaining in the diagonal then represents the size of the largest non-zero minor in the matrix and is accordingly the rank of the matrix.

Example. For

$$A = \begin{bmatrix} 3 & 6 & 5 & 2 \\ 6 & 16 & 18 & 7 \\ 3 & 6 & 5 & 2 \end{bmatrix}, \qquad P = P_{21}(-2) = \begin{bmatrix} 1 & 0 & 0 \\ -2 & 1 & 0 \\ 0 & 0 & 1 \end{bmatrix}$$

and

$$\mathrm{P}^* = P_{31}(-1) = \begin{bmatrix} 1 & 0 & 0 \\ 0 & 1 & 0 \\ -1 & 0 & 1 \end{bmatrix},$$

we have

$$PA = \begin{bmatrix} 3 & 6 & 5 & 2 \\ 0 & 4 & 8 & 3 \\ 3 & 6 & 5 & 2 \end{bmatrix} \quad \text{and} \quad P^*PA = \begin{bmatrix} 3 & 6 & 5 & 2 \\ 0 & 4 & 8 & 3 \\ 0 & 0 & 0 & 0 \end{bmatrix}.$$

This matrix P^*PA has two "diagonal" elements that are non-zero. Its largest order minor that is nonsingular is therefore a 2×2 and so the rank of the matrix is 2. Consequently the rank of A is also 2, because P^*PA is simply A pre-multiplied by elementary operators, a procedure that does not affect rank.

Finding the rank of a matrix by this means is very simple—merely manipulate the rows of the matrix by adding multiples of them to each other to reduce the sub-diagonal elements to zero. Initially the first row is used to reduce the sub-diagonal elements of the first column to zeros, basing the calculations on the first term of the first row. Then the elements below the diagonal in the second column are reduced to zeros using the second row, based on its diagonal term. Because the sub-diagonal elements of the first column have already been made zero in the first step of this procedure this second step does not affect their values. And so the process is continued, using each row this way in turn, until all remaining rows are zero or until the last row is reached. The number of non-zero diagonal elements is then the rank.

Matrices related as are the successive products in this procedure are said to be *equivalent*. Thus two matrices A and B are equivalent if B can be obtained from A by multiplying A by elementary operators. We write $A \cong B$ for A equivalent to B.

Examples. Consider

$$A = \begin{bmatrix} 1 & 2 & 3 & 4 \\ 6 & 8 & 3 & 2 \\ 2 & 1 & 0 & 1 \end{bmatrix}.$$

Leaving the first row as it is we can reduce the sub-diagonal elements of the first column to zero by subtracting six times the first row from the second and twice the first row from the third. Thus

$$A \cong \begin{bmatrix} 1 & 2 & 3 & 4 \\ 0 & -4 & -15 & -22 \\ 0 & -3 & -6 & -7 \end{bmatrix}.$$

If we now multiply the third row by 4 (equivalent to pre-multiplication by an R-type operator) and then subtract 3 times the second row we get

$$A \cong \begin{bmatrix} 1 & 2 & 3 & 4 \\ 0 & -4 & -15 & -22 \\ 0 & 0 & 21 & 38 \end{bmatrix}$$

and hence conclude that $r(A) = 3$. As a second example consider

$$B = \begin{bmatrix} 1 & 2 \\ 3 & 6 \\ 6 & 4 \\ 5 & 6 \end{bmatrix}.$$

If r_i denotes the ith row, then the operations $r_2 - 3r_1$, $r_3 - 6r_1$ and $r_4 - 5r_1$ yield

$$B \cong \begin{bmatrix} 1 & 2 \\ 0 & 0 \\ 0 & -8 \\ 0 & -4 \end{bmatrix}.$$

The operations of subtracting half row 3 from row 4 and of interchanging rows 2 and 3 then give

$$B \cong \begin{bmatrix} 1 & 2 \\ 0 & -8 \\ 0 & 0 \\ 0 & 0 \end{bmatrix}$$

so that $r(B) = 2$.

10. EQUIVALENCE

Equivalence has already been defined in the previous paragraph: two matrices are equivalent when one can be derived from the other by multiplying it by a series of elementary operators. That is, B is equivalent to A if

$$P_1 \ldots P_s A Q_1 \ldots Q_t = B$$

where $P_1, \ldots, P_s$ and $Q_1, \ldots, Q_t$ are elementary operators. If A is $r \times c$ each of the P's is square with order r, each Q is square also with order c, and B is the same order as A, namely $r \times c$. If we write P for the product of the P's and Q for the product of the Q's, P and Q are square, their inverses exist, and we have $PAQ = B$, and hence

$$A = P^{-1}BQ^{-1}. \qquad (16)$$

Since the inverse of an elementary operator is itself an elementary operator P^{-1} and Q^{-1}, which are products of the inverses of the individual P's and Q's, are also products of elementary operators. Therefore equation (16) shows that multiplying B by elementary operators can lead to A. Consequently B is equivalent to A and we write $B \cong A$. Hence we have shown that if A is equivalent to B, B is equivalent to A. Furthermore, since multiplication by elementary operators does not affect rank, $r(A) = r(B)$.

11. REDUCTION TO EQUIVALENT CANONICAL FORM

We have just seen that if A is equivalent to B there exist matrices P and Q such that $PAQ = B$ where P and Q are products of elementary operators. P represents operations on rows and Q represents operations on columns. An interesting and useful application of this result is that any matrix A having rank r is equivalent to a matrix of the form $C = \begin{bmatrix} I_r & 0 \\ 0 & 0 \end{bmatrix}$, where C is the same order as A, I_r is the identity matrix of order r and the zeros are null matrices of appropriate order. We first give an example of this result and then state and prove it as a formal theorem.

Example. The matrix

$$A = \begin{bmatrix} 2 & 6 & 4 & 2 \\ 4 & 15 & 14 & 7 \\ 2 & 9 & 10 & 5 \end{bmatrix},$$

is clearly of rank 2 (its second row equals the sum of the other two). To develop C from A we carry out the following operations on A.

Operations	Effect on A
i. row 2 − 2(row 1) row 3 − row 1	$\begin{bmatrix} 2 & 6 & 4 & 2 \\ 0 & 3 & 6 & 3 \\ 0 & 3 & 6 & 3 \end{bmatrix}$
ii. row 3 − row 2	$\begin{bmatrix} 2 & 6 & 4 & 2 \\ 0 & 3 & 6 & 3 \\ 0 & 0 & 0 & 0 \end{bmatrix}$
iii. column 2 − 3(column 1) column 3 − 2(column 1) column 4 − column 1	$\begin{bmatrix} 2 & 0 & 0 & 0 \\ 0 & 3 & 6 & 3 \\ 0 & 0 & 0 & 0 \end{bmatrix}$
iv. column 3 − 2(column 2) column 4 − column 2	$\begin{bmatrix} 2 & 0 & 0 & 0 \\ 0 & 3 & 0 & 0 \\ 0 & 0 & 0 & 0 \end{bmatrix}$

Notice that at this stage A has been amended to be of the form

$$\Delta = \begin{bmatrix} D_r & 0 \\ 0 & 0 \end{bmatrix}$$

where D_r is a diagonal matrix with r non-zero elements. This is often referred to as the *diagonal form* of A. It can be reduced to the C form in several ways. Suppose we operate on columns and

v. multiply column 1 by $\frac{1}{2}$ and column 2 by $\frac{1}{3}$.

This gives

$$A \cong C = \begin{bmatrix} 1 & 0 & 0 & 0 \\ 0 & 1 & 0 & 0 \\ 0 & 0 & 0 & 0 \end{bmatrix} = \begin{bmatrix} I_2 & 0 \\ 0 & 0 \end{bmatrix},$$

in concert with A being of rank 2.

For a matrix of order $m \times n$ and rank r, operations of this nature lead to

$$A_{m\times n} \cong \begin{bmatrix} I_r & 0_{r\times n-r} \\ 0_{m-r\times r} & 0_{m-r\times n-r} \end{bmatrix}$$

the zeros being null matrices of the orders shown. If $r = m, < n$ the form is $[I_m \quad 0]$, it is $\begin{bmatrix} I_n \\ 0 \end{bmatrix}$ if $r = n, < m$ and I_n if $r = m = n$. The formal theorem concerning this result is as follows.

Theorem. Any non-null matrix of rank r is equivalent to a matrix

$$C = \begin{bmatrix} I_r & \cdot \\ \cdot & \cdot \end{bmatrix}$$

where I_r is the identity matrix of order r and where the dots represent null matrices of appropriate order.

Implied in the theorem is the existence of matrices P and Q such that $PAQ = C$ where P and Q are products of matrices of elementary operators. If A is $m \times n$, P is square of order m and Q is square of order n, and

$$P_m A_{m\times n} Q_n = C = \begin{bmatrix} I_r & \cdot \\ \cdot & \cdot \end{bmatrix}.$$

The matrix C is usually known as the *equivalent canonical form* of A, or the *canonical form under equivalence*. The procedure for obtaining it is often called *reduction to canonical form*.

To find P and Q we note that $PAQ = (PI)A(IQ)$ and hence performing on the identity matrix the operations performed on A to reduce it to canonical form produces the matrices P and Q, grouping row operations together to get P and column operations to get Q. Thus in the example we obtain P by applying operations (i) and (ii) to I_3. Hence

$$I_3 = \begin{bmatrix} 1 & 0 & 0 \\ 0 & 1 & 0 \\ 0 & 0 & 1 \end{bmatrix} \cong \begin{bmatrix} 1 & 0 & 0 \\ -2 & 1 & 0 \\ -1 & 0 & 1 \end{bmatrix} \cong \begin{bmatrix} 1 & 0 & 0 \\ -2 & 1 & 0 \\ 1 & -1 & 1 \end{bmatrix} = P;$$

and carrying out operations (iii), (iv) and (v) on I_4 gives Q:

$$I_4 = \begin{bmatrix} 1 & 0 & 0 & 0 \\ 0 & 1 & 0 & 0 \\ 0 & 0 & 1 & 0 \\ 0 & 0 & 0 & 1 \end{bmatrix} \cong \begin{bmatrix} 1 & -3 & -2 & -1 \\ 0 & 1 & 0 & 0 \\ 0 & 0 & 1 & 0 \\ 0 & 0 & 0 & 1 \end{bmatrix}$$

$$\cong \begin{bmatrix} 1 & -3 & 4 & 2 \\ 0 & 1 & -2 & -1 \\ 0 & 0 & 1 & 0 \\ 0 & 0 & 0 & 1 \end{bmatrix} \cong \begin{bmatrix} \frac{1}{2} & -1 & 4 & 2 \\ 0 & \frac{1}{3} & -2 & -1 \\ 0 & 0 & 1 & 0 \\ 0 & 0 & 0 & 1 \end{bmatrix} = Q.$$

To check on the results we find that

$$PA = \begin{bmatrix} 1 & 0 & 0 \\ -2 & 1 & 0 \\ 1 & -1 & 1 \end{bmatrix} \begin{bmatrix} 2 & 6 & 4 & 2 \\ 4 & 15 & 14 & 7 \\ 2 & 9 & 10 & 5 \end{bmatrix} = \begin{bmatrix} 2 & 6 & 4 & 2 \\ 0 & 3 & 6 & 3 \\ 0 & 0 & 0 & 0 \end{bmatrix}$$

and

$$PAQ = \begin{bmatrix} 2 & 6 & 4 & 2 \\ 0 & 3 & 6 & 3 \\ 0 & 0 & 0 & 0 \end{bmatrix} \begin{bmatrix} \frac{1}{2} & -1 & 4 & 2 \\ 0 & \frac{1}{3} & -2 & -1 \\ 0 & 0 & 1 & 0 \\ 0 & 0 & 0 & 1 \end{bmatrix} = \begin{bmatrix} 1 & 0 & 0 & 0 \\ 0 & 1 & 0 & 0 \\ 0 & 0 & 0 & 0 \end{bmatrix}$$

as required.

***Proof of theorem.**

Since A is non-null there is at least one element that is non-zero. Suppose it is a_{pq}. Then by successively multiplying A by E-type operators (pre- and post-multiplying), the element a_{pq} can be brought to the leading position. Call the resulting matrix B_1. Then $B_1 \cong A$ and $b_{11} = a_{pq} \neq 0$. Now pre-multiply B_1 by P-type operators, $P_{i1}(-b_{i1}/b_{11})$ for $i = 2, 3, \ldots, m$, to reduce to zero all elements (except the first) of the first column of B_1; and likewise post-multiply B_1 by $P_{1j}(-b_{1j}/b_{11})$ for $j = 2, 3, \ldots, n$, to reduce the first row (except the first element) to zeros. Call the resulting matrix B_2. Then

$$A_{m\times n} \cong B_1 \cong B_2 = \begin{bmatrix} b_{11} & 0 & \ldots & 0 \\ 0 & c_{22} & \ldots & c_{2n} \\ \cdot & & & \cdot \\ \cdot & & & \cdot \\ \cdot & & & \cdot \\ 0 & c_{m2} & \ldots & c_{mn} \end{bmatrix}.$$

This process can now be repeated operating on the c_{ij}'s, provided that at least one of them is non-zero; and repeated until, in a form similar to this, there are only zero terms remaining apart from the non-zero terms that have been brought to the diagonal at each step in the process. Furthermore, at each step the resulting matrix is equivalent to A and has rank r, the same as A; and by the nature of the operations (of bringing non-zero elements into the diagonal and zeroizing off-diagonal elements) we know the final form is diagonal, as far as non-zero elements are concerned; and because the rank is r there must be exactly r of these elements. Multiplying each non-null column by the reciprocal of the non-zero term therein yields the final form as $\begin{bmatrix} I_r & \cdot \\ \cdot & \cdot \end{bmatrix}$ and

$$A_{m\times n} \cong \begin{bmatrix} I_r & \cdot \\ \cdot & \cdot \end{bmatrix},$$

with

$$PAQ = \begin{bmatrix} I_r & \cdot \\ \cdot & \cdot \end{bmatrix} \tag{17}$$

where P and Q are the products of the elementary operators. It is clear that the proof of the theorem is largely a formal statement of the procedures used in the example.

The matrices P and Q are not unique in this process. For instance, in the example we might perform an operation equivalent to (v) on rows instead of columns. PAQ would still be C although P and Q would be

$$P = \begin{bmatrix} \frac{1}{2} & 0 & 0 \\ -\frac{2}{3} & \frac{1}{3} & 0 \\ 1 & -1 & 1 \end{bmatrix} \quad \text{and} \quad Q = \begin{bmatrix} 1 & -3 & 4 & 2 \\ 0 & 1 & -2 & -1 \\ 0 & 0 & 1 & 0 \\ 0 & 0 & 0 & 1 \end{bmatrix}.$$

The order in which the individual elementary operations are carried out conditions the final forms of P and Q; whatever they may be, C will always be the same for any particular matrix. It will be the same order as A, and for all matrices of rank r it can be partitioned as $C = \begin{bmatrix} I_r & 0 \\ 0 & 0 \end{bmatrix}$ where the zeros are null matrices of appropriate order.

12. CONGRUENT REDUCTION OF SYMMETRIC MATRICES

The procedure just discussed is applicable to any matrix. With symmetric matrices it is more specialized, arising from the property of symmetry. Notice in the example that at stage (iv) the matrix has been reduced to what is called the diagonal form,

$$\Delta = \begin{bmatrix} D_r & 0 \\ 0 & 0 \end{bmatrix},$$

where D_r is a diagonal matrix of r non-zero elements. The canonical form C was derived from this by multiplying each non-null column by the reciprocal of the non-zero term therein. This is step (v) in the example, and is also the last step in the proof of the theorem.

Now consider similar procedures for a symmetric matrix. Keeping in mind the nature of such a matrix, that each column is the same as the corresponding row, we realize that the same operations as are made on the rows of A to reduce the sub-diagonal elements to zero will, if performed on the columns of A, reduce the elements above the diagonal to zeros also. Therefore, when A is symmetric, performing the same operations on columns as on rows reduces A to the diagonal form Δ. By the nature of the matrices of elementary operators this means that in PAQ the matrix Q equals the transpose of P, i.e., $Q = P'$. Hence we have $PAP' = \Delta$.

Let us suppose all diagonal elements of D_r are positive. Then we can form the diagonal matrix whose elements are the reciprocals of the square roots of the diagonal elements of D_r. Call this matrix R_r: $R_r^2 = D_r^{-1}$. If A is of order n and if F is formed from R_r as

$$F = \begin{bmatrix} R_r & 0 \\ 0 & I_{n-r} \end{bmatrix},$$

then

$$FPAP'F = C = \begin{bmatrix} I_r & 0 \\ 0 & 0 \end{bmatrix},$$

and writing $FP = P^*$ gives $P^*AP^{*\prime} = C$. This means that, provided all elements of D_r are positive, a symmetric matrix can be reduced to canonical form under equivalence by pre-multiplying by a matrix P^* and post-multiplying by its transpose $P^{*\prime}$. This is called the *congruent reduction* of the symmetric matrix A, and C is known as the *canonical form under congruence*. The reader requiring a more formal proof of this result is referred, for example, to Aitken (1948), Section 29.

Suppose now that not all diagonal elements of D_r are positive, but that q of them are negative. Then pre- and post-multiply PAP' appropriately so that the first $r - q$ diagonal elements of D_r are positive and the last q are negative. Now define F as before except that R_r has as elements the reciprocals of the square roots of the elements of D_r disregarding sign. With $FP = P^*$ we then have

$$P^*AP^{*\prime} = \begin{bmatrix} I_{r-q} & 0 & 0 \\ 0 & -I_q & 0 \\ 0 & 0 & 0 \end{bmatrix}.$$

This is the general canonical form under congruence, for the symmetric matrix A. When $q = 0$ it reduces to C. The difference between the order of I_{r-q} and that of I_q is known as the *signature* of A, i.e., signature $= r - 2q$. Retaining the negative signs in the diagonal means that the reduction to this form is entirely in terms of real numbers. If we are prepared to use imaginary numbers, involving $i = \sqrt{-1}$, then reduction to C can be made, using an F that involves imaginary numbers.

Example. For the symmetric matrix

$$A = \begin{bmatrix} 4 & 12 & 20 \\ 12 & 45 & 78 \\ 20 & 78 & 136 \end{bmatrix}$$

the following operations reduce A to diagonal form.

Operations	**Reduction of A**
i. row 2 − 3(row 1) row 3 − 5(row 1)	$A \cong \begin{bmatrix} 4 & 12 & 20 \\ 0 & 9 & 18 \\ 0 & 18 & 36 \end{bmatrix}$
ii. column 2 − 3(column 1) column 3 − 5(column 1)	$A \cong \begin{bmatrix} 4 & 0 & 0 \\ 0 & 9 & 18 \\ 0 & 18 & 36 \end{bmatrix}$
iii. row 3 − 2(row 2)	$A \cong \begin{bmatrix} 4 & 0 & 0 \\ 0 & 9 & 18 \\ 0 & 0 & 0 \end{bmatrix}$
iv. column 3 − 2(column 2)	$A \cong \begin{bmatrix} 4 & 0 & 0 \\ 0 & 9 & 0 \\ 0 & 0 & 0 \end{bmatrix}$

P is obtained by carrying out operations (i) and (iii) on I_3:

$$I_3 = \begin{bmatrix} 1 & 0 & 0 \\ 0 & 1 & 0 \\ 0 & 0 & 1 \end{bmatrix} \cong \begin{bmatrix} 1 & 0 & 0 \\ -3 & 1 & 0 \\ -5 & 0 & 1 \end{bmatrix} \cong \begin{bmatrix} 1 & 0 & 0 \\ -3 & 1 & 0 \\ 1 & -2 & 1 \end{bmatrix} = P.$$

Operations (ii) and (iv) yield P', and the product PAP' equals $\begin{bmatrix} 4 & 0 & 0 \\ 0 & 9 & 0 \\ 0 & 0 & 0 \end{bmatrix}$, as the reader can verify for himself. The canonical form is then obtained as $P^*AP^{*\prime}$ where

$$P^* = \begin{bmatrix} 1/\sqrt{4} & 0 & 0 \\ 0 & 1/\sqrt{9} & 0 \\ 0 & 0 & 1 \end{bmatrix} P = \begin{bmatrix} 1/2 & 0 & 0 \\ 0 & 1/3 & 0 \\ 0 & 0 & 1 \end{bmatrix} \begin{bmatrix} 1 & 0 & 0 \\ -3 & 1 & 0 \\ 1 & -2 & 1 \end{bmatrix}$$

$$= \begin{bmatrix} 1/2 & 0 & 0 \\ -1 & 1/3 & 0 \\ 1 & -2 & 1 \end{bmatrix}.$$

The reader can verify that $P^*AP^{*\prime}$ is the same form as C.

13. THE RANK OF A PRODUCT MATRIX

A result that is useful in subsequent work concerns the relationship between the rank of a matrix product AB and the ranks of A and B.

Theorem. The rank of a product, $A_{m\times q}B_{q\times n}$, cannot exceed the rank of either A or B.

**Proof.*

If $r(A) = r$ there exist matrices P and Q such that

$$PAQ = \begin{bmatrix} I_r & \cdot \\ \cdot & \cdot \end{bmatrix}.$$

Therefore

$$PA = \begin{bmatrix} I_r & \cdot \\ \cdot & \cdot \end{bmatrix} Q^{-1}$$

and

$$PAB = \begin{bmatrix} I_r & \cdot \\ \cdot & \cdot \end{bmatrix} Q^{-1}B = \begin{bmatrix} G \\ \cdot \end{bmatrix} \text{ say,}$$

where G is $r \times n$. Therefore $r(PAB)$ cannot exceed r; and because P is a product of elementary operators $r(PAB) = r(AB)$, so that $r(AB) \not> r = r(A)$. Similarly it may be shown that $r(AB) \not> r(B)$. Hence $r(AB)$ cannot exceed the rank of either A or B.

It may be noted in addition that $r(A)$ equals or is less than the smaller of m and q and $r(B)$ equals or is less than the smaller of q and n. Therefore, when A is $m \times q$ and B is $q \times n$, $r(AB)$ cannot exceed the smallest of the values m, n, q, $r(A)$ and $r(B)$.

*Brief mention has already been made of left and right inverses of rectangular matrices in Section 4.7. It was shown there that for $A_{r\times c}$ no matrix $B_{c\times r}$ exists for which both $AB = I_r$ and $BA = I_c$. We now consider the conditions under which one or other of these equations can be true. With the aid of the theorem just given the situation is simple. The identity matrix I_r has rank r, and therefore the equation $A_{r\times c}B_{c\times r} = I_r$ can be true only if both A and B have rank at least equal to r. Likewise $D_{c\times r}$ for $D_{c\times r}A_{r\times c} = I_c$ exists only if the rank of A is at least equal to c. But for $r \neq c$, A has rank equal or less than the smaller of both r and c so that it cannot be equal or greater than both of them. Therefore $B_{c\times r}$ and $D_{c\times r}$ for which $AB = I_r$ and $DA = I_c$ cannot both exist. Only one of them can, and then only if A is of full rank. If this occurs and $r < c$, B exists, but if $r > c$, D exists. Thus a rectangular matrix has neither a left nor a right inverse if it is not of full rank. When it is of full rank it has only a right inverse if there are less rows than columns, and only a left inverse if there are more rows than columns.[1]

14. EXERCISES

1 For $A = \begin{bmatrix} 1 & 2 & 3 \\ 4 & 5 & 7 \\ 9 & 8 & 6 \end{bmatrix}$

and $E = \begin{bmatrix} 0 & 1 & 0 \\ 1 & 0 & 0 \\ 0 & 0 & 1 \end{bmatrix}$, $R = \begin{bmatrix} 1 & 0 & 0 \\ 0 & 1 & 0 \\ 0 & 0 & 2 \end{bmatrix}$ and $P = \begin{bmatrix} 1 & 3 & 0 \\ 0 & 1 & 0 \\ 0 & 0 & 1 \end{bmatrix}$,

show that

(*a*) $EA = \begin{bmatrix} 4 & 5 & 7 \\ 1 & 2 & 3 \\ 9 & 8 & 6 \end{bmatrix}$, $AE = \begin{bmatrix} 2 & 1 & 3 \\ 5 & 4 & 7 \\ 8 & 9 & 6 \end{bmatrix}$, $EAE = \begin{bmatrix} 5 & 4 & 7 \\ 2 & 1 & 3 \\ 8 & 9 & 6 \end{bmatrix}$;

(*b*) $RA = \begin{bmatrix} 1 & 2 & 3 \\ 4 & 5 & 7 \\ 18 & 16 & 12 \end{bmatrix}$, $AR = \begin{bmatrix} 1 & 2 & 6 \\ 4 & 5 & 14 \\ 9 & 8 & 12 \end{bmatrix}$, $RAR = \begin{bmatrix} 1 & 2 & 6 \\ 4 & 5 & 14 \\ 18 & 16 & 24 \end{bmatrix}$;

(*c*) $PA = \begin{bmatrix} 13 & 17 & 24 \\ 4 & 5 & 7 \\ 9 & 8 & 6 \end{bmatrix}$, $AP = \begin{bmatrix} 1 & 5 & 3 \\ 4 & 17 & 7 \\ 9 & 35 & 6 \end{bmatrix}$, $PAP = \begin{bmatrix} 13 & 56 & 24 \\ 4 & 17 & 7 \\ 9 & 35 & 6 \end{bmatrix}$;

(*d*) $REA = \begin{bmatrix} 4 & 5 & 7 \\ 1 & 2 & 3 \\ 18 & 16 & 12 \end{bmatrix}$, $REAP = \begin{bmatrix} 4 & 17 & 7 \\ 1 & 5 & 3 \\ 18 & 70 & 12 \end{bmatrix}$;

(*e*) $-|E| = 1 = |P| = \frac{1}{2}|R|$;

(*f*) $E = E^{-1} = E'$; and $R = R'$;

(*g*) $R^{-1} = \begin{bmatrix} 1 & 0 & 0 \\ 0 & 1 & 0 \\ 0 & 0 & \frac{1}{2} \end{bmatrix}$ and $P^{-1} = \begin{bmatrix} 1 & -3 & 0 \\ 0 & 1 & 0 \\ 0 & 0 & 1 \end{bmatrix}$.

2. Show that $\begin{bmatrix} 1 \\ 2 \\ 1 \end{bmatrix}$, $\begin{bmatrix} -1 \\ 3 \\ 2 \end{bmatrix}$ and $\begin{bmatrix} -13 \\ -1 \\ 2 \end{bmatrix}$ is not a set of LINN vectors and find the linear relationship existing between the three vectors.

[1] See page 142 for discussion of full rank relative to rectangular matrices.

3. Prove that $\begin{bmatrix}1\\2\\1\end{bmatrix}$, $\begin{bmatrix}-1\\3\\2\end{bmatrix}$ and $\begin{bmatrix}1\\1\\0\end{bmatrix}$ is a set of LINN vectors and find the linear combination of them that equals the vector $\begin{bmatrix}a\\b\\c\end{bmatrix}$.

4. Find the rank of

$$A = \begin{bmatrix}1 & 6\\2 & 9\\4 & 3\end{bmatrix}, \qquad B = \begin{bmatrix}6 & 4 & -1 & 2 & 5\\3 & 0 & -1 & 2 & 7\\18 & 3 & 4 & -2 & 0\\6 & 8 & 0 & 0 & -4\end{bmatrix},$$

$$C = \begin{bmatrix}3 & 7 & 6\\2 & 1 & 7\\4 & 6 & 3\\-1 & -1 & 0\\6 & 8 & 3\end{bmatrix}, \qquad D = \begin{bmatrix}1 & 1 & 0 & 2\\-1 & 3 & 6 & 0\\1 & 5 & 6 & 4\\6 & 4 & -3 & 11\end{bmatrix},$$

$$E = \begin{bmatrix}1 & -1 & 0\\3 & 2 & -4\\5 & 0 & -4\\1 & -6 & 4\end{bmatrix}, \quad \text{and} \quad F = \begin{bmatrix}21 & 6 & 3\\12 & 16 & 36\\-63 & 13 & 18\\0 & 93 & 81\end{bmatrix}.$$

5. Express each of the previous matrices, A, D and E, as a product XY, where the number of rows in Y is the rank of the matrix.

6. For $A = \begin{bmatrix}1 & 2 & 4 & 0\\-2 & -3 & -1 & 1\\0 & 1 & 7 & 1\\-2 & -2 & 6 & 2\\-3 & -6 & -12 & 0\end{bmatrix}$

show the following:

(*a*) All minors of A, of order 3 or more, are zero.
(*b*) The rank of A is 2.
(*c*) For any pair of linearly independent rows each of the other three rows is a linear function thereof.
(*d*) A can be expressed as XY, where X is 5×2 and Y is 2×4, in the following forms:

$$A = \begin{bmatrix}1 & 0\\0 & 1\\2 & 1\\2 & 2\\-3 & 0\end{bmatrix}\begin{bmatrix}1 & 2 & 4 & 0\\-2 & -3 & -1 & 1\end{bmatrix}$$

$$= \begin{bmatrix} -1 & 1/2 \\ 1 & 0 \\ -1 & 1 \\ 0 & 1 \\ 3 & -3/2 \end{bmatrix} \begin{bmatrix} -2 & -3 & -1 & 1 \\ -2 & -2 & 6 & 2 \end{bmatrix}.$$

7. Reduce the following matrices to diagonal form, $PAQ = \begin{bmatrix} D_r & \cdot \\ \cdot & \cdot \end{bmatrix}$ and thence to canonical form $PAQ^* = \begin{bmatrix} I_r & \cdot \\ \cdot & \cdot \end{bmatrix}$ where $Q^* = Q \begin{bmatrix} D_r^{-1} & \cdot \\ \cdot & I \end{bmatrix}$:

$$A_1 = \begin{bmatrix} 7 & 13 \\ 2 & 9 \end{bmatrix} \qquad A_2 = \begin{bmatrix} 3 & 6 & 48 \\ 1 & 9 & 2 \\ 4 & 1 & 3 \end{bmatrix}$$

$$A_3 = \begin{bmatrix} 1 & 0 & -1 \\ 3 & -4 & 2 \\ 5 & -4 & 0 \\ 1 & 4 & -6 \end{bmatrix} \qquad A_4 = \begin{bmatrix} 3 & 6 & 2 & 4 \\ 9 & 1 & 3 & 2 \\ 6 & -5 & 1 & -2 \end{bmatrix}.$$

8. Reduce the following symmetric matrices to diagonal form and thence to canonical form:

$$B_1 = \begin{bmatrix} 4 & 4 & 10 \\ 4 & 20 & 18 \\ 10 & 18 & 29 \end{bmatrix} \qquad B_2 = \begin{bmatrix} 4 & 6 & 12 \\ 6 & 8 & 1 \\ 12 & 1 & 5 \end{bmatrix}$$

$$B_3 = \begin{bmatrix} 4 & -2 & 0 \\ -2 & 3 & -2 \\ 0 & -2 & 2 \end{bmatrix} \qquad B_4 = \begin{bmatrix} 1 & 8 & 6 & 7 \\ 8 & 65 & 99 & 40 \\ 6 & 99 & 81 & 78 \\ 7 & 40 & 78 & 21 \end{bmatrix}.$$

REFERENCE

Aitken A. C. (1948). *Determinants and Matrices*, Fifth Edition, Oliver and Boyd, Edinburgh.

CHAPTER 6

LINEAR EQUATIONS AND GENERALIZED INVERSES

Linear equations in several unknowns can be represented in matrix form as $Ax = y$ where x is the vector of unknowns, y is a vector of known values and A is the matrix of coefficients. Equations of this nature have been mentioned on a number of occasions in previous chapters. So long as A has an inverse the solution is $x = A^{-1}y$, a situation demanding that A be square and that its rank equal its order. The more general case is now considered, that of having p equations in q unknowns,

$$A_{p\times q}x_{q\times 1} = y_{p\times 1},$$

with the rank of A being r, $r(A) = r$. The instances when the solution is $x = A^{-1}y$ are then particular cases of this, namely when $p = q = r$.

Throughout this chapter we deal with linear equations only, of the kind $3a + 7b + 2c + d = 13$ that is linear in scalar unknowns, which shall be denoted by the letters a, b, c and d. At no time are equations involving powers of the unknowns considered. Thus by "equations" we will always mean "linear equations". The style of presentation established in the preceding chapter is continued, with certain results being stated formally as theorems.

1. EQUATIONS HAVING MANY SOLUTIONS

The equation $3a = 12$ has but a single solution, $a = 4$. Likewise the simultaneous equations $2a + b = 7$ and $a + 3b = 16$ have the single solution $a = 1$ and $b = 5$. But consider the equations $Ax = y$:

$$\begin{aligned} 2a + 3b + c &= 14 \\ a + b + c &= 6 \\ 3a + 5b + c &= 22. \end{aligned} \tag{1}$$

By direct substitution it will be found that the values $a = 4$, $b = 2$ and $c = 0$ are a solution, and so is the set of values $a = 6$, $b = 1$ and $c = -1$. Indeed, it may be shown that each of the following sets of values is a solution.

TABLE 1. FIVE SOLUTIONS TO EQUATIONS (1)

Unknowns	Solutions*				
	$\tilde{x}_1$	$\tilde{x}_2$	$\tilde{x}_3$	$\tilde{x}_4$	$\tilde{x}_5$
a	4	6	0	−6	1.2
b	2	1	4	7	3.4
c	0	−1	2	5	1.4

* The notation $\tilde{x}$ is used for a solution, to distinguish it from x, the vector of unknowns.

Readers accustomed to thinking of a set of linear equations as having only a single solution may find this idea of many solutions novel. Nevertheless it occurs quite frequently, especially in certain statistical analyses that are widely applicable to biological data (see Chapter 10).

2. CONSISTENT EQUATIONS

a. Definition

Consider the two equations

$$a + b = 5$$

and

$$2a + 2b = 11.$$

Obviously if one of them is true the other cannot be, for the second is incompatible, or inconsistent, with the first. Considered as a set of linear equations they are said to be *inconsistent*. Similarly, consider

$$\begin{aligned} 2a + 3b + c &= 14 \\ a + b + c &= 6 \\ 3a + 5b + c &= 19; \end{aligned} \tag{1a}$$

subtracting twice the first from the sum of the second and third leads to the meaningless result $0 = -3$. This is because the left-hand side of the third equation is twice that of the first minus that of the second, but without the same relationship being true of the right-hand sides. This incompatibility leads to the equations being described as inconsistent. Now consider equations (1) again. They differ from (1a) only in the right-hand side of the last equation, which is 22 instead of 19. This means, however, that in (1) the right-hand side of the last equation is twice that of the first minus that of the second, the same relationship that holds for the left-hand sides. Now consider these equations in matrix form $Ax = y$:

$$\begin{bmatrix} 2 & 3 & 1 \\ 1 & 1 & 1 \\ 3 & 5 & 1 \end{bmatrix} \begin{bmatrix} a \\ b \\ c \end{bmatrix} = \begin{bmatrix} 14 \\ 6 \\ 22 \end{bmatrix}. \tag{2}$$

The relationship which exists among the individual equations of (1) applies to both the rows of A and the corresponding elements of y, namely the last row of A (element of y) is twice the first minus the second. Equations of this nature, in which linear relationships existing among the rows of A also hold among the corresponding elements of y, are said to be *consistent*. A formal definition follows.

Definition. The linear equations $Ax = y$ are defined as consistent if linear relationships existing among the rows of A also exist among the corresponding elements of y.

This definition might appear to apply only when the rank of A is less than the number of rows, for only then is it possible to have linear relationships among the rows of A. But, in a formal way, it also applies when the rank of A is the same as the number of rows, for by definition the equations $Ax = y$ are then consistent with no relationships existing among the rows of A. The definition is therefore generally applicable.

b. Existence of solutions

It is convenient at this point to initiate a sequence of theorems that may be established from the definition of consistency. The first concerns the occasions on which linear equations can be solved.

Theorem 1: A set of linear equations can be solved if, and only if, they are consistent.

Proof. If $r(A) = r$, r of the p rows in A are linearly independent (LINN). Assume they are the first r rows and denote them by the matrix A^*. (If this assumption is not true $Ax = y$ is not affected by interchanging equations in order to redefine A and y so that the assumption is true. We therefore retain it.) The remaining rows of A are therefore linear combinations of the first r LINN rows now denoted by A^*, and can be written as CA^* where C represents the matrix of coefficients of these linear combinations. Hence A can be expressed in partitioned form as $A = \begin{bmatrix} A^* \\ CA^* \end{bmatrix}$. Conformable with this the vector of right-hand sides, y, can also be partitioned, as $y = \begin{bmatrix} y_1 \\ y_2 \end{bmatrix}$, where y_1 has r elements and y_2 has $p - r$. The equations $Ax = y$ then become

$$\begin{bmatrix} A^* \\ CA^* \end{bmatrix} x = \begin{bmatrix} y_1 \\ y_2 \end{bmatrix}, \tag{3}$$

or

$$A^*x = y_1 \tag{4}$$

and

$$CA^*x = y_2. \tag{5}$$

Suppose that $x = \tilde{x}$ is a solution of equations (3). Then from (4) $A^*\tilde{x} = y_1$, and pre-multiplication by C gives $CA^*\tilde{x} = Cy_1$. But putting $x = \tilde{x}$ in (5) also gives $CA^*\tilde{x} = y_2$. Therefore $y_2 = Cy_1$, so giving equations (3) as

$$\begin{bmatrix} A^* \\ CA^* \end{bmatrix} x = \begin{bmatrix} y_1 \\ Cy_1 \end{bmatrix}. \tag{6}$$

This equation shows that linear relationships existing among the rows of A also exist among the corresponding elements of y. Therefore, by definition, the equations are consistent. Hence we have proved that if equations $Ax = y$ have a solution they are consistent. To prove the converse note that if the equations are consistent they can be written as (6), in which case (5) is $CA^*x = Cy_1$. This is certainly satisfied by any vector $\tilde{x}$ that satisfies (4). Therefore the problem of solving consistent equations reduces to solving (4).

Having assumed, with $r(A) = r$, that the first r rows of A are independent and denoted by A^*, it is now further assumed that the first r columns of A^* are independent, so that A^* can be partitioned as $A^* = [A_1 \quad A_2]$ where A_1 is $r \times r$ and non-singular and A_2 is $r \times (q - r)$. (If this assumption is not true A^* can be redefined by appropriately interchanging columns of A^* and elements of x to make it so, and without affecting equations (4) in any way. The assumption is therefore retained.) Conformable with partitioning A^* the vector x can also be partitioned, as $x = \begin{bmatrix} x_1 \\ x_2 \end{bmatrix}$ where x_1 has r elements and x_2 has $q - r$. Equation (4), $A^*x = y_1$, then becomes

$$[A_1 \quad A_2]\begin{bmatrix} x_1 \\ x_2 \end{bmatrix} = y_1 \tag{7}$$

or

$$A_1x_1 + A_2x_2 = y_1, \tag{8}$$

and since A_1 is nonsingular this is the same as

$$x_1 = A_1^{-1}y_1 - A_1^{-1}A_2x_2. \tag{9}$$

Solving $Ax = y$, earlier shown equivalent to solving (4), is now seen to be identical to satisfying (9). To obtain a solution of consistent equations $Ax = y$ we therefore need only find values of x_1 and x_2 that satisfy (9), and then by the partitioning of x,

$$x = \begin{bmatrix} x_1 \\ x_2 \end{bmatrix} = \begin{bmatrix} A_1^{-1}y_1 - A_1^{-1}A_2x_2 \\ x_2 \end{bmatrix} \tag{10}$$

is a solution $Ax = y$. Nothing prevents our computing the terms of (10), so it represents a legitimate solution of the equations $Ax = y$ provided they are consistent. Thus we have shown that consistent equations have a solution and Theorem 1 is proved.

Note that although the proof is described on the basis of r being less than both p and q it holds true even when it equals either or both these values. In particular, when $r = p = q$, A is square, A_2 and x_2 have no elements and the solution given in (10) becomes $x = A^{-1}y$, as would be expected.

Example. To solve equations (1) by expression (10) we start from the matrix form given in (2). Partitioned as (3) this is

$$\begin{bmatrix} 2 & 3 & 1 \\ 1 & 1 & 1 \\ \hdashline 3 & 5 & 1 \end{bmatrix} \begin{bmatrix} a \\ b \\ c \end{bmatrix} = \begin{bmatrix} 14 \\ 6 \\ \hdashline 22 \end{bmatrix}$$

so that (7) is

$$\left[\begin{array}{cc:c} 2 & 3 & 1 \\ 1 & 1 & 1 \end{array}\right] \begin{bmatrix} a \\ b \\ \hdashline c \end{bmatrix} = \begin{bmatrix} 14 \\ 6 \end{bmatrix}.$$

Consequently (9) is

$$\begin{bmatrix} a \\ b \end{bmatrix} = \begin{bmatrix} 2 & 3 \\ 1 & 1 \end{bmatrix}^{-1} \begin{bmatrix} 14 \\ 6 \end{bmatrix} - \begin{bmatrix} 2 & 3 \\ 1 & 1 \end{bmatrix}^{-1} \begin{bmatrix} 1 \\ 1 \end{bmatrix} c$$

which reduces to

$$\begin{bmatrix} a \\ b \end{bmatrix} = \begin{bmatrix} 4 - 2c \\ 2 + c \end{bmatrix}. \tag{11}$$

In this case x_1 is the vector $\begin{bmatrix} a \\ b \end{bmatrix}$ and x_2 is the scalar c. Applying (11) to expression (10) now gives the solution as

$$\tilde{x} = \begin{bmatrix} 4 - 2c \\ 2 + c \\ c \end{bmatrix}. \tag{12}$$

In particular if $c = 0$, $\tilde{x} = \begin{bmatrix} 4 \\ 2 \\ 0 \end{bmatrix}$ is a solution, as is $\tilde{x} = \begin{bmatrix} 6 \\ 1 \\ -1 \end{bmatrix}$ for $c = -1$. These are two of the solutions shown in Table 1. The reader should satisfy himself that the other solutions given there can also, for appropriate values of c, be derived from (12).

That equations (1) have many solutions is clearly evident from the form of the general solution given in (12), for, by direct substitution, we find that

$$\begin{bmatrix} 2 & 3 & 1 \\ 1 & 1 & 1 \\ 3 & 5 & 1 \end{bmatrix} \begin{bmatrix} 4 - 2c \\ 2 + c \\ c \end{bmatrix} = \begin{bmatrix} 14 \\ 6 \\ 22 \end{bmatrix}$$

no matter what the value of c is. This means, of course, that there is an infinite number of solutions.

The arbitrariness of the solutions in the example just discussed is true also of the general form of solution, expression (10). There, for $\tilde{x}_1$ appropriately constructed from (9) using any $\tilde{x}_2$, $\tilde{x} = \begin{bmatrix} \tilde{x}_1 \\ \tilde{x}_2 \end{bmatrix}$ will be a solution for *any* $\tilde{x}_2$. Thus (10) not only represents a method of obtaining solutions to equations $Ax = y$, but it also demonstrates the possibility of there being many solutions. Depending on the nonexistence or existence of x_2 there will be either a single unique solution or an infinite number of solutions. When $r = q$, the number of unknowns, there is no vector x_2, A^* and A_1 are the same matrix and solution (10) reduces to $x = A_1^{-1}y_1$, which is the sole, unique solution. If, in addition, $p = r = q$, A, A^* and A_1 are all the same and the unique solution is the familiar $x = A^{-1}y$. When $r < q$, x_2 exists and there are many solutions, indeed an infinite number of them.

Examples. The only solution to

$$\begin{bmatrix} 2 & 3 \\ 1 & 1 \\ 3 & 5 \\ 4 & 1 \\ 1 & -2 \end{bmatrix} \begin{bmatrix} a \\ b \end{bmatrix} = \begin{bmatrix} 14 \\ 6 \\ 22 \\ 18 \\ 0 \end{bmatrix}$$

is

$$\begin{bmatrix} a \\ b \end{bmatrix} = \begin{bmatrix} 2 & 3 \\ 1 & 1 \end{bmatrix}^{-1} \begin{bmatrix} 14 \\ 6 \end{bmatrix} = \begin{bmatrix} 4 \\ 2 \end{bmatrix}.$$

But

$$\begin{bmatrix} 2 & 3 & 1 \\ 1 & 1 & 1 \\ 3 & 5 & 1 \end{bmatrix} \begin{bmatrix} a \\ b \\ c \end{bmatrix} = \begin{bmatrix} 14 \\ 6 \\ 22 \end{bmatrix}$$

has an infinite number of solutions:

$$\begin{bmatrix} a \\ b \\ c \end{bmatrix} = \begin{bmatrix} 4 - 2c \\ 2 + c \\ c \end{bmatrix}.$$

For A_{pxq} the equations $Ax = y$ are, when $p = q$ and A is square, usually described as being of full rank when $r = p = q$, their sole solution being $x = A^{-1}y$; and for $r < p = q$ the equations are described as "equations not of full rank", and have infinitely many solutions. These descriptions could be broadened to cover rectangular matrices as well, by referring to all equations for which $r = q$ as being equations of full rank: they have a unique solution. And all equations for which $r < q$ could be described as equations not of full rank: they have an infinite number of solutions. Whatever description is used the important fact remains: when the rank of A equals the number of unknowns in the equations $Ax = y$ they have a unique solution, otherwise they have infinitely many solutions.

c. Tests for consistency

It is clear from Theorem 1 that consistency is a property required of any set of equations before attempting to solve them, for otherwise they have no solution and it is pointless to try to look for one. Therefore, before discussing further procedures for solving consistent equations, we consider methods of testing whether or not equations are consistent. If they are consistent they can be solved, otherwise they cannot.

The most common test for consistency is one involving the partitioned matrix $[A \quad y]$ formed by adding y as an extra column to A. Such a matrix is often referred to (Aitken, 1948, for example) as an *augmented matrix*, A augmented by y. It is the basis of the following theorem, which provides the test.

Theorem 2a. The equations $Ax = y$ are consistent if, and only if, the rank of the augmented matrix $[A \quad y]$ equals the rank of A.

**Proof.*

If $Ax = y$ is consistent, the augmented matrix can, by equation (6), be written as

$$[A \quad y] = \begin{bmatrix} A^* & y_1 \\ CA^* & Cy_1 \end{bmatrix}.$$

This obviously has the same number of LINN rows as A, namely r, and hence the same rank as A. Conversely, if $[A \quad y]$ and A have the same rank, then from (3) so do $\begin{bmatrix} A^* & y_1 \\ CA^* & y_2 \end{bmatrix}$ and $\begin{bmatrix} A^* \\ CA^* \end{bmatrix}$. This can only be true if $y_2 = Cy_1$, i.e. if $Ax = y$ is consistent.

The test for consistency based on Theorem 2a requires ascertaining the rank of $[A \quad y]$ and the rank of A. If the ranks are equal the equations are consistent, otherwise they are inconsistent. Since in later sections of this chapter we derive methods for solving equations that utilize the diagonal form of A, we give an alternative test for consistency that can be used in conjunction with those methods.

Theorem 2b. If A is a matrix of p rows and rank r, and if PAQ is a diagonal form of A, then the equations $Ax = y$ are consistent if, and only if, the last $p - r$ elements of Py are zero.

**Proof.*

With PAQ being a diagonal form of A, PA can be written as $PA = \begin{bmatrix} A_r \\ 0 \end{bmatrix}$ where A_r is a matrix of r LINN rows, P being a product of elementary operators. Therefore, if the equations $Ax = y$ are consistent, so are $PAx = Py$; i.e., so are $\begin{bmatrix} A_r x \\ 0 \end{bmatrix} = Py$. Hence the last $p - r$ elements of Py must be zero. Conversely, if the last $p - r$ elements of Py are zero, the equations $\begin{bmatrix} A_r \\ 0 \end{bmatrix} x = Py$ are consistent, and since P^{-1} exists the equations $P^{-1}\begin{bmatrix} A_r \\ 0 \end{bmatrix} x = y$ are also; i.e. $Ax = y$ is a set of consistent equations. So the theorem is proved.

As part of the calculations for obtaining PAQ the test for consistency based on this theorem is patently simple: r, the rank of A, is obtained as the number of non-zero elements in PAQ, and if the last $p - r$ elements of Py are zero the equations are consistent, otherwise they are inconsistent.

Example. For equations (1) the diagonal form PAQ is

$$\begin{bmatrix} 1 & 0 & 0 \\ -\frac{1}{2} & 1 & 0 \\ -2 & 1 & 1 \end{bmatrix}\begin{bmatrix} 2 & 3 & 1 \\ 1 & 1 & 1 \\ 3 & 5 & 1 \end{bmatrix}\begin{bmatrix} 1 & -3/2 & -2 \\ 0 & 1 & 1 \\ 0 & 0 & 1 \end{bmatrix} = \begin{bmatrix} 2 & 0 & 0 \\ 0 & -\frac{1}{2} & 0 \\ 0 & 0 & 0 \end{bmatrix}.$$

Therefore the rank of A is 2, and for the equations to be consistent the last $3 - 2 = 1$ rows of Py must be zero, which they are, for

$$Py = \begin{bmatrix} 1 & 0 & 0 \\ -\frac{1}{2} & 1 & 0 \\ -2 & 1 & 1 \end{bmatrix}\begin{bmatrix} 14 \\ 6 \\ 22 \end{bmatrix} = \begin{bmatrix} 14 \\ -1 \\ 0 \end{bmatrix}.$$

3. MORE OR FEWER EQUATIONS THAN UNKNOWNS

We have just seen how consistent equations $Ax = y$ can be solved using expressions (9) and (10). These require selecting r LINN rows of A and partitioning them as $[A_1 \quad A_2]$ where A_1 is square, of order r and nonsingular. A_1^{-1} then exists and so (9) and (10) can be used. The ability to proceed in this way relies on first finding a suitable A_1, which involves (i) ascertaining the rank, r, of A, (ii) locating r linearly independent rows of A and r linearly independent columns among those rows, and (iii) interchanging, if need be, rows and columns of A and rows of x and y to recreate A in a form suited to partitioning.

There is usually no difficulty in carrying out these steps when only a few equations are involved, but when the number of equations is at all large any one of the steps may be quite tedious. Fortunately, however, they can be avoided by using methods that require neither ascertaining the rank of A nor searching for a nonsingular minor. Furthermore, all cases of either more or fewer equations than unknowns (rectangular A) are handled similarly, and in a manner only slightly different from that used when the number of equations is the same as the number of unknowns (square A). The methods involve matrices known as generalized inverses.

4. GENERALIZED INVERSE MATRICES

a. Definitions

Penrose (1955) shows that for any matrix A there is a unique matrix K which satisfies the following four conditions:

$$\text{i. } AKA = A \qquad \text{iii. } (KA)' = KA$$
$$\text{ii. } KAK = K \qquad \text{iv. } (AK)' = AK$$

He called K the generalized inverse of A and showed that it exists and is unique no matter what the form of A is, be it square (singular or nonsingular) or rectangular.

Example. Conditions (i) through (iv) are satisfied for

$$A = \begin{bmatrix} 1 & 0 & 2 \\ 0 & -1 & 1 \\ -1 & 0 & -2 \\ 1 & 2 & 0 \end{bmatrix} \quad \text{and} \quad K = \tfrac{1}{66}\begin{bmatrix} 6 & -2 & -6 & 10 \\ 0 & -11 & 0 & 22 \\ 12 & 7 & -12 & -2 \end{bmatrix}.$$

Not only did Penrose show the existence and uniqueness of a matrix satisfying conditions (i) through (iv), but he also demonstrated its use in solving linear equations. However, K is not easily computed when A is at all large in size, and variations on K exist that are just as useful in solving equations but not so difficult to compute. Several different names have been used for these variants. Rao (1955) and Greville (1957), for example, use the term "pseudo inverse", Wilkinson (1958) considers a matrix which he calls an "effective inverse" and Federer and Zelen (1964) refer to any matrix satisfying conditions (i), (ii) and (iii) as a "weak generalized inverse". The term generalized inverse has also been used by Rao (1962) for any matrix T for which $x = Ty$ is a solution to consistent equations $Ax = y$. This, as shown in the next theorem, implies that $ATA = A$, namely condition (i). For this reason, and also because most variants of Penrose's generalized inverse that have been proposed satisfy condition (i), we choose here to define a *generalized inverse* as being any matrix G for which $AGA = A$, that is to say, it is any matrix satisfying just the first of Penrose's four conditions.

The theorem already alluded to ensures that Rao's (1962) definition is included in the one just given, and most of the other definitions that have been used are also included because they incorporate condition (i). Such others of the conditions (ii), (iii) and (iv) as may be included in any definition do not negate the definition just given, they only restrict it, the greatest restriction being when all of them are included. In this case we may refer to the generalized inverse so defined, namely that considered by Penrose, as the "unique generalized inverse".

* The left and right inverses discussed in Section 4.7 each satisfy three of the conditions (i) through (iv). Thus if L is a left inverse of A, $LA = I$, in which case conditions (i), (ii) and (iii) are clearly satisfied for $K = L$. Similarly, right inverses satisfy (i), (ii) and (iv).

b. The product $H = GA$

Two simple properties of the product $H = GA$ arise from the definition of G as any matrix for which $AGA = A$. First, H and A have the same rank for, since $H = GA$, $r(H) \leqq r(A)$, and, because $A = AGA = AH$, $r(A) \leqq r(H)$. Therefore $r(H) = r(A)$. Secondly, $H^2 = H$: this is so because $H^2 = GAGA = G(AGA) = GA = H$. Matrices having this property are often described as *idempotent*: H is idempotent because $H^2 = H$ (see Chapter 8).

c. Computing a generalized inverse

A procedure for obtaining from A the unique generalized inverse K can be found in Penrose (1955), and several methods of obtaining variants of K are given by Rao (1962). The simplest of these is now presented. Although the method to be described applies equally as well to rectangular matrices as to square ones it will be discussed initially in terms of square matrices. The minor modifications needed for rectangular matrices are given at the end of the chapter.

Section 5.11 deals with the method of reducing any matrix A to diagonal form

$$PAQ = \Delta = \begin{bmatrix} D_r & 0 \\ 0 & 0 \end{bmatrix}$$

where Δ is the same order as A, r is the rank of A, D_r is a diagonal matrix having r non-zero elements, and the 0's are null matrices of appropriate order. The matrices P and Q are products of elementary operators.

Example. For the matrix of equations (1) considered earlier,

$$A = \begin{bmatrix} 2 & 3 & 1 \\ 1 & 1 & 1 \\ 3 & 5 & 1 \end{bmatrix},$$

a diagonal form is obtained using

$$P = \begin{bmatrix} 1 & 0 & 0 \\ -\frac{1}{2} & 1 & 0 \\ -2 & 1 & 1 \end{bmatrix} \quad \text{and} \quad Q = \begin{bmatrix} 1 & -3/2 & -2 \\ 0 & 1 & 1 \\ 0 & 0 & 1 \end{bmatrix}.$$

Thus

$$PAQ = \Delta = \begin{bmatrix} 2 & 0 & 0 \\ 0 & -\frac{1}{2} & 0 \\ 0 & 0 & 0 \end{bmatrix} = \begin{bmatrix} D_2 & 0 \\ 0 & 0 \end{bmatrix} \quad \text{where} \quad D_2 = \begin{bmatrix} 2 & 0 \\ 0 & -\frac{1}{2} \end{bmatrix}.$$

Let us now define a new matrix Δ^- as

$$\Delta^- = \begin{bmatrix} D^{-1} & 0 \\ 0 & 0 \end{bmatrix},$$

where the symbol Δ^- is read as "delta minus" and where D^{-1} is the inverse of D in the usual manner. Then the matrix G defined as

$$G = Q\Delta^- P$$

is a generalized inverse of A, i.e. $AGA = A$. Thus in the example

$$\Delta^- = \begin{bmatrix} \frac{1}{2} & 0 & 0 \\ 0 & -2 & 0 \\ 0 & 0 & 0 \end{bmatrix}$$

and

$$\begin{aligned} G &= Q\Delta^- P \\ &= \begin{bmatrix} 1 & -3/2 & -2 \\ 0 & 1 & 1 \\ 0 & 0 & 1 \end{bmatrix} \begin{bmatrix} \frac{1}{2} & 0 & 0 \\ 0 & -2 & 0 \\ 0 & 0 & 0 \end{bmatrix} \begin{bmatrix} 1 & 0 & 0 \\ -\frac{1}{2} & 1 & 0 \\ -2 & 1 & 1 \end{bmatrix} \\ &= \begin{bmatrix} -1 & 3 & 0 \\ 1 & -2 & 0 \\ 0 & 0 & 0 \end{bmatrix} \end{aligned}$$

and by carrying out the multiplications it will be found that $AGA = A$.

The equality $AGA = A$ for G defined in this way is based upon properties of Δ and Δ^-, namely that

$$\Delta\Delta^- = \begin{bmatrix} I_r & 0 \\ 0 & 0 \end{bmatrix} = \Delta^-\Delta$$

and that both $\Delta^-\Delta\Delta^- = \Delta^-$ and $\Delta\Delta^-\Delta = \Delta$. (The reader should satisfy himself that these relations are true, both directly from the definitions of Δ and Δ^- and in the example.) In addition, because P and Q are products of elementary operators their inverses exist, so that $A = P^{-1}\Delta Q^{-1}$. As a consequence, AGA reduces to being $P^{-1}\Delta Q^{-1}$ equal to A and so, by our definition, G is a generalized inverse of A. Notice that G is called "a" generalized inverse and not "the" generalized inverse of A because it is not unique. This is clear from the diagonal form $PAQ = \Delta$ in which none of the matrices P, Q and Δ is unique and so, therefore, neither is $G = Q\Delta^-P$.

Two further properties of G derived in this manner are of interest. First, $GAG = G$, because GAG reduces to $Q\Delta^-P = G$, and second, the rank of $H - I$, where $H = GA$, is $q - r$, q being the order of A. This is evident from using $G = Q\Delta^-P$ and $A = P^{-1}\Delta Q^{-1}$ to write

$$H = GA = Q\Delta^-\Delta Q^{-1} = Q\begin{bmatrix} I_r & 0 \\ 0 & 0 \end{bmatrix}Q^{-1}$$

so that

$$H - I_q = Q\left\{\begin{bmatrix} I_r & 0 \\ 0 & 0 \end{bmatrix} - I_q\right\}Q^{-1}.$$

Therefore, because Q is a product of elementary operators,

$$\text{rank } [H - I_q] = \text{rank}\left\{\begin{bmatrix} I_r & 0 \\ 0 & 0 \end{bmatrix} - I_q\right\} = q - r.$$

These two results are additional to those proved before for any G for which $AGA = A$, namely that $r(H) = r(A)$ and $H^2 = H$. All four results play a part in the application of G to the problem of solving the equations $Ax = y$ for x. The reader might satisfy himself that they hold true for G of the example.

5. SOLVING LINEAR EQUATIONS USING GENERALIZED INVERSES

a. Obtaining a solution

The relationship between a generalized inverse of A and the consistent equations $Ax = y$ is set out in the following theorem adapted from Rao (1962).

Theorem 3. The consistent equations $Ax = y$ have the solution $x = Gy$ if, and only if, $AGA = A$.

**Proof.*

Let a_j be the jth column of A, and consider the equations $Ax = a_j$. The vector x_0 defined as having all elements zero except for unity as the jth element is clearly a solution of $Ax = a_j$. Consequently, by Theorem 1, the equations $Ax = a_j$ are consistent. Therefore, if $x = Gy$ is a solution of $Ax = y$, $x = Ga_j$ is a solution of $Ax = a_j$ and $AGa_j = a_j$. This is true for all j, namely for all columns of A, $AGa_j = a_j$. Therefore $AGA = A$.

Conversely, if $AGA = A$, $AGAx = Ax$, and if $Ax = y$ this becomes $AGy = y$, i.e., $A(Gy) = y$. Therefore $x = Gy$ is a solution of $Ax = y$. Its existence means the equations are consistent, and so the theorem is proved.

Example. Equations (1) continue to be used as an example. Thus

$$A = \begin{bmatrix} 2 & 3 & 1 \\ 1 & 1 & 1 \\ 3 & 5 & 1 \end{bmatrix} \qquad \text{has} \quad G = \begin{bmatrix} -1 & 3 & 0 \\ 1 & -2 & 0 \\ 0 & 0 & 0 \end{bmatrix}$$

as a generalized inverse and hence

$$\tilde{x} = Gy = \begin{bmatrix} -1 & 3 & 0 \\ 1 & -2 & 0 \\ 0 & 0 & 0 \end{bmatrix} \begin{bmatrix} 14 \\ 6 \\ 22 \end{bmatrix} = \begin{bmatrix} 4 \\ 2 \\ 0 \end{bmatrix}$$

is a solution, as may be verified by direct substitution.

Theorem 3 indicates how a solution of equations $Ax = y$ may be obtained: find G for which $AGA = A$ and Gy is a solution. But, as shown earlier, when the rank of A is less than the number of unknowns there is a multitude of solutions. The following theorem provides a method of deriving this multitude from the one solution given by Theorem 3.

Theorem 4. If A is a matrix of q columns, and if $AGA = A$ and $H = GA$, then $\tilde{x} = Gy + (H - I)z$ is a solution of consistent equations $Ax = y$ for z being any arbitrary vector of order q.

Proof. If

$$\begin{aligned} \tilde{x} &= Gy + (H - I)z, \\ A\tilde{x} &= AGy + A(H - I)z \\ &= AGy + (AGA - A)z \\ &= AGy, \text{ because } AGA = A, \\ &= y \text{ by Theorem 3.} \end{aligned}$$

Hence $\tilde{x}$ is a solution of $Ax = y$, for any q-order vector z.

Example (*continued*). In equations (1)

$$Gy = \begin{bmatrix} 4 \\ 2 \\ 0 \end{bmatrix} \quad \text{and} \quad H = GA = \begin{bmatrix} 1 & 0 & 2 \\ 0 & 1 & -1 \\ 0 & 0 & 0 \end{bmatrix}.$$

Therefore

$$\tilde{x} = Gy + (H - I)z = \begin{bmatrix} 4 \\ 2 \\ 0 \end{bmatrix} + \begin{bmatrix} 0 & 0 & 2 \\ 0 & 0 & -1 \\ 0 & 0 & -1 \end{bmatrix} z$$

is a solution. Taking $z = \begin{bmatrix} u_1 \\ u_2 \\ u_3 \end{bmatrix}$ as a vector of arbitrary elements u_1, u_2, u_3, the solution becomes

$$\tilde{x} = \begin{bmatrix} 4 \\ 2 \\ 0 \end{bmatrix} + \begin{bmatrix} 2u_3 \\ -u_3 \\ -u_3 \end{bmatrix} = \begin{bmatrix} 4 + 2u_3 \\ 2 - u_3 \\ -u_3 \end{bmatrix}$$

which is exactly the same form as solution (12) obtained earlier by partitioning. With $u_3 = 0$, 1 and -2 we get the same explicit solutions as before:

$$\tilde{x}_1 = \begin{bmatrix} 4 \\ 2 \\ 0 \end{bmatrix}, \quad \tilde{x}_2 = \begin{bmatrix} 6 \\ 1 \\ -1 \end{bmatrix} \quad \text{and} \quad \tilde{x}_3 = \begin{bmatrix} 0 \\ 4 \\ 2 \end{bmatrix}.$$

The solution $\tilde{x} = Gy + (H - I)z$ given by Theorem 4 is essentially the same form as that obtained in equation (10) through partitioning A:

$$\tilde{x} = \begin{bmatrix} A_1^{-1}y_1 \\ 0 \end{bmatrix} + \begin{bmatrix} -A_1^{-1}A_2 \\ I \end{bmatrix} x_2.$$

Both forms will generate, through allocating different sets of values to the arbitrary vectors z and x_2, the same series of vectors that satisfy $Ax = y$. Since the rank of $H - I$ is $q - r$ the vector $(H - I)z$ contains only $q - r$ independent elements, the same number as x_2. Hence both forms of solution generate the same number of LINN vectors that are solutions of $Ax = y$. Vectors of this nature are referred to as LINN solutions.

The important difference between the two solutions given by Theorem 4 and equation (10) lies in the relative effort involved in calculating them. The earlier one, (10), requires choosing a nonsingular matrix A_1 from among the rows and columns of A whereas Theorem 4 only demands finding a generalized inverse. This, we can now readily appreciate, is a considerably easier task computationally than selecting the necessary linearly independent rows and columns required for choosing A_1. And it is especially easier when handling large numbers of equations; the calculations for obtaining a generalized inverse are easily processed by a high-speed computer whereas those needed for choosing A_1 would be quite cumbersome.

b. Independent solutions

Having established a method for solving linear equations and having shown in doing so that it is possible for them to have an infinite number of solutions, we ask two questions. (i) To what extent are the solutions linearly independent? (ii) What relationships exist among the solutions? Since each solution is a vector of order q there can of course, be no more than q LINN solutions. In fact there are fewer, as the following theorem shows.

Theorem 5. If A is a matrix of q columns and rank r, and if y is a non-null vector, the number of LINN solutions to the consistent equations $Ax = y$ is $q - r + 1$.

**Proof.*

For $PAQ = \Delta$, $G = Q\Delta^- P$ and $H = GA$, the rank of $H - I$ is $q - r$ and $\tilde{x} = Gy + (H - I)z$ is a solution of $Ax = y$. Because $H - I$ has rank $q - r$ there are only $q - r$ independent elements in $(H - I)z$. Therefore there are only $q - r$ LINN vectors $(H - I)z$, and using these in $\tilde{x}$ gives $q - r$ LINN solutions to $Ax = y$.

But $\tilde{x} = Gy$ is also a solution. Let us assume that for $i = 1, 2, \ldots, q - r$, $\tilde{x}_i = Gy + (H - I)z_i$ are the $q - r$ LINN solutions corresponding to $q - r$ LINN vectors $(H - I)z_i$ and that $\tilde{x} = Gy$ is linearly dependent on the $\tilde{x}_i$. Then, for a series of scalars λ_i, $i = 1, 2, \ldots, q - r$, not all of which are zero,

$$\begin{aligned} Gy &= \Sigma\lambda_i\tilde{x}_i \\ &= \Sigma\lambda_i[Gy + (H - I)z_i] \\ &= Gy(\Sigma\lambda_i) + \Sigma\lambda_i(H - I)z_i. \end{aligned} \tag{13}$$

Now because the vectors $(H - I)z_i$ are linearly independent this expression can be free of the arbitrary vectors z_i only if the second summation is zero, namely if every λ_i is zero. This means (13) is no longer true for some λ_i non-zero. Therefore Gy is independent of the $\tilde{x}_i$ and so Gy and $\tilde{x}_i$ for $i = 1, 2, \ldots, q - r$ form a set of $q - r + 1$ LINN solutions.

This theorem means that $\tilde{x} = Gy$ and $\tilde{x} = Gy + (H - I)z$ for $q - r$ LINN vectors z are LINN solutions of $Ax = y$. All other solutions will be linear combinations of those forming a set of LINN solutions. Note that if y is a null vector, Gy is not a non-null vector, and so there are then only $q - r$ LINN solutions.

Example (*continued*).

$$\tilde{x}_1 = \begin{bmatrix} 4 \\ 2 \\ 0 \end{bmatrix}, \quad \tilde{x}_2 = \begin{bmatrix} 6 \\ 1 \\ -1 \end{bmatrix} \quad \text{and} \quad \tilde{x}_3 = \begin{bmatrix} 0 \\ 4 \\ 2 \end{bmatrix}$$

are solutions to equations (1) in which $q = 3$ and $r = 2$. There are therefore $3 - 2 + 1 = 2$ LINN solutions. Hence $\tilde{x}_3$ is a linear combination of $\tilde{x}_1$ and $\tilde{x}_2$: $\tilde{x}_3 = 3\tilde{x}_1 - 2\tilde{x}_2$.

A means of constructing solutions as linear combinations of other solutions is contained in the following theorem.

Theorem 6. If $\tilde{x}_1, \tilde{x}_2, \ldots, \tilde{x}_s$ are any s solutions of consistent equations $Ax = y$ for which $y \neq 0$, then any linear combination of these solutions, $x^* = \sum_{i=1}^{s} \lambda_i \tilde{x}_i$, is also a solution of the equations if, and only if,

$$\sum_{i=1}^{s} \lambda_i = 1.$$

**Proof.*

(To simplify notation, the limits of summation are omitted.) Because

$$x^* = \Sigma\lambda_i\tilde{x}_i,$$
$$Ax^* = A\Sigma\lambda_i\tilde{x}_i = \Sigma\lambda_i A\tilde{x}_i.$$

And because $\tilde{x}_i$ is a solution $A\tilde{x}_i = y$ for all i, so giving

$$Ax^* = \Sigma\lambda_i y \tag{14}$$
$$= y(\Sigma\lambda_i). \tag{15}$$

Now if x^* is a solution of $Ax = y$, $Ax^* = y$, and by comparison with (15) this means, y being non-null, that $\Sigma\lambda_i = 1$. Conversely, if $\Sigma\lambda_i = 1$, equation (15) implies that $Ax^* = y$, namely that x^* is a solution. So the theorem is proved.

Notice that although the theorem is specified in terms of s solutions it refers to *any* s solutions. Hence, for any number of solutions, whether linearly independent or not, any linear combination of them is itself a solution provided the coefficients in that combination sum to unity.

Example (*continued*). We have already seen that $\tilde{x}_3 = 3\tilde{x}_1 - 2\tilde{x}_2$ is a solution of equations (1) and the sum of the coefficients in $\tilde{x}_3$ is $3 - 2 = 1$. The same is true of $\tilde{x}_4 = 0.73\tilde{x}_2 + 0.27\tilde{x}_1$,

and of $\tilde{x}_5 = 0.23\tilde{x}_1 + 0.45\tilde{x}_2 + 0.32\tilde{x}_3$.

The reader should calculate these vectors explicitly and satisfy himself that they are solutions of equations (1).

Example. For equations

$$Ax = \begin{bmatrix} 1 & 2 & -1 & 9 \\ 2 & 4 & 3 & 3 \\ -1 & -2 & 6 & -24 \\ 1 & 2 & 4 & -6 \end{bmatrix} x = \begin{bmatrix} 4 \\ 13 \\ 1 \\ 9 \end{bmatrix} = y \tag{16}$$

a diagonal form of A comes from using

$$P = \begin{bmatrix} 1 & 0 & 0 & 0 \\ -2 & 1 & 0 & 0 \\ 3 & -1 & 1 & 0 \\ 1 & -1 & 0 & 1 \end{bmatrix} \quad \text{and} \quad Q = \begin{bmatrix} 1 & 1 & -2 & -6 \\ 0 & 0 & 1 & 0 \\ 0 & 1 & 0 & 3 \\ 0 & 0 & 0 & 1 \end{bmatrix},$$

giving
$$PAQ = \Delta = \begin{bmatrix} 1 & 0 & 0 & 0 \\ 0 & 5 & 0 & 0 \\ 0 & 0 & 0 & 0 \\ 0 & 0 & 0 & 0 \end{bmatrix}.$$

The rank of A is therefore 2, and because

$$Py = \begin{bmatrix} 4 \\ 5 \\ 0 \\ 0 \end{bmatrix}$$

the equations are consistent. The generalized inverse $G = Q\Delta^- P$ is found to be

$$G = \begin{bmatrix} 0.6 & 0.2 & 0 & 0 \\ 0 & 0 & 0 & 0 \\ -0.4 & 0.2 & 0 & 0 \\ 0 & 0 & 0 & 0 \end{bmatrix}, \quad \text{with} \quad H = GA = \begin{bmatrix} 1 & 2 & 0 & 6 \\ 0 & 0 & 0 & 0 \\ 0 & 0 & 1 & -3 \\ 0 & 0 & 0 & 0 \end{bmatrix}.$$

Hence solutions to the equations are obtained as

$$\tilde{x} = Gy + (H - I)z$$

$$= \begin{bmatrix} 5 \\ 0 \\ 1 \\ 0 \end{bmatrix} + \begin{bmatrix} 0 & 2 & 0 & 6 \\ 0 & -1 & 0 & 0 \\ 0 & 0 & 0 & -3 \\ 0 & 0 & 0 & -1 \end{bmatrix} \begin{bmatrix} u_1 \\ u_2 \\ u_3 \\ u_4 \end{bmatrix} \tag{17}$$

$$= \begin{bmatrix} 5 + 2u_2 + 6u_4 \\ -u_2 \\ 1 - 3u_4 \\ -u_4 \end{bmatrix}, \tag{18}$$

where the u's are arbitrary scalars. For example, with $u_2 = 1$ and $u_4 = 2$

the explicit solution is $\tilde{x}_1 = \begin{bmatrix} 19 \\ -1 \\ -5 \\ -2 \end{bmatrix}$ and for $u_2 = 0$ and $u_4 = 1$,

$\tilde{x}_2 = \begin{bmatrix} 11 \\ 0 \\ -2 \\ -1 \end{bmatrix}$ is another solution.

c. The equation $Ax = 0$

The two preceding theorems apply only to cases where y is a non-null vector. Consideration is now given to equations $Ax = 0$, namely $Ax = y$ when y is null.

First note that in the scalar counterpart of $Ax = 0$, namely $ax = 0$ where a and x are scalars, the conclusion is that a and/or x are zero. But this is not so for $Ax = 0$, for with A being non-null there are often non-null vectors x for which $Ax = 0$.

Example. Consider equations (16) with $y = 0$:

$$\begin{aligned} a + 2b - c + 9d &= 0 \\ 2a + 4b + 3c + 3d &= 0 \\ -a - 2b + 6c - 24d &= 0 \\ a + 2b + 4c - 6d &= 0. \end{aligned} \tag{19}$$

An obvious solution is $a = 0 = b = c = d$, corresponding to the null vector $x = 0$ in $Ax = 0$. But non-null solutions can also be found. For example, $a = 8$, $b = -1$, $c = -3$ and $d = -1$ is a solution and so is $a = 2$, $b = -4$, $c = 3$ and $d = 1$.

By the definition of consistency it is clear that equations $Ax = 0$ are always consistent, and with $y = 0$ their general solution from Theorem 4 is $\tilde{x} = (H - I)z$ for arbitrary z of order q. Now the rank of $H - I$ is $q - r$, so that when $r = q$, $H - I$ is null and so is $\tilde{x}$; and this is the only solution for $r = q$. Otherwise, for $r < q$, the solution is $\tilde{x} = (H - I)z$ which can be non-null. This means that the only condition under which $Ax = 0$ has non-null solutions is when the rank of A is less than its number of columns, and under this condition there will always be non-null solutions. Furthermore, because $r(H - I) = q - r$ there are only $q - r$ LINN vectors $(H - I)z$ and so only $q - r$ LINN solutions to $Ax = 0$, as mentioned following Theorem 5. In addition, in proving Theorem 6 (14) reduces to $Ax^* = 0$ because $y = 0$, indicating that for solutions $\tilde{x}_i$, $x^* = \sum \lambda_i \tilde{x}_i$ is also a solution no matter what values the λ_i's take. Hence any linear combination of any solutions of $Ax = 0$ is itself a solution.

Example (*continued*). Equations (19) can be written as

$$\begin{bmatrix} 1 & 2 & -1 & 9 \\ 2 & 4 & 3 & 3 \\ -1 & -2 & 6 & -24 \\ 1 & 2 & 4 & -6 \end{bmatrix} x = 0.$$

Their non-null solutions $\tilde{x} = (H - I)z$, where $H - I$ is the same as in (17), are

$$\tilde{x} = \begin{bmatrix} 2u_2 + 6u_4 \\ -u_2 \\ -3u_4 \\ -u_4 \end{bmatrix}.$$

Thus for $u_2 = 1$ and $u_4 = 1$, and for $u_2 = 4$ and $u_4 = -1$ the respective solutions are $\tilde{x}_1 = \begin{bmatrix} 8 \\ -1 \\ -3 \\ -1 \end{bmatrix}$ and $\tilde{x}_2 = \begin{bmatrix} 2 \\ -4 \\ 3 \\ 1 \end{bmatrix}$, and any linear combination of these is also a solution, for example,

$$\begin{bmatrix} 62.24 \\ -19.48 \\ -11.64 \\ -3.88 \end{bmatrix} = 7\tilde{x}_1 + 3.12\tilde{x}_2$$

is a solution.

d. Linear combinations of elements of a solution

We have one final theorem concerning solutions of linear equations. It relates to linear combinations of the scalar values which are elements of any vector that is a solution. The theorem is as follows, again due to Rao (1962).

Theorem 7. For consistent equations $Ax = y$ for which $AGA = A$ and $H = GA$, the linear combination $k'\tilde{x}$ of elements of any solution $\tilde{x}$ is unique if, and only if, $k'H = k'$.

**Proof.*

For a solution $\tilde{x}$ given by Theorem 4,

$$k'\tilde{x} = k'Gy + k'(H - I)z.$$

This is independent of the arbitrary z if $k'H = k'$; and since *any* solution can be put in the form $\tilde{x}$ by appropriate choice of z, the value of $k'\tilde{x}$ for any $\tilde{x}$ is $k'Gy$ provided that $k'H = k'$.

It may not be entirely clear that when $k'H = k'$ the value of $k'\tilde{x} = k'Gy$ is unique no matter which of the many generalized inverses is used for the matrix G. We therefore clarify this point. First, by Theorem 5 there are $q - r + 1$ LINN solutions of the form $\tilde{x} = Gy + (H - I)z$. Let these solutions be $\tilde{x}_i$ for $i = 1, 2, \ldots, q - r + 1$. Suppose for some other generalized inverse G^* say, we have a solution

$$x^* = G^*y + (H^* - I)z.$$

Then, since the $\tilde{x}_i$'s are a LINN set of $q - r + 1$ solutions, x^* must be a linear combination of them; that is, there is a set of scalars λ_i, for $i = 1, 2, \ldots, q - r + 1$, such that

$$x^* = \sum_{i=1}^{q-r+1} \lambda_i \tilde{x}_i$$

where not all the λ_i's are zero and for which, by Theorem 6, $\Sigma\lambda_i = 1$. Further, if $k'H = k'$, $k'\tilde{x}_i = k'Gy$ for all i, so that the same linear combination of the elements of x^* is

$$\begin{aligned} k'x^* &= k'\Sigma\lambda_i\tilde{x}_i = \Sigma\lambda_i k'\tilde{x}_i = \Sigma\lambda_i k'Gy \\ &= k'Gy(\Sigma\lambda_i) = k'Gy. \end{aligned}$$

Hence $k'x^*$ for *any* solution x^* equals $k'Gy$ if $k'H = k'$. This proves the theorem conclusively.

Example (*continued*). In the solution of equations (16) derived earlier, it will be found that $k'H = k'$ for

$$k' = (1 \quad 2 \quad 1 \quad 3).$$

Hence

$$k'\tilde{x}_1 = k'\begin{bmatrix} 19 \\ -1 \\ -5 \\ -2 \end{bmatrix} = k'\tilde{x}_2 = k'\begin{bmatrix} 11 \\ 0 \\ -2 \\ -1 \end{bmatrix},$$

the common value being 6. In fact, as can be seen from (18), this is true of all solutions because

$$k'\tilde{x} = [1 \quad 2 \quad 1 \quad 3]\begin{bmatrix} 5 + 2u_2 + 6u_4 \\ - u_2 \\ 1 - 3u_4 \\ - u_4 \end{bmatrix} = 6,$$

regardless of the values of u_2 and u_4.

The values of k' for which $k'\tilde{x}$ is unique for all solutions $\tilde{x}$ are those given by the equation $k'H = k'$. And by the idempotency property of H it is apparent that $k' = w'H$ is a solution of $k'H = k'$, no matter what the elements of w' are. Hence, for any vector w', $k' = w'H$ gives $k'\tilde{x}$ as

$$\begin{aligned} k'\tilde{x} &= w'H\tilde{x} = w'HGy + w'H(H - I)z \\ &= w'HGy = w'GAGy, \end{aligned}$$

and if G is such that $GAG = G$ as well as AGA equaling A,

$$k'\tilde{x} = w'Gy. \tag{20}$$

Thus for any vector w' and $k' = w'H$, $k'\tilde{x} = w'Gy$ is the same for all solutions $\tilde{x}$. Since $r(H) = r$, $k' = w'H$ provides a set of r LINN vectors k' that have this property; for example, for two such vectors k'_1 and k'_2 say, $k'_1\tilde{x}$ and $k'_2\tilde{x}$ are different but each has a value that is the same for all solutions $\tilde{x}$.

Example (*continued*). Using $w' = [w_0 \quad w_1 \quad w_2 \quad w_3]$ and the H of (17) gives

$$k' = w'H = [w_0 \quad w_1 \quad w_2 \quad w_3]\begin{bmatrix} 1 & 2 & 0 & 6 \\ 0 & 0 & 0 & 0 \\ 0 & 0 & 1 & -3 \\ 0 & 0 & 0 & 0 \end{bmatrix}$$

$$= [w_0 \quad 2w_0 \quad w_2 \quad 6w_0 - 3w_2]$$

and with Gy as in (17)

$$k'\tilde{x} = w'Gy = [w_0 \quad w_1 \quad w_2 \quad w_3]\begin{bmatrix} 5 \\ 0 \\ 1 \\ 0 \end{bmatrix}$$

$$= 5w_0 + w_2. \tag{21}$$

Hence any expression $k'\tilde{x}$ that is unique no matter what solution is used will have the value $5w_0 + w_2$ where $k' = [w_0 \quad 2w_0 \quad w_2 \quad 6w_0 - 3w_2]$. For example, the $k' = [1 \quad 2 \quad 1 \quad 3]$ used before is derived by putting $w_0 = 1$ and $w_2 = 1$; and from (21), $k'\tilde{x}$ then equals 6, as was obtained.

These results concerning unique linear functions of elements of a solution have important application in the statistical analyses of linear models—a topic that is devoted to Chapter 10.

e. Complete example

We here indicate the salient points of a complete example. Suppose the equations $Ax = y$ are

$$\begin{bmatrix} 5 & 2 & -1 & 2 \\ 2 & 2 & 3 & 1 \\ 1 & 1 & 4 & -1 \\ 2 & -1 & -3 & -1 \end{bmatrix} x = \begin{bmatrix} 7 \\ 9 \\ 5 \\ -6 \end{bmatrix}.$$

For row and column operations represented respectively by

$$P = \begin{bmatrix} 0 & 0 & 1 & 0 \\ 1 & 0 & -5 & 0 \\ 0 & 1 & -2 & 0 \\ -1 & 2 & -1 & 1 \end{bmatrix} \quad \text{and} \quad Q = \begin{bmatrix} 1 & -1 & 3 & 7/15 \\ 0 & 1 & -7 & -28/15 \\ 0 & 0 & 1 & 9/15 \\ 0 & 0 & 0 & 1 \end{bmatrix},$$

we have

$$PAQ = \Delta = \begin{bmatrix} 1 & 0 & 0 & 0 \\ 0 & -3 & 0 & 0 \\ 0 & 0 & -5 & 0 \\ 0 & 0 & 0 & 0 \end{bmatrix}$$

and

$$G = Q\Delta^{-}P = \tfrac{1}{15}\begin{bmatrix} 5 & -9 & 8 & 0 \\ -5 & 21 & -17 & 0 \\ 0 & -3 & 6 & 0 \\ 0 & 0 & 0 & 0 \end{bmatrix}.$$

From Δ we see that $r(A) = 3$ and in Py we find that the last element is zero. Hence the equations are consistent and can be solved using G with

$$H = GA = \begin{bmatrix} 1 & 0 & 0 & -7/15 \\ 0 & 1 & 0 & 28/15 \\ 0 & 0 & 1 & -9/15 \\ 0 & 0 & 0 & 0 \end{bmatrix}. \tag{22}$$

It may be noted in passing that H is idempotent and $AH = AGA = A$, as can be verified by direct multiplication.

Solutions to the equations are given by $\tilde{x} = Gy + (H - I)z$ where z is arbitrary, and if z is taken as $z' = [u_1 \quad u_2 \quad u_3 \quad u_4]$ where the u's are scalar the solution is

$$\tilde{x} = \tfrac{1}{15}\begin{bmatrix} -6 - 7u_4 \\ 69 + 28u_4 \\ 3 - 9u_4 \\ -15u_4 \end{bmatrix}, \tag{23}$$

u_4 being any arbitrary scalar value.

With the arbitrary vector $w' = [w_0 \quad w_1 \quad w_2 \quad w_3]$, unique values of $k'\tilde{x}$ are available, using (22), for any k' of the form

$$k' = w'H = [w_0 \quad w_1 \quad w_2 \quad (-7w_0 + 28w_1 - 9w_2)/15], \tag{24}$$

and from (23) the corresponding value of $k'\tilde{x}$ is

$$k'\tilde{x} = w'Gy = (-6w_0 + 69w_1 + 3w_2)/15. \tag{25}$$

Because $n = 4$ and $r(A) = 3 = r(H)$, $4 - 3 + 1 = 2$ LINN solutions can be found. For example, putting $u_4 = 0$ in (23) gives

$$\tilde{x}_1 = \begin{bmatrix} -0.4 \\ 4.6 \\ 0.2 \\ 0 \end{bmatrix}, \quad \text{and} \quad u_4 = -3 \text{ gives } \tilde{x}_2 = \begin{bmatrix} 1 \\ -1 \\ 2 \\ 3 \end{bmatrix}.$$

Also, there will be three LINN vectors k' satisfying (24); for example, $k_1' = [3 \quad 0 \quad 1 \quad -2]$, $k_2' = [4 \quad 1 \quad 5 \quad -3]$ and $k_3' = [1 \quad 4 \quad 0 \quad 7]$, for which the corresponding values of $k'\tilde{x}$ are, by (25), $k_1'\tilde{x} = -1$, $k_2'\tilde{x} = 4$ and $k_3'\tilde{x} = 18$.

Any solution other than $\tilde{x}_1$ and $\tilde{x}_2$ will be a linear combination of these two. For example,

$$\tilde{x}_3 = \begin{bmatrix} -1.8 \\ 10.2 \\ -1.6 \\ -3.0 \end{bmatrix}$$

is a solution, and $\tilde{x}_3 = 2\tilde{x}_1 - \tilde{x}_2$. And any other k' satisfying (24) will be a linear combination of k_1', k_2' and k_3', and the corresponding value of $k'\tilde{x}$ will be the same linear combination of $k_1'\tilde{x}$, $k_2'\tilde{x}$ and $k_3'\tilde{x}$. For example, for $k_4' = [11 \quad -1 \quad 30 \quad -25]$,

$$k_4' = -5k_1' + 7k_2' - 2k_3'$$

and

$$\begin{aligned} k_4'\tilde{x} &= -5(k_1'\tilde{x}) + 7(k_2'\tilde{x}) - 2(k_3'\tilde{x}) \\ &= -5(-1) + 7(4) - 2(18) = -3. \end{aligned}$$

The reader should verify that, for appropriate choice of w', k_4' and $k_4'\tilde{x}$ satisfy (24) and (25) respectively.

This procedure may appear somewhat lengthy for such a small example, but its advantages are clearly apparent if they are thought of in terms of a large number of equations, say 80 to 100. Each step in the procedure involves matrix operations that are easily carried out no matter what the size or rank of the matrix is and each step is readily amenable to computer processing. As a result, the problem of solving linear equations can be reduced to a straightforward computing procedure, available either as a desk-calculator operation for cases invoving relatively few variables or as a computer operation where large numbers of variables are involved.

The last topic to be dealt with is the generalized inverse of a rectangular matrix, which is needed when there are more or fewer equations than unknowns. This is now considered.

6. RECTANGULAR MATRICES

a. A simple procedure

A method of solving a set of equations which has a different number of equations from unknowns is illustrated in the following example.

Example. Suppose the equations $Ax = y$ are

$$\begin{bmatrix} 1 & 4 \\ 2 & 7 \\ 3 & 10 \end{bmatrix} x = \begin{bmatrix} 7 \\ 13 \\ 19 \end{bmatrix}.$$

Let us redefine A by adding a column of zeros to it to make it square. A generalized inverse of the redefined A can then be obtained. Thus if

$$A = \begin{bmatrix} 1 & 4 & 0 \\ 2 & 7 & 0 \\ 3 & 10 & 0 \end{bmatrix}, \quad \text{and} \quad P = \begin{bmatrix} 1 & 0 & 0 \\ -2 & 1 & 0 \\ 1 & -2 & 1 \end{bmatrix}$$

and

$$Q = \begin{bmatrix} 1 & -4 & 0 \\ 0 & 1 & 0 \\ 0 & 0 & 1 \end{bmatrix},$$

it will be found that

$$PAQ = \Delta = \begin{bmatrix} 1 & 0 & 0 \\ 0 & -1 & 0 \\ 0 & 0 & 0 \end{bmatrix} \quad \text{and} \quad G = Q\Delta^{-}P = \begin{bmatrix} -7 & 4 & 0 \\ 2 & -1 & 0 \\ 0 & 0 & 0 \end{bmatrix}.$$

Let us now redefine G by dropping off the same number of rows of zeros as columns were added to A, namely one, at the same time reverting to the original form of A. Then

$$G = \begin{bmatrix} -7 & 4 & 0 \\ 2 & -1 & 0 \end{bmatrix}$$

is a generalized inverse of

$$A = \begin{bmatrix} 1 & 4 \\ 2 & 7 \\ 3 & 10 \end{bmatrix}, \quad \text{with} \quad H = GA = \begin{bmatrix} 1 & 0 \\ 0 & 1 \end{bmatrix}.$$

The solution to the equations has the same form as previously,

$$\tilde{x} = Gy + (H - I)z, = \begin{bmatrix} 3 \\ 1 \end{bmatrix}$$

in this case, there being no arbitrariness about the solution in this example because $H - I = 0$.

The key to this method is at once apparent: if A is rectangular it is redefined by adding sufficient rows (columns) of zeros to make it square. A generalized inverse is then derived in the usual fashion, and is redefined by dropping off the same number of columns (rows) of zeros as had rows (columns) been added to A. If A as originally given has order $p \times q$, the order of G is $q \times p$, and the same properties of G and procedures for solving equations $Ax = y$ hold as previously. There is nothing more to the procedure than this, just add rows (columns) of zeros to A, obtain G and drop off columns (rows) of zeros. Note the uniformity of this procedure: it is the same whether there are more or fewer equations than unknowns.

b. An alternative

An alternative procedure involves a minor modification to the definition of Δ^-, but no redefining of A and G. It is based on the fact that reduction to canonical form under equivalence can be carried out for rectangular matrices as readily as it can for square ones. The matrices P and Q are then of different orders and the process of finding G is as exemplified below.

Example. The A of the previous example is

$$A = \begin{bmatrix} 1 & 4 \\ 2 & 7 \\ 3 & 10 \end{bmatrix},$$

and for

$$P = \begin{bmatrix} 1 & 0 & 0 \\ -2 & 1 & 0 \\ 1 & -2 & 1 \end{bmatrix} \quad \text{and} \quad Q = \begin{bmatrix} 1 & -4 \\ 0 & 1 \end{bmatrix}$$

$$PAQ = \Delta = \begin{bmatrix} 1 & 0 \\ 0 & -1 \\ 0 & 0 \end{bmatrix}.$$

From this, Δ^- is defined as

$$\Delta^- = \begin{bmatrix} 1 & 0 & 0 \\ 0 & -1 & 0 \end{bmatrix}$$

and the method proceeds as before.

The only difference between this method and the one used for square matrices is that Δ^- is defined as the transpose of Δ with its non-zero elements replaced by their reciprocals. If A is $p \times q$ so is Δ, but Δ^- will be $q \times p$.

7. EXERCISES

1. Find a generalized inverse of each of the following matrices.

(*a*) $\begin{bmatrix} 2 & 1 & 4 \\ 6 & 9 & 3 \\ 4 & 4 & 5 \end{bmatrix}$ (*b*) $\begin{bmatrix} 1 & 0 & -1 & 2 \\ 3 & 1 & 2 & 1 \\ 4 & 3 & -2 & 1 \\ 13 & 11 & 3 & -5 \end{bmatrix}$

(c) $\begin{bmatrix} 1 & 2 & 1 & 2 \\ 1 & 3 & 2 & 1 \\ 0 & 1 & 1 & 1 \\ -1 & 2 & 3 & 1 \end{bmatrix}$ (d) $\begin{bmatrix} 4 & 3 \\ 1 & 2 \end{bmatrix}$

2. Find a generalized inverse of each of the matrices given in Exercise 2 of Chapter 5.

3. Find a set of linearly independent non-null solutions for each of the following sets of equations and show that any other solution is a linear combination of them.

(a) $\begin{bmatrix} 6 & 2 & 0 \\ -1 & 0 & 3 \\ 3 & 2 & 9 \end{bmatrix} x = \begin{bmatrix} 8 \\ 2 \\ 14 \end{bmatrix}$ (b) $\begin{bmatrix} 4 & -9 & -1 & 2 \\ 3 & 1 & 0 & 1 \\ 10 & -7 & -1 & 4 \\ 25 & -2 & -1 & 9 \end{bmatrix} x = \begin{bmatrix} 7 \\ 5 \\ 17 \\ 42 \end{bmatrix}$

(c) $\begin{bmatrix} 1 & 1 & 0 & 1 \\ -1 & -1 & 1 & 1 \\ 1 & 0 & 0 & 1 \\ -1 & 0 & 1 & 1 \end{bmatrix} x = \begin{bmatrix} 8 \\ -1 \\ 6 \\ 1 \end{bmatrix}$ (d) $\begin{bmatrix} 1 & 2 & 3 & 4 \\ 5 & 6 & 7 & 8 \end{bmatrix} x = \begin{bmatrix} 10 \\ 26 \end{bmatrix}$

(e) $\begin{bmatrix} 1 & 0 & 2 \\ 4 & 1 & 7 \\ 3 & 2 & 3 \end{bmatrix} x = \begin{bmatrix} 12 \\ 46 \\ 27 \end{bmatrix}$ (f) $\begin{bmatrix} 6 & 1 & 4 & 2 & 1 \\ 3 & 0 & 1 & 4 & 2 \\ -3 & -2 & -5 & 8 & 4 \end{bmatrix} x = \begin{bmatrix} 11 \\ 6 \\ -4 \end{bmatrix}$

4. Show that the equations of Exercise 3 are consistent.

5. If G is the generalized inverse of A show that $\tilde{x} = (G + GA - I)y$ is a solution of $Ax = y$, and that $\tilde{x} = [G(I + \lambda A) - \lambda I]y$ is also a solution, where λ is any scalar.

6. For the generalized inverse based on P, Q and Δ as used in this chapter, show that (a) $GAQP = QPAG$; and (b) $H = Q\Delta\Delta^{-}Q^{-1}$.

7. If $A = A'$, $U'U = I$ and $U'AU = \Delta$, the diagonal form used in this chapter, show that for $R = U\Delta^{-}U'$,

(a) $RAR = R$
(b) $ARA = A$
(c) $(RA)' = RA$
(d) $(AR)' = AR$
(e) $(RA)^2 = RA$
(f) $RA = AR$
(g) $A^2 = U\Delta^2 U'$
(h) $(I - RA)^2 = I - RA$.

8. If G is any generalized inverse of $X'X$, prove that
(a) GX' is a generalized inverse of X, and (b) XGX' is unique.

9. If G is a symmetric generalized inverse of $X'X$ satisfying the first two of Penrose's four conditions, and if

(i) $a = GX'y$, (ii) $s = (y - Xa)'(y - Xa)$,

and (iii) $b = a - GQ(Q'GQ)^{-1}(Q'a - m)$,

show that

(iv) $s = y'y - a'X'y$,

(v) $Q'b = m$,

and (vi) $(y - Xb)'(y - Xb) = s + (Q'a - m)'(Q'GQ)^{-1}(Q'a - m)$.

REFERENCES

Aitken, A. C. (1948). *Determinants and Matrices*, Fifth Edition, Oliver and Boyd, Edinburgh.

Federer, W. T. and M. Zelen (1964). Application of the calculus for factorial arrangements. III. Analysis of factorials with unequal numbers of observations. (Unpublished) Biometrics Unit, Cornell University, and The National Cancer Institute.

Greville, T. N. E. (1957). The pseudo inverse of a rectangular or singular matrix and its application to the solution of systems of linear equations. *S.I.A.M. Newsletter*, **5**, 3–6.

Penrose, R. A. (1955). A generalized inverse for matrices. *Proc. Camb. Phil. Soc.*, **51**, 406–413.

Rao, C. R. (1955). Analysis of dispersion for multiple classified data with unequal numbers of cells. *Sankhya*, **15**, 253–280.

Rao, C. R. (1962). A note on a generalized inverse of a matrix with applications to problems in mathematical statistics. *J. Roy. Stat. Soc.* (B), **24**, 152–158.

Wilkinson, G. N. (1958). Estimation of missing values for the analysis of incomplete data. *Biometrics*, **14**, 257–286.

CHAPTER 7

LATENT ROOTS AND VECTORS

1. AGE DISTRIBUTION VECTORS

Studies of a living population often include investigating the distribution of individuals according to their age. For any appropriately chosen unit of time, depending on the species concerned, this distribution can be summarized as a vector n_t:

$$n'_t = [v_{0t} \quad v_{1t} \quad v_{2t} \quad \dots \quad v_{it} \quad \dots \quad v_{kt}]$$

where v_{it} is the number of individuals that are aged i at time t, i and t being measured in the selected unit of time. Such individuals, although described as being of age i, are all those whose exact age at time t is at least i units of time but less than $i + 1$. In this way the vector n_t represents the distribution of the population according to age at time t, for all ages up to but not including $k + 1$. We might therefore refer to n_t as the age distribution vector at time t.

An immediate problem of interest is to find the value of the age distribution vector at time $t + 1$, n_{t+1}, and to investigate its relationship to n_t. This is achieved by means of a matrix containing age-specific fertility and survival rates as elements. For the sake of simplicity the development is confined to females and uses the following two parameters:

p_i = the probability that a female aged i at time t survives to time $t + 1$,

and

f_i = the number of daughters alive at time $t + 1$ that were born during time t to $t + 1$ to a female aged i at time t.

Now consider the elements of n_{t+1}, $v_{i,t+1}$ for $i = 0, 1, \ldots, k$. By definition

$$\begin{aligned} v_{0,t+1} &= \text{number of females aged 0 to 1 at time } t+1 \\ &= \text{number of females alive at time } t+1 \\ &\quad\ \text{that were born during time } t \text{ to } t+1. \end{aligned}$$

Hence, from the definition of $f_0, f_1, \ldots, f_k$,

$$v_{0,t+1} = f_0 v_{0t} + f_1 v_{1t} + \ldots + f_k v_{kt}.$$

Similarly, the definition of $v_{1,t+1}$ is

$$\begin{aligned} v_{1,t+1} &= \text{number of females aged 1 to 2 at time } t+1 \\ &= \text{survivors of those females aged 0 to 1 at time } t \end{aligned}$$

and from the manner of defining p_0 this means $v_{1,t+1} = p_0 v_{0t}$. Likewise $v_{2,t+1} = p_1 v_{1,t}$, and in general $v_{i,t+1} = p_{i-1} v_{i-1,t}$ for $i = 1, 2, \ldots, k$. These results are brought together in the matrix equation

$$n_{t+1} = \begin{bmatrix} v_{0,t+1} \\ v_{1,t+1} \\ v_{2,t+1} \\ \cdot \\ \cdot \\ \cdot \\ v_{k,t+1} \end{bmatrix} = \begin{bmatrix} f_0 & f_1 & f_2 & \cdot & \cdot & \cdot & f_k \\ p_0 & 0 & 0 & \cdot & \cdot & \cdot & 0 \\ 0 & p_1 & 0 & \cdot & \cdot & \cdot & 0 \\ \cdot & & \cdot & \cdot & & & \cdot \\ \cdot & & & \cdot & \cdot & & \cdot \\ \cdot & & & & \cdot & \cdot & \cdot \\ 0 & 0 & 0 & & & p_{k-1} & 0 \end{bmatrix} \begin{bmatrix} v_{0,t} \\ v_{1,t} \\ v_{2,t} \\ \cdot \\ \cdot \\ \cdot \\ v_{k,t} \end{bmatrix} = Mn_t,$$

where M represents the $(k + 1) \times (k + 1)$ matrix whose non-zero elements are the f's and p's. The first row of M contains the f's, and the elements immediately below the diagonal elements are the p's. The latter, by their position in M, are referred to as *sub-diagonal elements*. Thus the relationship between n_{t+1} and n_t is that of a linear transformation

$$n_{t+1} = Mn_t. \tag{1}$$

Bernadelli (1941) and Leslie (1945 and 1948), to mention but two of the earliest references, discuss many implications of this relationship, especially Leslie who treats the topic at great length. For the moment one question only is pertinent: does a time ever come when the age distribution becomes stable? Stability is here taken to mean that corresponding elements of the age distribution vectors would be the same, apart from a constant factor, from one unit of time to the next. Should this state of affairs occur at time

$t = s$, say, $v_{i,s+1}/v_{i,s}$ would equal $v_{i',s+1}/v_{i',s}$ for $i \neq i'$, and were this common ratio to equal λ the consequent relationship between n_{s+1} and n_s would be $n_{s+1} = \lambda n_s$. Utilizing (1) this would mean $Mn_s = \lambda n_s$. The question of the existence of a stable age distribution therefore reduces itself to asking if there exists a vector n_s and a scalar λ such that $Mn_s = \lambda n_s$. The general nature of this question is now discussed; other implications of equation (1) relative to age-distribution vectors are considered subsequently.

2. DERIVATION OF LATENT ROOTS AND VECTORS

Generalization of the question just posed is the following: does there exist, for a square matrix A, a vector u and a scalar λ such that

$$Au = \lambda u\ ? \tag{2}$$

If a scalar and vector exist that satisfy this equation it can be rewritten as $Au - \lambda u = 0$, equivalent to

$$(A - \lambda I)u = 0.$$

(The symbol 0 here represents a null vector, and henceforth is used interchangeably as a null vector or null matrix, and also as the scalar zero.) Now, as seen in Chapter 6, an equation of this form has a non-null solution for the vector u only if the rank of $(A - \lambda I)$ is less than its order, in which case its determinant is zero; i.e.,

$$|A - \lambda I| = 0. \tag{3}$$

Accordingly, this equation establishes conditions under which equation (2) is true, namely, values of λ which satisfy (3) are such that (2) is also satisfied. Equation (3) is known as the *characteristic equation.* When A is of order n the characteristic equation is a polynomial in λ of degree n, and hence has n solutions $\lambda_1, \lambda_2, \ldots, \lambda_n$ say. For each of them (2) holds true, and so in general we would expect to find n vectors $u_1, u_2, \ldots, u_n$ corresponding to the n λ's. It is shown later that this is not always the case, but let us assume for the moment that it is. Then for each λ_i that is a solution to (3), (2) is true, i.e.,

$$Au_i = \lambda_i u_i, \qquad \text{for} \quad i = 1, 2, \ldots, n. \tag{4}$$

The λ_i's are known as the *latent roots* of the matrix A, and the corresponding vectors u_i are the *latent vectors* of A. The word "latent" is not the only term that can be used in this context; the roots and vectors are referred to variously as latent roots, characteristic roots, λ-roots or eigenvalues, and the corresponding vectors as latent vectors, characteristic vectors, λ-vectors or eigen vectors. The terminology "latent roots" and "latent vectors" is used exclusively in this book.

Example. The matrix

$$A = \begin{bmatrix} 1 & 4 \\ 9 & 1 \end{bmatrix}$$

has characteristic equation

$$\left| \begin{bmatrix} 1 & 4 \\ 9 & 1 \end{bmatrix} - \begin{bmatrix} \lambda & 0 \\ 0 & \lambda \end{bmatrix} \right| = 0;$$

i.e.,

$$\begin{vmatrix} 1 - \lambda & 4 \\ 9 & 1 - \lambda \end{vmatrix} = 0. \qquad (5)$$

Note that a characteristic equation is always of this form: equated to zero is the determinant of A amended by subtracting λ from every diagonal element. Expanding (5) gives

$$(1 - \lambda)^2 - 36 = 0,$$

and hence $\lambda = -5$ or 7. Methods for obtaining the latent vector corresponding to each solution of the characteristic equation are discussed later, but meanwhile it can be seen here that

$$\begin{bmatrix} 1 & 4 \\ 9 & 1 \end{bmatrix} \begin{bmatrix} 2 \\ -3 \end{bmatrix} = -5 \begin{bmatrix} 2 \\ -3 \end{bmatrix} \qquad (6)$$

and

$$\begin{bmatrix} 1 & 4 \\ 9 & 1 \end{bmatrix} \begin{bmatrix} 2 \\ 3 \end{bmatrix} = 7 \begin{bmatrix} 2 \\ 3 \end{bmatrix}, \qquad (7)$$

these being examples of equation (2). Thus $\begin{bmatrix} 2 \\ -3 \end{bmatrix}$ is the latent vector corresponding to the latent root -5, and $\begin{bmatrix} 2 \\ 3 \end{bmatrix}$ is the vector for the root 7.

Illustration. Bernadelli (1941) hypothesizes a species of beetle "which lives three years only, and which propagates in its third year of life. Let the survival rate of the first age group be $\frac{1}{2}$, of the second $\frac{1}{3}$, and assume that each female in the age group 2 to 3 produces, in the average, 6 new living females". Then the M matrix of fertility and survival rates of the preceding section is

$$M = \begin{bmatrix} 0 & 0 & 6 \\ \frac{1}{2} & 0 & 0 \\ 0 & \frac{1}{3} & 0 \end{bmatrix}.$$

Corresponding to the latent root of unity is the latent vector $\begin{bmatrix}6\\3\\1\end{bmatrix}$ so that

$$\begin{bmatrix}6\\3\\1\end{bmatrix} = \begin{bmatrix}0 & 0 & 6\\ \frac{1}{2} & 0 & 0\\ 0 & \frac{1}{3} & 0\end{bmatrix}\begin{bmatrix}6\\3\\1\end{bmatrix}.$$

Hence any population of these beetles starting out with an age distribution in the proportion 6:3:1 will always stay in that proportion.

Equation (4) leads to a simple result that is found useful in subsequent development. If λ_i is a solution of the characteristic equation of A and if u_i is the corresponding latent vector, then by (4) $Au_i = \lambda_i u_i$. Pre-multiplication by A gives

$$A^2 u_i = A\lambda_i u_i = \lambda_i A u_i$$

because λ_i is a scalar, and using (4) again this becomes $A^2 u_i = \lambda_i^2 u_i$. This is easily extended so that in general

$$A^k u_i = \lambda_i^k u_i, \tag{8}$$

where k is any integer, positive only should A be singular, otherwise positive or negative.

The first step involved in calculating the latent vectors of a matrix is to obtain its latent roots by solving the characteristic equation. Methods for determining the corresponding latent vectors then depend on certain conditions relative to the latent roots. Two situations are initially distinguishable: when the latent roots are all different, and when they are not all different.

3. LATENT ROOTS ALL DIFFERENT

a. Linear independence of latent vectors

When the latent roots of a matrix are all different the corresponding latent vectors form a linearly independent set of vectors.

**Proof.*

Assume the latent vectors u_i corresponding to the latent roots λ_i are dependent, the λ_i being all different. Then, by the definition of dependence, there exists a set of scalars, $c_1, c_2, \ldots, c_n$, not all zero, such that

$$c_1 u_1 + c_2 u_2 + \ldots + c_n u_n = 0.$$

Pre-multiplying this equation in turn by $A, A^2, \ldots, A^{n-1}$ and using results implicit in (8) leads to the following set of equations:

$$\begin{bmatrix} 1 & 1 & \dots & 1 \\ \lambda_1 & \lambda_2 & & \lambda_n \\ \lambda_1^2 & \lambda_2^2 & & \lambda_n^2 \\ \cdot & & & \cdot \\ \cdot & & & \cdot \\ \cdot & & & \cdot \\ \lambda_1^{n-1} & \lambda_2^{n-1} & \dots & \lambda_n^{n-1} \end{bmatrix} \begin{bmatrix} c_1 u'_1 \\ c_2 u'_2 \\ c_3 u'_3 \\ \cdot \\ \cdot \\ \cdot \\ c_n u'_n \end{bmatrix} = 0. \tag{9}$$

Denote the n-order matrix on the left, the matrix of the λ's and their powers, by B. Then B is an alternant matrix (see Section 3.3), and because of assuming the λ's are all different its determinant is non-zero. Therefore B^{-1} exists and pre-multiplying (9) by B^{-1} gives

$$c_1 u_1 = c_2 u_2 = \dots = c_n u_n = 0.$$

But this implies $c_1 = c_2 = \dots = c_n = 0$, which contradicts the assumption of linear dependence of the u's. Therefore the u's are linearly independent.

An important consequence of the latent vectors being independent is that if they are used as columns of a matrix

$$U = [u_1 \quad u_2 \quad \dots \quad u_n],$$

U is square of order n and, because its columns are independent, $r(U) = n$, U is nonsingular and U^{-1} exists. Therefore, by writing

$$Au_i = \lambda_i u_i, \qquad \text{for} \quad i = 1, 2, \dots, n,$$

as

$$A[u_1 \quad u_2 \quad \dots \quad u_n] = [u_1 \quad u_2 \quad \dots \quad u_n] \begin{bmatrix} \lambda_1 & 0 & & \dots & 0 \\ 0 & \lambda_2 & & & \cdot \\ & & & & \cdot \\ & & & & \cdot \\ \cdot & & & \cdot & \\ \cdot & & & \cdot & 0 \\ \cdot & & & & \cdot \\ 0 & \dots & & 0 & \lambda_n \end{bmatrix}$$

we get

$$AU = UD$$

where D is the n-order diagonal matrix of the latent roots $\lambda_1, \lambda_2, \dots, \lambda_n$. Hence

$$A = UDU^{-1} \qquad \text{and} \qquad D = U^{-1}AU. \tag{10}$$

D is often referred to as the *canonical form of A under similarity*.

Example (*continued*). It has already been shown that the matrix $A = \begin{bmatrix} 1 & 4 \\ 9 & 1 \end{bmatrix}$ has latent roots -5 and 7, the corresponding latent vectors being $\begin{bmatrix} 2 \\ -3 \end{bmatrix}$ and $\begin{bmatrix} 2 \\ 3 \end{bmatrix}$. The matrix U which has these vectors as columns is

$$U = \begin{bmatrix} 2 & 2 \\ -3 & 3 \end{bmatrix} \quad \text{with} \quad U^{-1} = \tfrac{1}{12}\begin{bmatrix} 3 & -2 \\ 3 & 2 \end{bmatrix},$$

and

$$\begin{aligned} U^{-1}AU &= \tfrac{1}{12}\begin{bmatrix} 3 & -2 \\ 3 & 2 \end{bmatrix}\begin{bmatrix} 1 & 4 \\ 9 & 1 \end{bmatrix}\begin{bmatrix} 2 & 2 \\ -3 & 3 \end{bmatrix} \\ &= \begin{bmatrix} -5 & 0 \\ 0 & 7 \end{bmatrix} \\ &= D, \end{aligned}$$

the diagonal matrix of the latent roots.

Equations (10) provide a useful method of deriving the powers of A once U and U^{-1} have been obtained. For then only the powers of D are required, which is an easy task because D is diagonal. Thus from (10)

$$A^2 = AA = UDU^{-1}UDU^{-1} = UD^2U^{-1}$$

and in general, for p being an integer,

$$A^p = UD^pU^{-1}$$

including, if it exists, $(A^{-1})^p = A^{-p} = UD^{-p}U^{-1}$.

Illustrations. Instances in which the integer powers of a matrix are required abound in biology. One arises from the illustration used at the beginning of the chapter, where the equation $n_{t+1} = Mn_t$ was established for age-distribution vectors. Suppose time is measured from some zero point at which the age-distribution vector is known, n_0 say. Successive values of the vector are then

$$\begin{aligned} n_1 &= Mn_0, \\ n_2 &= Mn_1 = M^2n_0, \\ n_3 &= Mn_2 = M^3n_0, \\ &\cdot \\ &\cdot \\ &\cdot \end{aligned}$$

and

$$n_r = Mn_{r-1} = M^rn_0,$$

and if M has been reduced to canonical form under similarity, $M = UDU^{-1}$, these values are readily obtained as

$$n_r = UD^rU^{-1}n_0, \qquad \text{for} \quad r = 1, 2, \ldots .$$

Since D is diagonal the computational task is considerably reduced compared to obtaining $M^r n_0$ as it stands. Because D^r has only its diagonal elements non-zero UD^r is U with its columns multiplied by respective elements of D^r, and thus $UD^rU^{-1}n_0$ can be derived explicitly in terms of r. This is illustrated in detail shortly.

The situation of requiring powers of a matrix also occurs in some of the other illustrations that have been mentioned. For example, the probability transition matrix for a species of flour beetle referred to in Section 1.6 is analogous to the matrix M just discussed. So also is the matrix P of the pulmonary tuberculosis illustration in Section 4.8, and in the discussion of inbreeding in Section 2.5*b* the generation matrix plays a similar role. In all these cases the integer powers of a matrix are required; the procedure just indicated provides a useful technique for obtaining them.

b. Calculating latent vectors

For each λ_i that is a solution of the characteristic equation we have, by (4), $Au_i = \lambda_i u_i$, and hence

$$(A - \lambda_i I)u_i = 0. \tag{11}$$

Now λ_i has been derived as a solution to the equation $|A - \lambda I| = 0$. Therefore $|A - \lambda_i I| = 0$ and consequently (see Section 6.5*c*) equation (11) has $n - r_i$ LINN solutions for u_i, where r_i is the rank of $A - \lambda_i I$. With the latent roots being all different $r_i = n - 1$ for all i (this being a special case of the theorem in the next section). Thus (11) has just a single LINN solution for u_i, which can be obtained by giving an arbitrary value to one element of u_i in (11) and solving the resultant equations for the other $n - 1$ elements. This is done for each λ_i in turn, to obtain the respective u_i vectors.

Example. For

$$A = \begin{bmatrix} 1 & 4 & 1 \\ 2 & 1 & 0 \\ -1 & 3 & 1 \end{bmatrix}$$

the characteristic equation is

$$\begin{vmatrix} 1-\lambda & 4 & 1 \\ 2 & 1-\lambda & 0 \\ -1 & 3 & 1-\lambda \end{vmatrix} = 0$$

which expands, by diagonal expansion, as

$$-\lambda^3 + \lambda^2(1+1+1) - \lambda\left\{\begin{vmatrix} 1 & 4 \\ 2 & 1 \end{vmatrix} + \begin{vmatrix} 1 & 1 \\ -1 & 1 \end{vmatrix} + \begin{vmatrix} 1 & 0 \\ 3 & 1 \end{vmatrix}\right\} + \begin{vmatrix} 1 & 4 & 1 \\ 2 & 1 & 0 \\ -1 & 3 & 1 \end{vmatrix} = 0,$$

and reduces to $\lambda(\lambda^2 - 3\lambda - 4) = 0$. Hence the latent roots of A are $\lambda = 0$, -1 and 4. Denoting the elements of u_i by α_i, β_i and γ_i, equation (11) is

$$\begin{bmatrix} 1-\lambda_i & 4 & 1 \\ 2 & 1-\lambda_i & 0 \\ -1 & 3 & 1-\lambda_i \end{bmatrix}\begin{bmatrix} \alpha_i \\ \beta_i \\ \gamma_i \end{bmatrix} = 0.$$

For $\lambda_1 = 0$ this becomes

$$\begin{aligned} \alpha_1 + 4\beta_1 + \gamma_1 &= 0 \\ 2\alpha_1 + \beta_1 \quad\quad &= 0 \\ -\alpha_1 + 3\beta_1 + \gamma_1 &= 0. \end{aligned}$$

Taking the arbitrary value as $\alpha_1 = 1$ leads to solutions $\beta_1 = -2$ and $\gamma_1 = 7$, and the corresponding latent vector is $u_1 = \begin{bmatrix} 1 \\ -2 \\ 7 \end{bmatrix}$. Similarly for the second solution to the characteristic equation, $\lambda_2 = -1$, the equations are

$$\begin{aligned} 2\alpha_2 + 4\beta_2 + \gamma_2 &= 0 \\ 2\alpha_2 + 2\beta_2 \quad\quad &= 0 \\ -\alpha_2 + 3\beta_2 + 2\gamma_2 &= 0, \end{aligned}$$

and taking $\alpha_2 = 2$ as the arbitrary value gives $\beta_2 = -2$ and $\gamma_2 = 4$. Hence the second latent vector is $u_2 = \begin{bmatrix} 2 \\ -2 \\ 4 \end{bmatrix}$. Finally, for $\lambda_3 = 4$, the equations have solutions associated with the arbitrary value $\alpha_3 = 3$ of $\beta_3 = 2$ and $\gamma_3 = 1$. Hence $u_3 = \begin{bmatrix} 3 \\ 2 \\ 1 \end{bmatrix}$ and the matrix of the latent vectors is

$$U = [u_1 \quad u_2 \quad u_3] = \begin{bmatrix} 1 & 2 & 3 \\ -2 & -2 & 2 \\ 7 & 4 & 1 \end{bmatrix}.$$

It may be verified that $|U| = 40$, that

$$U^{-1} = \tfrac{1}{20}\begin{bmatrix} -5 & 5 & 5 \\ 8 & -10 & -4 \\ 3 & 5 & 1 \end{bmatrix}$$

and hence, as in (10),

$$U^{-1}AU = \begin{bmatrix} 0 & 0 & 0 \\ 0 & -1 & 0 \\ 0 & 0 & 4 \end{bmatrix} = \begin{bmatrix} \lambda_1 & 0 & 0 \\ 0 & \lambda_2 & 0 \\ 0 & 0 & \lambda_3 \end{bmatrix} = D.$$

Likewise $A = UDU^{-1}$ and

$$A^p = UD^pU^{-1} = \frac{1}{20}\begin{bmatrix} 1 & 2 & 3 \\ -2 & -2 & 2 \\ 7 & 4 & 1 \end{bmatrix}\begin{bmatrix} 0 & 0 & 0 \\ 0 & (-1)^p & 0 \\ 0 & 0 & 4^p \end{bmatrix}\begin{bmatrix} -5 & 5 & 5 \\ 8 & -10 & -4 \\ 3 & 5 & 1 \end{bmatrix}.$$

The reader will find it worthwhile to carry out these multiplications and satisfy himself that the results are as shown. He should also apply the same methods to the first example of this Section.

Illustration. Li (1954, page 107) discusses the five genotypes of an autotetraploid based on two alleles, *a* and *A*, at a single locus. Grouping the genotypes into three different types, T_0 for *aaaa* or *AAAA*, T_1 for *aAAA* or *Aaaa*, and T_2 for *aaAA*, the probability under selfing that an individual of type T_i produces a progeny of type T_j in unit time can be represented by p_{ij}. The nine values of this probability can be arrayed as a probability transition matrix

$$P = \begin{bmatrix} p_{00} & p_{01} & p_{02} \\ p_{10} & p_{11} & p_{12} \\ p_{20} & p_{21} & p_{22} \end{bmatrix} = \begin{bmatrix} 1 & 0 & 0 \\ \frac{1}{4} & \frac{1}{2} & \frac{1}{4} \\ \frac{2}{36} & \frac{16}{36} & \frac{18}{36} \end{bmatrix}.$$

One question of interest to geneticists is "what is the probability that an individual of type T_i will have descendants of type T_j in the kth generation of selfing?" An extension of the question is to find the value of this probability when k is infinitely large. The whole problem involves deriving the kth power of P because its elements are, for k generations of selfing, the probabilities required. As already indicated, equations (10) are ideally suited to this problem. The calculations are summarized below.

The characteristic equation of P has solutions for the latent roots λ as 1, 1/6 and 5/6. The equations to find the latent vectors are

$$\begin{bmatrix} 1-\lambda_i & 0 & 0 \\ \frac{1}{4} & \frac{1}{2}-\lambda_i & \frac{1}{4} \\ \frac{1}{18} & \frac{8}{18} & \frac{9}{18}-\lambda_i \end{bmatrix}\begin{bmatrix} \alpha_i \\ \beta_i \\ \gamma_i \end{bmatrix} = 0.$$

For $\lambda_1 = 1$, and taking $\alpha_1 = 1$ as the necessary arbitrary value, it will be found that the latent vector is $u_1 = \begin{bmatrix} 1 \\ 1 \\ 1 \end{bmatrix}$; with $\lambda_2 = 1/6$ and $\alpha_2 = 0$ the vector is $u_2 = \begin{bmatrix} 0 \\ 3 \\ -4 \end{bmatrix}$; and for $\lambda_3 = 5/6$ and $\alpha_3 = 0$ the vector is $u_3 = \begin{bmatrix} 0 \\ 3 \\ 4 \end{bmatrix}$.

Hence

$$U = \begin{bmatrix} 1 & 0 & 0 \\ 1 & 3 & 3 \\ 1 & -4 & 4 \end{bmatrix} \quad \text{with} \quad U^{-1} = \tfrac{1}{24}\begin{bmatrix} 24 & 0 & 0 \\ -1 & 4 & -3 \\ -7 & 4 & 3 \end{bmatrix}.$$

Thus,

$$P^k = \tfrac{1}{24}\begin{bmatrix} 1 & 0 & 0 \\ 1 & 3 & 3 \\ 1 & -4 & 4 \end{bmatrix}\begin{bmatrix} 1 & 0 & 0 \\ 0 & (1/6)^k & 0 \\ 0 & 0 & (5/6)^k \end{bmatrix}\begin{bmatrix} 24 & 0 & 0 \\ -1 & 4 & -3 \\ -7 & 4 & 3 \end{bmatrix},$$

which reduces to

$$P^k = \begin{bmatrix} 1 & 0 & 0 \\ 1 - r_k/2 - 3s_k/8 & r_k/2 & 3s_k/8 \\ 1 - r_k/2 - 2s_k/3 & 2s_k/3 & r_k/2 \end{bmatrix}$$

where $r_k = (5/6)^k + (1/6)^k$ and $s_k = (5/6)^k - (1/6)^k$.

From these expressions the elements of P^k are readily obtainable for any value of k. The probability of a genotype of type T_i having a descendant of type T_j after k generations of selfing is then given by the appropriate element of P^k. In particular, when k is infinitely large r_k and s_k tend to zero and the value of P^k becomes $\begin{bmatrix} 1 & 0 & 0 \\ 1 & 0 & 0 \\ 1 & 0 & 0 \end{bmatrix}$, showing that ultimately all genotypes become of type T_0.

Not all matrices used in this manner have a limiting form.

Illustration. The matrix appropriate to Bernadelli's beetles (*loc. cit.*) is

$$A = \begin{bmatrix} 0 & 0 & 6 \\ \frac{1}{2} & 0 & 0 \\ 0 & \frac{1}{3} & 0 \end{bmatrix}$$

and without even considering its latent roots and vectors it is easily verified that

$$A^2 = \begin{bmatrix} 0 & 2 & 0 \\ 0 & 0 & 3 \\ \frac{1}{6} & 0 & 0 \end{bmatrix}$$

and $A^3 = I$. Hence $A^4 = A$, $A^5 = A^2$, $A^6 = I$, $A^7 = A$ and so on, with regular periodicity. Thus, as Bernadelli shows, if the beetle population is initially 3600 beetles, divided equally between its three stages of life, the sequence of age-distribution vectors from then on is as shown in Table 1.

TABLE 1. AGE-DISTRIBUTION VECTORS FOR BERNADELLI'S BEETLES

Year	0	1	2	3	4	...
Vector	n_0	$n_1 = An_0$	$n_2 = A^2 n_0$	$n_3 = A^3 n_0$	$n_4 = A^4 n_0$	...
Elements	1200	7200	2400	1200	7200	...
of	1200	600	3600	1200	600	...
Vector	1200	400	200	1200	400	...

The regular periodicity, with a three-year cycle, is clearly evident. As shown earlier, the population is stable only if its initial age distribution is in the proportion 6:3:1. In general, if the initial proportion is $6x:3y:z$ the series of age-distribution vectors will be

$$\begin{bmatrix} 6x \\ 3y \\ z \end{bmatrix}, \begin{bmatrix} 6z \\ 3x \\ y \end{bmatrix}, \begin{bmatrix} 6y \\ 3z \\ x \end{bmatrix}, \begin{bmatrix} 6x \\ 3y \\ z \end{bmatrix}, \quad \ldots \quad \text{and so on.}$$

4. MULTIPLE LATENT ROOTS

Example. Suppose an attempt is made to express

$$A = \begin{bmatrix} 2 & -1 & 1 \\ 3 & 3 & -2 \\ 4 & 1 & 0 \end{bmatrix}$$

in the form $A = UDU^{-1}$. The characteristic equation reduces to

$$\lambda^3 - 5\lambda^2 + 7\lambda - 3 = 0,$$

yielding roots $\lambda = 3$, 1 and 1. Accordingly the equations $(A - \lambda_i I)u_i = 0$ for finding the latent vectors are, in the case of $\lambda_1 = 3$,

$$\begin{aligned} -\alpha_1 - \beta_1 + \gamma_1 &= 0 \\ 3\alpha_1 \quad\quad - 2\gamma_1 &= 0 \\ 4\alpha_1 + \beta_1 - 3\gamma_1 &= 0 \end{aligned}$$

for which a solution is $u_1 = \begin{bmatrix} \alpha_1 \\ \beta_1 \\ \gamma_1 \end{bmatrix} = \begin{bmatrix} 2 \\ 1 \\ 3 \end{bmatrix}$. Similarly for $\lambda_2 = 1$ the vector is obtained from $(A - \lambda_2 I)u_2 = 0$:

$$\begin{aligned} \alpha_2 - \beta_2 + \gamma_2 &= 0 \\ 3\alpha_2 + 2\beta_2 - 2\gamma_2 &= 0 \\ 4\alpha_2 + \beta_2 - \gamma_2 &= 0. \end{aligned} \tag{12}$$

A solution is $u_2 = \begin{bmatrix} \alpha_2 \\ \beta_2 \\ \gamma_2 \end{bmatrix} = \begin{bmatrix} 0 \\ 1 \\ 1 \end{bmatrix}$. The third latent root must now be considered, $\lambda_3 = 1$. Since it equals λ_2 the equations $(A - \lambda_3 I)u_3 = 0$ for deriving u_3 will be the same as (12). Now, two solutions u_2 and u_3 of (12) that are not independent of each other would be of no use, for then $U = [u_1 \quad u_2 \quad u_3]$ would have no inverse and the form $U^{-1}AU$ would be meaningless. But from Section 6.5*c* it is known that (12) has two LINN solutions if $r(A - \lambda_2 I) = 1$. Therefore we conclude that a third-order square matrix which has two equal latent roots can be expressed as $U^{-1}AU = D$ only if $r(A - \lambda^* I)$ is 1, where λ^* is the value of the two equal roots. This is a particular case of the general result concerning the existence of the form $U^{-1}AU = D$ when two or more latent roots have the same value. This general result, which is of great value, is embodied in the following theorem.

Theorem. For a square matrix A, of order n, whose latent roots are $\lambda_1, \lambda_2, \ldots, \lambda_s$ with multiplicities $m_1, m_2, \ldots, m_s$, where $\sum_{k=1}^{s} m_k = n$, the necessary and sufficient condition under which A can be expressed in the form $U^{-1}AU = D$, where D is a diagonal matrix of all the n latent roots, is that the rank of $(A - \lambda_k I)$ be equal to $n - m_k$ for all $k = 1, 2, \ldots, s$.

By λ_k being a latent root with multiplicity m_k we mean that the characteristic equation $|A - \lambda I| = 0$ has m_k roots all equal to λ_k. This means that $(\lambda - \lambda_k)^{m_k}$ is a factor of $|A - \lambda I|$, and hence, because the characteristic equation is a polynomial of degree n in λ, $\sum_{k=1}^{s} m_k = n$. Naturally this includes the case when $m_k = 1$ for all values of k, this being the situation already discussed, that of all the latent roots being different.

**Proof.*[1]

i. Given that $U^{-1}AU = D$ we show that $r(A - \lambda_k I) = n - m_k$. From the nature of D it follows that $(D - \lambda_k I)$ has exactly m_k zeros in the diagonal and hence

$$r(D - \lambda_k I) = n - m_k.$$

But $U^{-1}AU = D$ so that $A = UDU^{-1}$. Therefore

$$\begin{aligned} (A - \lambda_k I) &= (UDU^{-1} - \lambda_k I) \\ &= U(D - \lambda_k I)U^{-1}, \end{aligned}$$

[1] I am indebted to Dr. B. L. Raktoe for the bulk of this proof.

and since the rank of a matrix is unaffected by multiplying the matrix by nonsingular matrices (because they are products of elementary operators)

$$r(A - \lambda_k I) = r(D - \lambda_k I) = n - m_k.$$

ii. Conversely, given that $r(A - \lambda_k I) = n - m_k$, we show that A can be expressed in the form $U^{-1}AU = D$.

Since $r(A - \lambda_k I) = n - m_k$, the equation $(A - \lambda_k I)x = 0$ has exactly $n - (n - m_k) = m_k$ linearly independent non-null solutions. But by definition these solutions are latent vectors of A. Hence associated with each λ_k there is a set of m_k independent latent vectors. We now show that the sets are independent of each other. Suppose they are not, and that one vector of the second set of vectors, y_2 say, is a linear combination of the vectors of the first set, $z_1, z_2, \ldots, z_{m_1}$. Then

$$y_2 = \sum_{i=1}^{m_1} c_i z_i$$

for some scalars c_i that are not all zero. Multiplying this equation by A leads to $Ay_2 = \Sigma c_i A z_i$. Because y_2 and the z's are latent vectors corresponding respectively to the different latent roots λ_1 and λ_2,

$$\lambda_2 y_2 = \Sigma c_i \lambda_1 z_i = \lambda_1 \Sigma c_i z_i = \lambda_1 y_2,$$

which cannot be true because $\lambda_2 \neq \lambda_1$ and they are not both zero. Therefore the initial supposition is wrong, and we conclude that all s sets of m_k latent vectors, for $k = 1, 2, \ldots, s$, are independent. If these vectors are used as columns to form a matrix U, U is nonsingular and U^{-1} exists. Furthermore, since its columns are latent vectors of A, $AU = UD$ and hence $U^{-1}AU = D$ exists. Thus the sufficiency condition has been proved and the proof of the theorem is complete.

To summarize the theorem we can say that for any root λ having multiplicity m the form $U^{-1}AU = D$ exists if and only if

$$r(A - \lambda I) = n - m;$$

and this includes the case of single roots, for which $m = 1$.

Example. For

$$A = \begin{bmatrix} -1 & -2 & -2 \\ 1 & 2 & 1 \\ -1 & -1 & 0 \end{bmatrix}$$

the characteristic equation reduces to $(\lambda - 1)(\lambda^2 - 1) = 0$ giving roots 1, 1 and -1. Hence $\lambda = 1$ is a multiple root with multiplicity 2 for which the rank of $A - \lambda I$ must be investigated. Thus

$$A - I = \begin{bmatrix} -2 & -2 & -2 \\ 1 & 1 & 1 \\ -1 & -1 & -1 \end{bmatrix}$$

and its rank is patently $1 = 3 - 2 = n - m$. Therefore the form $U^{-1}AU = D$ exists.

To find U, independent vectors u must be obtained such that $(A - I)u = 0$. Clearly two possibilities are $u_1 = \begin{bmatrix} 1 \\ -1 \\ 0 \end{bmatrix}$ and $u_2 = \begin{bmatrix} 1 \\ 0 \\ -1 \end{bmatrix}$. And for $\lambda_3 = -1$ the equations $(A - \lambda_3 I)u_3 = 0$ have solution $u_3 = \begin{bmatrix} 2 \\ -1 \\ 1 \end{bmatrix}$ and hence $U = [u_1 \quad u_2 \quad u_3] = \begin{bmatrix} 1 & 1 & 2 \\ -1 & 0 & -1 \\ 0 & -1 & 1 \end{bmatrix}$. The reader should satisfy himself that $U^{-1}AU$ is the appropriate diagonal matrix.

Example. With

$$A = \begin{bmatrix} 2 & -1 & 1 \\ 3 & 3 & -2 \\ 4 & 1 & 0 \end{bmatrix}$$

the characteristic equation has roots $\lambda = 1$, 1 and 3. For the multiple root of 1, with multiplicity $m = 2$, $A - \lambda I$ is

$$(A - I) = \begin{bmatrix} 1 & -1 & 1 \\ 3 & 2 & -2 \\ 4 & 1 & -1 \end{bmatrix},$$

and has rank 2 (row 3 equals row 1 plus row 2). Thus with $n = 3$ and $m = 2$, $r(A - I) = 2 \neq 3 - 2$, so the form $U^{-1}AU = D$ does not exist. This example is the one considered immediately before the theorem.

5. SOME PROPERTIES OF LATENT ROOTS

Properties of latent roots and vectors are utilized in many ways. A few are now discussed.

a. Powers of latent roots

On several occasions use has been made of equation (4), that if λ is a latent root of A and u is the corresponding vector, then $Au = \lambda u$. As developed in (8), repeated applications of this lead to $A^k u = \lambda^k u$. By comparison it is apparent that if $Au = \lambda u$ because λ is a latent root and u the corresponding vector, then λ^k in $A^k u = \lambda^k u$ must be a latent root of A^k. Hence it is concluded that if λ is a latent root of A, λ^k is a latent root of A^k where k is positive if A is singular and positive or negative if A is nonsingular. In particular, when A is nonsingular with latent root λ, $1/\lambda$ is a latent root of A^{-1}.

Example. If

$$A = \begin{bmatrix} 4 & 3 \\ 4 & 8 \end{bmatrix}, \quad A^2 = \begin{bmatrix} 28 & 36 \\ 48 & 76 \end{bmatrix} \quad \text{and} \quad A^{-1} = \tfrac{1}{20}\begin{bmatrix} 8 & -3 \\ -4 & 4 \end{bmatrix}.$$

The characteristic equations are as follows:

for A: $\lambda^2 - 12\lambda + 20 = 0$, with roots 2 and 10;
for A^2: $\lambda^2 - 104\lambda + 400 = 0$, with roots 4 and 100;
for A^{-1}: $20\lambda^2 - 12\lambda + 1 = 0$, with roots 1/2 and 1/10.

Clearly, the latent roots of A^2 are the squares of those of A; and the latent roots of A^{-1} are the reciprocals of those of A.

b. Sum and product of latent roots

By diagonal expansion (Section 3.5) the equation

$$\begin{vmatrix} a_{11} - \lambda & a_{12} & a_{13} \\ a_{21} & a_{22} - \lambda & a_{23} \\ a_{31} & a_{32} & a_{33} - \lambda \end{vmatrix} = 0$$

reduces to

$$-\lambda^3 + (-\lambda)^2 tr_1(A) + (-\lambda) tr_2(A) + |A| = 0$$

using the $tr_i(A)$ notation defined in Section 3.5. For A of order n this takes the form

$$(-\lambda)^n + (-\lambda)^{n-1} tr_1(A) + (-\lambda)^{n-2} tr_2(A) + \cdots + (-\lambda) tr_{n-1}(A) + |A| = 0.$$

Applying elementary rules of the theory of equations therefore gives

$$\sum_{i=1}^{n} \lambda_i = tr(A) \qquad \text{and} \qquad \prod_{i=1}^{n} \lambda_i = |A|.$$

These results are quite straightforward: the sum of the latent roots of a matrix equals the trace of the matrix, namely the sum of the diagonal elements, and the product of the latent roots is the determinant of the matrix.

Example. The latent roots of $A = \begin{bmatrix} 1 & 4 & 1 \\ 2 & 1 & 0 \\ -1 & 3 & 1 \end{bmatrix}$ are 0, -1 and 4.

Their sum is 3, which equals the trace of A, $1 + 1 + 1$, and their product is zero, equal to the determinant of A. (The first row equals the sum of the second and third rows.)

c. Non-zero latent roots

The conditions under which a matrix having multiple latent roots can be expressed in the form $U^{-1}AU = D$ have already been discussed. When it can be so expressed U is nonsingular and therefore

$$\text{rank}(D) = \text{rank}(A) = r.$$

But the rank of D is identical to the number of non-zero latent roots; hence the number of such roots is r, and we can write

$$U^{-1}AU = D = \begin{bmatrix} D_r^* & 0 \\ 0 & 0 \end{bmatrix}$$

where D_r^* is a diagonal, of order r, its diagonal elements being the r non-zero latent roots of A.

Example. The latent roots of $\begin{bmatrix} 2 & 3 & 7 \\ 1 & 2 & 4 \\ 1 & 1 & 3 \end{bmatrix}$ are given by

$$\lambda(\lambda^2 - 7\lambda + 2) = 0.$$

Thus the number of non-zero latent roots is 2, the rank of the matrix.

d. Latent roots of a scalar product

If λ is a latent root of A and u is a corresponding latent vector, $Au = \lambda u$. Therefore, for a scalar c, $cAu = c\lambda u$, showing that $c\lambda$ is a latent root of cA; i.e. if λ is a latent root of A, $c\lambda$ is a latent root of cA. The equation $cAu = c\lambda u$ can also be written as $A(cu) = \lambda(cu)$, showing that if u is a latent vector of A so is cu.

Example. $A = \begin{bmatrix} 1 & 1 \\ 4 & 1 \end{bmatrix}$ has characteristic equation $(1 - \lambda)^2 = 4$ and hence its latent roots are 3 and -1. But the characteristic equation of $5A$ is $(5 - \lambda)^2 = 100$, giving latent roots 15 and -5 which are 5 times those of A. The vector $\begin{bmatrix} 1 \\ 2 \end{bmatrix}$ is a latent vector of A corresponding to the latent root $\lambda = 3$; as is also the vector $\begin{bmatrix} 4 \\ 8 \end{bmatrix}$.

e. The Cayley-Hamilton theorem

This is stated very simply: a matrix satisfies its own characteristic equation.

**Proof.*

This proof relies upon the property of linear independence of the latent vectors of a matrix. Other proofs are available, but the following one is satisfactory for our purposes.

The characteristic equation for the matrix A is $|A - \lambda I| = 0$. Writing t_1 for $-tr(A)$,

t_2 for $tr_2(A)$, t_3 for $-tr_3(A)$ and so on, the characteristic equation can be expanded (for A of order n) as

$$\lambda^n + t_1\lambda^{n-1} + t_2\lambda^{n-2} + \ldots + t_{n-1}\lambda + t_n = 0. \qquad (13)$$

This is a scalar equation and can be multiplied by any vector, u say, to give

$$\lambda^n u + t_1\lambda^{n-1}u + \ldots + t_{n-1}\lambda u + t_n u = 0.$$

If u is the latent vector corresponding to a latent root λ which satisfies (13), use can be made of equations (4) and (8) to rewrite the foregoing as

$$A^n u + t_1 A^{n-1}u + \ldots + t_{n-1}Au + t_n Iu = 0. \qquad (14)$$

This will be true for all latent vectors. But since they are of order n, are linearly independent and are n in number, any other n-order vector can be expressed as a linear combination of them. Therefore (14) holds for all n-order vectors and so is true in general. Hence

$$A^n + t_1A^{n-1} + \ldots + t_{n-1}A + t_n I = 0;$$

i.e. A satisfies its own characteristic equation.

A useful application of this result is that when the characteristic equation is known, even if the latent roots are not, the nth and successive powers of A can be obtained as polynomials of A of order $n - 1$. This is also true for the inverse.

Example. For $A = \begin{bmatrix} 4 & 3 \\ 2 & 5 \end{bmatrix}$, $A^2 = \begin{bmatrix} 22 & 27 \\ 18 & 31 \end{bmatrix}$, $A^{-1} = \frac{1}{14}\begin{bmatrix} 5 & -3 \\ -2 & 4 \end{bmatrix}$

and the characteristic equation of A is

$$\lambda^2 - 9\lambda + 14 = 0.$$

The Cayley-Hamilton theorem is satisfied because

$$A^2 - 9A + 14I = \begin{bmatrix} 22 & 27 \\ 18 & 31 \end{bmatrix} - 9\begin{bmatrix} 4 & 3 \\ 2 & 5 \end{bmatrix} + 14\begin{bmatrix} 1 & 0 \\ 0 & 1 \end{bmatrix} = \begin{bmatrix} 0 & 0 \\ 0 & 0 \end{bmatrix}.$$

Polynomials for obtaining powers of A of order 2 or more come from rewriting $A^2 - 9A + 14I = 0$ as

$$A^2 = 9A - 14I.$$

Hence

$$\begin{aligned} A^3 &= 9A^2 - 14A \\ &= 9(9A - 14I) - 14A \\ &= 67A - 126I \end{aligned}$$

and

$$\begin{aligned} A^4 &= 67A^2 - 126A \\ &= 67(9A - 14I) - 126A \\ &= 477A - 938I. \end{aligned}$$

* By this means recurrence relations can be established between the coefficients in successive powers of A, so enabling A^k, for $k \geqq n$, to be written as a polynomial in A of order $n - 1$, the coefficients being functions of k. Thus for the above example it can be shown that, for $k \geqq 2$,

$$A^k = (\tfrac{1}{5})(7^k - 2^k)A - (\tfrac{14}{5})(7^{k-1} - 2^{k-1})I.$$

Similar expressions can be derived for the inverse A^{-1} if it exists. For $A^2 - 9A + 14I = 0$, is equivalent to

$$14I = 9A - A^2$$

so giving

$$A^{-1} = \tfrac{1}{14}(9I - A).$$

Thus

$$\begin{aligned}(A^{-1})^2 &= \tfrac{1}{196}(81I + A^2 - 18A)\\ &= \tfrac{1}{196}(67I - 9A),\end{aligned}$$

and so on. The reader should verify these results explicitly.

6. DOMINANT LATENT ROOTS

Various theorems are to be found concerning the existence of a largest, or dominant, latent root of a matrix (see, for example, Part II of Chapter 4 in Frazer, Duncan and Collar, 1952). These theorems provide methods of finding the largest latent root of a matrix without having to solve the characteristic equation, dealing, *inter alia*, with matters such as multiple roots, multiple dominant roots and roots involving complex numbers as well as considering the problem of the actual existence of a dominant root. We shall not discuss these theorems here but will demonstrate a useful method for calculating the largest latent root *when it exists*. The existence of a real-valued dominant root is assumed throughout; on this basis a method for obtaining it is discussed.

The equation $A = UDU^{-1}$ has been established in (10). Writing

$$U = \{u_{ij}\}, \qquad U^{-1} = \{v_{ij}\} \qquad \text{and} \quad D = \{\lambda_i\}$$

for $i, j = 1, 2 \ldots, n$, where D is the diagonal matrix of the latent roots of A, the consequent result $A^k = UD^kU^{-1}$ can be written as

$$A^k = \begin{bmatrix} u_{11} & \cdots & u_{1n} \\ \cdot & & \cdot \\ \cdot & & \cdot \\ \cdot & & \cdot \\ u_{n1} & \cdots & u_{nn} \end{bmatrix} \begin{bmatrix} \lambda_1^k & & & \\ & \cdot & & \\ & & \cdot & \\ & & & \cdot \\ & & & & \lambda_n^k \end{bmatrix} \begin{bmatrix} v_{11} & \cdots & v_{1n} \\ \cdot & & \cdot \\ \cdot & & \cdot \\ \cdot & & \cdot \\ v_{n1} & \cdots & v_{nn} \end{bmatrix}$$

$$= \begin{bmatrix} \lambda_1^k u_{11} & \cdots & \lambda_n^k u_{11} \\ \cdot & & \cdot \\ \cdot & & \cdot \\ \cdot & & \cdot \\ \lambda_1^k \mathrm{u}_{n1} & \cdots & \lambda_n^k u_{nn} \end{bmatrix} \begin{bmatrix} v_{11} & \cdots & v_{1n} \\ \cdot & & \cdot \\ \cdot & & \cdot \\ \cdot & & \cdot \\ v_{n1} & \cdots & v_{nn} \end{bmatrix}. \tag{15}$$

Suppose that λ_1 is the dominant root, such that λ_1^k for some sufficiently large value of k is numerically so much greater than any of the values $\lambda_2^k, \ldots, \lambda_n^k$ that the latter may be taken as zero. Then (15) becomes

$$A^k = \begin{bmatrix} \lambda_1^k u_{11} & 0 & \ldots & 0 \\ \cdot & & & \cdot \\ \cdot & & & \cdot \\ \cdot & & & \cdot \\ \lambda_1^k u_{n1} & 0 & \ldots & 0 \end{bmatrix} \begin{bmatrix} v_{11} & \ldots & v_{1n} \\ \cdot & & \cdot \\ \cdot & & \cdot \\ \cdot & & \cdot \\ v_{n1} & \ldots & v_{nn} \end{bmatrix} \tag{16}$$

$$= \lambda_1^k \begin{bmatrix} u_{11} \\ \cdot \\ \cdot \\ \cdot \\ u_{n1} \end{bmatrix} [v_{11} \quad \ldots \quad v_{1n}].$$

Post-multiplying this by a non-null column vector x gives

$$A^k x = \lambda_1^k \begin{bmatrix} u_{11} \\ \cdot \\ \cdot \\ \cdot \\ u_{n1} \end{bmatrix} [v_{11} \quad \ldots \quad v_{1n}] \begin{bmatrix} x_1 \\ \cdot \\ \cdot \\ \cdot \\ x_n \end{bmatrix} = \mu\lambda_1^k \begin{bmatrix} u_{11} \\ \cdot \\ \cdot \\ \cdot \\ u_{n1} \end{bmatrix}$$

where μ is the scalar $\mu = \sum_{j=1}^{n} v_{1j}x_j$.

Defining
$$w_k = \begin{bmatrix} w_{1k} \\ \cdot \\ \cdot \\ \cdot \\ w_{nk} \end{bmatrix} = \mu\lambda_1^k \begin{bmatrix} u_{11} \\ \cdot \\ \cdot \\ \cdot \\ u_{n1} \end{bmatrix}$$

then gives $A^k x = w_k$ and similarly $A^{k-1}x = w_{k-1}$. Now, from the definition of w_k, the ratio of the ith element of w_k to the ith element of w_{k-1} is

$$\frac{w_{i,k}}{w_{i,k-1}} = \frac{\mu\lambda_1^k u_{i1}}{\mu\lambda_1^{k-1} u_{i1}} = \lambda_1. \tag{17}$$

Hence, provided that λ_1 is the largest latent root and is such that for some sufficiently large value of k, λ_1^k predominates to a great enough extent over $\lambda_2^k, \ldots, \lambda_n^k$ that equation (16) is approximately true—provided, that these conditions are satisfied, we have shown that for some arbitrary non-null vector x, $A^k x$ and $A^{k-1}x$ are vectors w_k and w_{k-1} respectively, such that the ratio of their elements is λ_1. This means that repeated pre-multiplication of x by A will, after k multiplications, lead to a vector w_k such that the ratio of *each* of its elements to the corresponding element of w_{k-1} is the same for all elements, this ratio being the largest latent root of A. And furthermore, w_{k-1} is the corresponding latent vector, because

$$Aw_{k-1} = A^k x = w_k = \lambda_1 w_{k-1}.$$

Although these results have been obtained by writing (16) as an exact equality, they are really only approximations. The degree of approximation depends upon k and upon the extent to which λ_1^k exceeds $\lambda_2^k, \ldots, \lambda_n^k$ that the latter can be assumed zero in (15) in the presence of λ_1^k. Increasing k increases the accuracy of the approximation, so that in carrying out this process it is possible to find the largest latent root to any required degree of accuracy by increasing k. The essential result, (17), is true for all values of i, that is, for all elements of w_k and w_{k-1}, namely of $A^k x$ and $A^{k-1}x$. In any numerical case it is unlikely that these ratios will be exactly equal for any particular value of k; they will differ, but the differences between them will decrease as the value of k is increased.

Example.

For the matrix $A = \begin{bmatrix} 1 & 1 \\ 4 & 1 \end{bmatrix}$ and the arbitrary vector $x = \begin{bmatrix} 1 \\ 1 \end{bmatrix}$ Table 2 shows the vector $w_k = A^k x$ for $k = 0, 1, \ldots, 6$, together with the ratios of corresponding elements, $w_{i,k}/w_{i,k-1}$.

TABLE 2. EXAMPLE OF ITERATIVE VALUES OF A DOMINANT LATENT ROOT

A	k: 0	1	2	3	4	5	6
	$w_k = A^k x$						
$\begin{bmatrix} 1 & 1 \\ 4 & 1 \end{bmatrix}$	$\begin{bmatrix} 1 \\ 1 \end{bmatrix}$	$\begin{bmatrix} 2 \\ 5 \end{bmatrix}$	$\begin{bmatrix} 7 \\ 13 \end{bmatrix}$	$\begin{bmatrix} 20 \\ 41 \end{bmatrix}$	$\begin{bmatrix} 61 \\ 121 \end{bmatrix}$	$\begin{bmatrix} 182 \\ 365 \end{bmatrix}$	$\begin{bmatrix} 547 \\ 1093 \end{bmatrix}$
	$w_{i,k}/w_{i,k-1}$						
$i = 1$	—	2	3.5	2.8	3.05	2.98	3.004
$i = 2$	—	5	2.6	3.2	2.95	3.02	2.995

We see at once the decrease in the differences between the ratios of corresponding elements as k increases—and at $k = 6$ it is concluded that the largest latent root is 3. This is confirmed by solving the characteristic equation.

Demonstrating this method with a 2×2 matrix has kept the arithmetic to a minimum, but it can be applied to matrices of any order. The calculations become quite substantial for large-sized matrices, but with high-speed computers they are no longer a limiting problem. Although the true test for ending the iterative procedure is that $w_{i,k}/w_{i,k-1}$ should be the same for all i, within the limits of accuracy desired, two alternative procedures can be used that are not correct theoretically but are frequently satisfactory in practice. They demand continuing the iterative procedure until either

$$\sum_{i=1}^{n} \frac{w_{i,k}}{w_{i,k-1}} = \sum_{i=1}^{n} \frac{w_{i,k-1}}{w_{i,k-2}}$$

or until

$$\frac{\sum_{i=1}^{n} w_{i,k}}{\sum_{i=1}^{n} w_{i,k-1}} = \frac{\sum_{i=1}^{n} w_{i,k-1}}{\sum_{i=1}^{n} w_{i,k-2}}.$$

Choice of which of these alternatives to use depends on the problem at hand and on matters of computing efficiency and numerical analysis which are beyond our scope here. Nevertheless, as alternative stopping procedures they are worth noting: one involves equating sums of ratios, the other involves ratios of sums.

Illustrations. Dominant latent roots have direct application in problems relating to age-distribution vectors and to probability transition matrices discussed in Sections 2 and 3. It was seen there that if n_0 is an age-distribution vector at time $t = 0$, the corresponding vector at time $t = k$ is $n_k = M^k n_0$. Suppose that λ_1 is the dominant latent root of M and that k is sufficiently large that λ_1^k is so much greater than the corresponding powers of other latent roots that they may be overlooked. Then, from the development given in (15) and (16), n_k can be expressed as

$$\begin{aligned} n_k &= M^k n_0 \\ &= \lambda_1^k \begin{bmatrix} u_{11} \\ \cdot \\ \cdot \\ \cdot \\ u_{n1} \end{bmatrix} [v_{11} \quad \cdots \quad v_{1n}] n_0. \end{aligned}$$

Writing u_1 for the column of u-terms (the first column of U), and v_1' for the row of v-terms (the first row of U^{-1}), n_k can be written as

$$n_k = \lambda_1^k u_1 v_1' n_0 = \lambda_1^k (v_1' n_0) u_1.$$

Because u_1 is the first column of U it is the latent vector of M corresponding to the latent root λ_1; and because $v_1'n_0$ is a scalar $(v_1'n_0)u_1 = u^*$ say, is also a latent vector of M corresponding to λ_1. Hence

$$n_k = M^k n_0 = \lambda_1^k u^*,$$

Likewise

$$n_{k+1} = M^{k+1} n_0 = \lambda_1^{k+1} u^*,$$

and so

$$n_{k+1} = \lambda_1 n_k.$$

But this is the very situation discussed at the beginning of the chapter, that of the age distribution being stable from one point in time to the next, apart from a constant, which is now seen to be λ_1, the dominant latent root of M. Hence, if and when stability in the age distribution is reached, the ratio of successive age-distribution vectors is the dominant latent root of M, with the vectors being proportional to the corresponding latent vector of M.

These results mean that if stability is the only matter for investigation in a problem relating to age distributions, probability transition matrices or similar situations, the ratio of successive vectors and a vector proportional to them may be found, provided that a dominant latent root exists, by carrying out the iterative process for deriving the dominant latent root and an associated latent vector. For example, in Section 3 the probability transition matrix for the selfing of autotetraploids is

$$P = \begin{bmatrix} 1 & 0 & 0 \\ \frac{1}{4} & \frac{1}{2} & \frac{1}{4} \\ \frac{2}{36} & \frac{16}{36} & \frac{18}{36} \end{bmatrix}.$$

With the arbitrary vector $x' = [1 \quad 1 \quad 1]$ it is easily seen that $Px = x$, and at once we conclude that the largest latent root is unity, with x being the associated latent vector.

If there is no latent root that can be termed dominant in the sense used here, the iterative procedure breaks down and tends to no limit. The matrix involved in Bernadelli's beetles is just such a case.

A large-scale use of this procedure was recently undertaken by Darwin and Williams (1964), for studying different hunting policies to be directed against the rabbit pest on New Zealand farmlands. There, for a time interval of four weeks, M-matrices of order 39 were used, of exactly the form described at the beginning of this chapter. Thirteen such matrices were available, representing the fertility and survival rates in each of the four-week periods of a calendar year. They differed because breeding habits and natural mortality vary with seasonal changes throughout the year. From these thirteen matrices, $M_1, M_2, \ldots, M_{13}$ say, the age-distribution vector at the end of a year, n_E, is obtained from that at the beginning of the year, n_0, as the product of the thirteen M's multiplied by n_0: $n_E = M_1 M_2 \ldots M_{13} n_0$. And, assuming a stable age distribution, the dominant latent root of the product $M_1 M_2 \ldots M_{13}$ gives the effective rate of annual increase in the rabbit population under natural conditions. But the point of interest was to compare the effect of different hunting policies, each designed to reduce the rabbit population. Comparison was made on the basis of minimum effect, namely that were any policy to be applied it would be used only once a year. Each policy was represented by a diagonal matrix of order 39, the diagonal elements being the percentages of rabbits of each age (measured in four-week units) that might be expected to survive the hunting. Interspersing this matrix between any two of the M matrices in the product of all thirteen of them (or at the beginning or end of the product) represents in this scheme of things the fourteen possible times of year at which the hunting policy could be waged on the rabbits. The dominant latent root for each of these fourteen possible products was then obtained on a high-speed computer, this value representing the effective rate of annual increase in the rabbit population taking into account age-differential fertility and natural death (survival) rates, with hunting policy superimposed. By this means, the effectiveness of eleven different hunting policies was compared.

The results showed that it is better to use a hunting policy that kills more old than young rabbits just before the old start a new breeding season, and to use the opposite sort of hunting policy when the young have had a chance to be killed (naturally) but not much of a chance to breed, that is, about six months away from the other optimum. While this result qualitatively agrees with general reasoning, its value was in showing the *size* of effect that can be achieved. Differences of three to four per cent kill annually showed appreciable differences in terms of overall effect in a population that is naturally increasing but at not too fast a rate.

7. FACTORIZING THE CHARACTERISTIC EQUATION

The examples of obtaining latent roots that have been used so far have all involved solving either a cubic or a quadratic equation in λ. This is not difficult, but in finding latent roots of large matrices there arises the problem of having to solve polynomials of large degree, for example, a 20×20 matrix leads to a polynomial of degree 20. Should the dominant latent root be obtainable by the procedure just discussed, the second largest root can be found by a similar process after factorizing out the dominant root by a method now to be described.

Operationally the procedure is as follows. Suppose λ_1 is the dominant latent root of A, and suppose u_1 is a latent vector corresponding to λ_1. Choose u_1 such that $u_1'u_1 = \lambda_1$. (This is achieved by obtaining any latent vector corresponding to λ_1, t_1 say, and deriving u_1 from t_1 as

$$u_1 = (\sqrt{\lambda_1/t_1't_1})t_1.$$

Then $u_1'u_1 = \lambda_1$.) The second largest latent root of A is then obtained as the largest latent root of $A - u_1u_1'$. The following theorem is the basis of this procedure.

Theorem. If λ_1 is any latent root of A and u_1 is a corresponding latent vector with $u_1'u_1 = \lambda_1$, then the latent roots of $A - u_1u_1'$ except one root of zero are, with the exception of λ_1, those of A.

**Proof.*

To prove this theorem it need only be shown that the characteristic equation of $A - u_1u'_1$ is that of A with the factor $\lambda - \lambda_1$ replaced by λ; i.e., that

$$|A - u_1u'_1 - \lambda I| = \frac{\lambda}{\lambda - \lambda_1}|A - \lambda I|.$$

To do this, use is made of the determinantal expansions developed in Section 3.7 that for A and D being nonsingular square matrices the partitioned determinant

$$\begin{vmatrix} A & B \\ C & D \end{vmatrix} = |A|\,|D - CA^{-1}B| = |D|\,|A - BD^{-1}C|.$$

Therefore

$$\begin{vmatrix} A - \lambda I & u_1 \\ u'_1 & 1 \end{vmatrix} = |A - \lambda I|\{1 - u'_1(A - \lambda I)^{-1}u_1\} = |A - \lambda I - u_1u'_1|.$$

Hence

$$|A - \lambda I - u_1u'_1| = |A - \lambda I|\{1 - u'_1(A - \lambda I)^{-1}u_1\}. \tag{18}$$

But because u_1 is a latent vector corresponding to the latent root λ_1, $Au_1 = \lambda_1u_1$, and so $(A - \lambda I)u_1 = (\lambda_1 - \lambda)u_1$. Therefore

$$\left(\frac{1}{\lambda_1 - \lambda}\right)u_1 = (A - \lambda I)^{-1}u_1$$

and

$$u'_1(A - \lambda I)^{-1}u_1 = \frac{u'_1u_1}{\lambda_1 - \lambda} = \frac{\lambda_1}{\lambda_1 - \lambda}$$

by the choice of u_1. Substituting this in (18) gives

$$|A - u_1u'_1 - \lambda I| = \frac{\lambda}{\lambda - \lambda_1}|A - \lambda I|,$$

and so the theorem is proved.

Example. The characteristic equation of

$$A = \begin{bmatrix} 1 & 1 & 0 \\ 3 & 1 & 2 \\ -10 & 9 & 1 \end{bmatrix}$$

reduces to

$$(\lambda - 2)(\lambda - 5)(\lambda + 4) = 0.$$

For the root $\lambda_1 = 5$ it will be found that a latent vector is

$$t_1 = \begin{bmatrix} 2 \\ 8 \\ 13 \end{bmatrix}, \qquad \text{for which} \quad u_1 = (\sqrt{\lambda_1/t_1't_1})t_1 = \sqrt{5/237}\begin{bmatrix} 2 \\ 8 \\ 13 \end{bmatrix}.$$

The amended characteristic equation

$$|A - \lambda I - u_1u_1'| = 0$$

is accordingly

$$\left|\begin{bmatrix} 1-\lambda & 1 & 0 \\ 3 & 1-\lambda & 2 \\ -10 & 9 & 1-\lambda \end{bmatrix} - \tfrac{5}{237}\begin{bmatrix} 4 & 16 & 26 \\ 16 & 64 & 104 \\ 26 & 104 & 169 \end{bmatrix}\right| = 0$$

which simplifies to

$$\begin{vmatrix} 217\text{–}237\lambda & 157 & -130 \\ 631 & -83\text{–}237\lambda & -46 \\ -2500 & 1613 & -608-237\lambda \end{vmatrix} = 0.$$

On expansion this becomes

$$(237\lambda)^3 + (237\lambda)^2 474 - (237\lambda)449352 = 0$$

reducing to $$\lambda(\lambda - 2)(\lambda + 4) = 0$$

which is the characteristic equation of A with λ replacing $\lambda - 5$.

In applying this technique to finding the second largest latent root of a matrix one would usually employ the methods of the previous section to find the dominant roots of A and $A - u_1u_1'$.

8. SYMMETRIC MATRICES

Just as reduction to equivalent canonical form (Sections 5.11 and 5.12) is a little different for symmetric matrices than it is for nonsymmetric ones, so also is this the case with reduction to canonical form under similarity. Thus if A is symmetric, $A = A'$, the reduction $U^{-1}AU = D$ holds true for U being orthogonal (Section 4.6), $UU' = I$, which gives $U'AU = D$. It is called the *canonical form under orthogonal similarity.*

a. Latent roots all different

If λ_i and λ_k are two different latent roots of a matrix A, with u_i and u_k being corresponding latent vectors,

$$Au_i = \lambda_i u_i \quad \text{and} \quad Au_k = \lambda_k u_k.$$

Using these equalities and noting that transposing a scalar leaves it unchanged, the following development can be made for a symmetric matrix $A = A'$:

$$\begin{aligned} \lambda_i u_k' u_i &= u_k'(\lambda_i u_i) = u_k' A u_i = u_i' A u_k \\ &= u_i' \lambda_k u_k = \lambda_k u_i' u_k = \lambda_k u_k' u_i. \end{aligned}$$

For $\lambda_i \neq \lambda_k$ this is true only if $u_k' u_i = 0$, and if all latent roots are different this will be the case for every pair of roots. Hence, for distinct latent roots, columns of $U = [u_1 \quad u_2 \quad \ldots \quad u_n]$ can be found such that each inner product of one with another is zero. By dividing each column by the square root of the sum of squares of each of its elements, a process known as *normalizing* a vector, the columns become such that the inner product of each with itself is 1. Hence $u_i' u_k = 0$ for $i \neq k$ and $u_i' u_i = 1$. Thus $U'U = I$, U is an orthogonal matrix and the reduction $U^{-1}AU = D$ becomes $U'AU = D$.

Example. The matrix

$$A = \begin{bmatrix} 3 & -6 & -4 \\ -6 & 4 & 2 \\ -4 & 2 & -1 \end{bmatrix}$$

has latent roots $\lambda = -1, -4$ and 11. Solving the equations $(A - \lambda_i I)u_i = 0$ for each latent root in turn, it will be found that three latent vectors are $\begin{bmatrix} 1 \\ 2 \\ -2 \end{bmatrix}$, $\begin{bmatrix} 2 \\ 1 \\ 2 \end{bmatrix}$ and $\begin{bmatrix} 2 \\ -2 \\ -1 \end{bmatrix}$, so that the un-normalized form of U is

$$\begin{bmatrix} 1 & 2 & 2 \\ 2 & 1 & -2 \\ -2 & 2 & -1 \end{bmatrix}.$$

The process of normalizing each column requires dividing the first column by $\sqrt{1^2 + 2^2 + 2^2} = 3$. The second column is divided by $\sqrt{2^2 + 1^2 + 2^2} = 3$, and this is also the divisor for column 3. Hence the normalized form of U is

$$U = \tfrac{1}{3}\begin{bmatrix} 1 & 2 & 2 \\ 2 & 1 & -2 \\ -2 & 2 & -1 \end{bmatrix}, \quad U'U = I \text{ and } U'AU = \begin{bmatrix} -1 & 0 & 0 \\ 0 & -4 & 0 \\ 0 & 0 & 11 \end{bmatrix} = D.$$

b. Multiple latent roots

We have shown that for two distinct latent roots λ_i and λ_k of a symmetric matrix latent vectors u_i and u_k exist such that $u_i'u_k = 0$. It remains to be shown that for a multiple latent root we can find a set of vectors that are mutually orthogonal. Suppose that λ_q is a multiple root with multiplicity m_q. Then for any A that can be expressed in the form $U^{-1}AU = D$, $r(A - \lambda_q I) = n - m_q$ (Section 4), and $(A - \lambda_q I)u_q = 0$ has m_q LINN solutions u_q. Denote one solution by v_{q1}. Now consider solving

$$(A - \lambda_q I)u_q = 0$$

and

$$v_{q1}'u_q = 0$$

simultaneously for u_q. This set of equations has $m_q - 1$ LINN solutions for u_q and any one of them, v_{q2} say, is a latent vector of A and is orthogonal to the vector v_{q1}. It will also be orthogonal to the latent vector corresponding to any latent root different from λ_q, because of the orthogonality property of latent vectors corresponding to distinct latent roots of symmetric matrices. Similarly if $m_q > 2$ a third vector corresponding to λ_q can be obtained by solving

$$(A - \lambda_q I)u_q = 0$$

$$v_{q1}'u_q = 0$$

and

$$v_{q2}'u_q = 0$$

which have $m_q - 2$ LINN solutions for u_q. This process can be continued until m_q solutions are obtained, all orthogonal to each other and to the vectors corresponding to the other roots. Thus U is again orthogonal, and we have $U'AU = D$ for $U'U = I$.

Example.

$$A = \begin{bmatrix} -1 & -2 & 1 \\ -2 & 2 & -2 \\ 1 & -2 & -1 \end{bmatrix}$$

has latent roots $\lambda = 4$, -2 and -2. First we must consider the rank of $(A - \lambda_2 I)$ for the multiple root $\lambda_2 = -2$:

$$(A + 2I) = \begin{bmatrix} 1 & -2 & 1 \\ -2 & 4 & -2 \\ 1 & -2 & 1 \end{bmatrix}$$

has rank 1 which equals $3 - 2$, the order of A minus the multiplicity of the multiple root. Therefore latent roots of A exist such that $U'AU = D$. The latent vectors are found by solving $(A - \lambda_i I)u_i = 0$. For $\lambda_1 = 4$,

$(A - \lambda_1 I)u_1 = 0$ has the solution $u_1 = \begin{bmatrix} 1 \\ -2 \\ 1 \end{bmatrix}$, and for $\lambda_2 = -2$ the equations $(A - \lambda_2 I)u_2 = 0$ reduce to

$$\alpha_2 - 2\beta_2 + \gamma_2 = 0 \tag{19}$$

where

$$u_2 = \begin{bmatrix} \alpha_2 \\ \beta_2 \\ \gamma_2 \end{bmatrix} = \begin{bmatrix} 1 \\ 1 \\ 1 \end{bmatrix} \tag{20}$$

is one solution. But $\lambda = -2$ is a multiple root and consequently a second solution to (19) is required, one whose inner product with (20) is zero. Thus a second latent vector corresponding to $\lambda = -2$ is derived by solving

$$\alpha_3 - 2\beta_3 + \gamma_3 = 0$$

and

$$\alpha_3 + \beta_3 + \gamma_3 = 0.$$

A solution is

$$u_3 = \begin{bmatrix} \alpha_3 \\ \beta_3 \\ \gamma_3 \end{bmatrix} = \begin{bmatrix} 1 \\ 0 \\ -1 \end{bmatrix},$$

and the un-normalized form of U is

$$[u_1 \quad u_2 \quad u_3] = \begin{bmatrix} 1 & 1 & 1 \\ -2 & 1 & 0 \\ 1 & 1 & -1 \end{bmatrix}.$$

The normalized form comes from dividing the columns by $\sqrt{6}$, $\sqrt{3}$ and $\sqrt{2}$, respectively, to give

$$U = \frac{1}{\sqrt{6}} \begin{bmatrix} 1 & \sqrt{2} & \sqrt{3} \\ -2 & \sqrt{2} & 0 \\ 1 & \sqrt{2} & -\sqrt{3} \end{bmatrix}.$$

It will be found that $U'U = I$ (i.e., U is orthogonal), and that $U'AU = D$ where D is diagonal with non-zero elements 4, -2 and -2.

c. Real latent roots

Since the latent roots of a matrix of order n are the roots of the characteristic equation, a polynomial in λ of degree n, it is not true in general that all the roots are real numbers; some may be pairs of complex numbers such as $a + ib$ and $a - ib$ where $i = \sqrt{-1}$ and a and b are real numbers. But if A is symmetric and has no complex numbers as elements (i.e., all elements are real), there are no complex latent roots. This we now prove.

**Proof.*

Suppose the roots are not all real and that λ is a complex root with x being the corresponding vector. Let

$$x = \{x_j\} = \{\alpha_j + i\beta_j\} \qquad \text{and} \quad \bar{x} = \{\bar{x}_j\} = \{\alpha_j - i\beta_j\} \tag{21}$$

for $j = 1, 2, \ldots, n$ and where $\bar{x}_j$ is the complex conjugate of x_j. Because λ and x are a latent root and vector $Ax = \lambda x$ and therefore

$$\bar{x}'Ax = \bar{x}'\lambda x = \lambda\bar{x}'x.$$

Expanding $\bar{x}'x$ as $\sum_j \bar{x}_j x_j$ and applying the expansions given in Section 2.9 to $\bar{x}'Ax$ leads to

$$\sum_j a_{jj}\bar{x}_j x_j + \sum_{k>j}\sum a_{jk}(\bar{x}_j x_k + \bar{x}_k x_j) = \lambda \sum_j \bar{x}_j x_j.$$

Substitution of (21) into this gives, after a little reduction,

$$\sum_j a_{jj}(\alpha_j^2 + \beta_j^2) + 2\sum_{j>k}\sum a_{jk}(\alpha_j\alpha_k + \beta_j\beta_k) = \lambda \sum_j (\alpha_j^2 + \beta_j^2).$$

All terms except λ in this equation are real, and therefore λ must be real also. Hence all latent roots of A are real.

d. Positive definite symmetric matrices

Consider the general quadratic form discussed in Section 2.9, $z = x'Ax$, where A is symmetric of rank r. The latent roots of A are real, and provided that A can be expressed in canonical form, this will be

$$U'AU = D = \begin{bmatrix} D_r^* & 0 \\ 0 & 0 \end{bmatrix}$$

where U is orthogonal and D_r^* is a diagonal matrix of the r non-zero latent roots of A. Suppose we make a transformation of variables (Section 2.6) from x to y, $y = U'x$, equivalent to $x = U'^{-1}y = Uy$. Then

$$\begin{aligned} z &= x'Ax \\ &= y'U'AUy \\ &= y'\begin{bmatrix} D_r^* & 0 \\ 0 & 0 \end{bmatrix} y \\ &= \sum_{i=1}^{r} \lambda_i y_i^2. \end{aligned}$$

Now if z is positive semi-definite (Section 2.9), meaning that it is never negative, then $\sum_{i=1}^{r} \lambda_i y_i^2$ must be likewise, and hence all λ_i, for $i = 1, \ldots, r$, are positive; that is, the non-zero latent roots of a positive semi-definite symmetric matrix are positive real numbers.

Illustration. If the vector x represents a set of n normally and independently distributed random variables which have zero mean and variance-covariance matrix $\sigma^2 I$, so does the vector $y = U'x$ where U is orthogonal. Furthermore, for $z = x'Ax$ where A is positive semi-definite symmetric, of rank r, z is distributed as $\sigma^2\chi^2$ with r degrees of freedom.

For, with $E(x) = 0$, $E(xx') = \sigma^2 I$, and $y = U'x$, the mean value of y is

$$E(y) = E(U'x) = U'E(x) = 0$$

and its variance-covariance matrix is

$$E(yy') = E(U'xx'U) = U'E(xx')U = \sigma^2 U'IU = \sigma^2 I.$$

And for $z = x'Ax$, with A being positive definite and symmetric of rank r, there exists a matrix U such that $U'U = I$ and $U'AU = D$, where D is diagonal with r real, positive non-zero elements in its diagonal, $\lambda_1, \ldots, \lambda_r$ say. Therefore for transformed variables

$$y = U'x, \qquad z = x'Ax = y'Dy = \sum_{i=1}^{r} \lambda_i y_i^2,$$

and the y's are a set of normally distributed random variables with zero mean and variance-covariance matrix $\sigma^2 I$. Thus each y_i^2 has the distribution $\sigma^2\chi_1^2$, a chi-square distribution with one degree of freedom. And because the λ's are positive, z is the sum of r independent variables $\lambda_i y_i^2$ each distributed as $\lambda_i\sigma^2\chi_1^2$. In cases where $\lambda_i\sigma^2 = 1$, $x'Ax$ is thus distributed as χ_r^2. An illustration of this is given in Section 8.3.

9. EXERCISES

1. Find the latent roots and latent vectors of the following matrices. In each case combine the latent vectors into a matrix U and verify that $U^{-1}AU = D$ where D is the diagonal matrix of the latent roots.

$$B = \begin{bmatrix} 1 & 4 & 1 \\ 2 & 1 & 0 \\ -1 & 3 & 1 \end{bmatrix} \qquad C = \begin{bmatrix} 2 & -2 & 3 \\ 10 & -4 & 5 \\ 5 & -4 & 6 \end{bmatrix}$$

$$E = \begin{bmatrix} 7 & 4 & -1 \\ 4 & 7 & -1 \\ -4 & -4 & 4 \end{bmatrix} \qquad F = \begin{bmatrix} 9 & 15 & 3 \\ 6 & 10 & 2 \\ 3 & 5 & 1 \end{bmatrix}$$

$$G = \begin{bmatrix} 4 & -2 & 0 \\ -2 & 3 & -2 \\ 0 & -2 & 2 \end{bmatrix} \qquad H = \begin{bmatrix} 1 & 2 & 1 \\ 1 & 1 & 4 \\ 2 & -4 & 1 \end{bmatrix}$$

$$K = \begin{bmatrix} -1 & -2 & 1 \\ -2 & 2 & -2 \\ 1 & -2 & -1 \end{bmatrix} \qquad L = \begin{bmatrix} 1 & 4 & 2 \\ 4 & 1 & 2 \\ 8 & 8 & 1 \end{bmatrix}$$

$$M = \begin{bmatrix} -9 & 2 & 6 \\ 2 & -9 & 6 \\ 6 & 6 & 7 \end{bmatrix} \qquad N = \begin{bmatrix} 2 & 7 & 1 \\ 2 & 3 & 8 \\ 9 & 4 & 4 \end{bmatrix}$$

2. Use the iterative procedure to establish the dominant latent root for each of the matrices in Exercise 1, and by using the factorization procedure of Section 7 find the other two latent roots.
3. Of the matrices in Exercise 1, show that $F = B^2 + B$, and that if λ is a latent root of B then $\lambda^2 + \lambda$ is a latent root of F. Show, generally, that if F is a polynomial of a matrix A, each latent root of F is that same polynomial of a latent root of A.
4. Show that the latent vectors of $\begin{bmatrix} a & b \\ c & d \end{bmatrix}$ are of the form $\begin{bmatrix} -b \\ a - \lambda_1 \end{bmatrix}$ and $\begin{bmatrix} -b \\ a - \lambda_2 \end{bmatrix}$ where λ_1 and λ_2 are the latent roots. Verify that $2a - \lambda_1 = \lambda_2$ when $a = d$.
5. Factorize the characteristic equation of

$$A = \begin{bmatrix} a & b & b \\ b & a & b \\ b & b & a \end{bmatrix}$$

using the latent root $\lambda = a + 2b$. Carry out the same process using the root $\lambda = a - b$.

6. Express $A = \begin{bmatrix} 1 & 0 & 0 \\ u & v & 0 \\ x & y & z \end{bmatrix}$ in the form $A = UDU^{-1}$ where $D = \begin{bmatrix} 1 & 0 & 0 \\ 0 & v & 0 \\ 0 & 0 & z \end{bmatrix}$.

In doing so, notice that $\begin{bmatrix} 1 & 0 & 0 \\ a & 1 & 0 \\ b & c & 1 \end{bmatrix}^{-1} = \begin{bmatrix} 1 & 0 & 0 \\ -a & 1 & 0 \\ ac - b & -c & 1 \end{bmatrix}$.

7. In a selfing series with a pair of linked genes, the probability transition matrix for the states of heterogeneity at 0, 1 and 2 loci is

$$P = \begin{bmatrix} 1 & 0 & 0 \\ \frac{1}{2} & \frac{1}{2} & 0 \\ \frac{1}{2} - c(1 - c) & 2c(1 - c) & \frac{1}{2} - c(1 - c) \end{bmatrix},$$

where c is the proportion of crossovers. Use the result obtained in Exercise 6 to show that for $z = \frac{1}{2} - c(1 - c)$

$$P^k = \begin{bmatrix} 1 & 0 & 0 \\ 1 - (1/2)^k & (1/2)^k & 0 \\ 1 + z^k - (1/2)^{k-1} & (1/2)^{k-1} - 2z^k & z^k \end{bmatrix}.$$

8. The genetic variances and covariances in a diploid series of recurrent back-crosses to the F_1's of two homozygotes is obtained from the power series of the matrix

$$T = \begin{bmatrix} 1/2 & 1/2 & 0 \\ 1/4 & 1/2 & 1/4 \\ 0 & 1/2 & 1/2 \end{bmatrix}.$$

Find the latent roots of T and show that

$$T^k = \begin{bmatrix} 1/4 + (1/2)^{k+1} & 1/2 & 1/4 - (1/2)^{k+1} \\ 1/4 & 1/2 & 1/4 \\ 1/4 - (1/2)^{k+1} & 1/2 & 1/4 + (1/2)^{k+1} \end{bmatrix}.$$

9. Investigation of the fate of a mutant recessive lethal gene under a series of brother-sister matings is based on the probability transition matrix

$$S = \begin{bmatrix} 1 & 0 & 0 \\ 1/4 & 1/2 & 1/4 \\ 1/9 & 4/9 & 4/9 \end{bmatrix}.$$

Show that

$$S^k = \begin{bmatrix} 1 & 0 & 0 \\ 1 - a_k - b_k & a_k & b_k \\ 1 - \alpha_k - \beta_k & \alpha_k & \beta_k \end{bmatrix},$$

where, with $a = \sqrt{145}$,

$$s_k = \frac{1}{2}\left[\left(\frac{17 + a}{36}\right)^k + \left(\frac{17 - a}{36}\right)^k\right]$$

and

$$d_k = \frac{1}{2a}\left[\left(\frac{17 + a}{36}\right)^k - \left(\frac{17 - a}{36}\right)^k\right],$$

the elements of S^k are given by

$$a_k = s_k + d_k, \qquad b_k = 9d_k,$$
$$\alpha_k = 16d_k \qquad \text{and} \qquad \beta_k = s_k - d_k.$$

REFERENCES

Bernadelli, H. (1941). Population waves. *J. Burma Res. Soc.*, **31**, 1–18.

Darwin, J. H. and R. M. Williams (1964). The effect of time of hunting on the size of a rabbit population. *New Zealand J. of Sci.*, **7**, 341–352.

Frazer, R. A., W. J. Duncan and A. R. Collar (1952). *Elementary Matrices*. Cambridge University Press, Cambridge.

Leslie, P. H. (1945). On the use of matrices in certain population mathematics. *Biometrika*, **33**, 183–212.

Leslie, P. H. (1948). Some further notes on the use of matrices in population mathematics. *Biometrika*, **35**, 213–245.

Li, C. C. (1955). *Population Genetics*. University of Chicago Press, Chicago.

CHAPTER 8

MISCELLANEA

This chapter contains a variety of miscellaneous topics that are found useful in applications of matrix algebra to biology and statistics.

1. ORTHOGONAL MATRICES

An orthogonal matrix was defined in Section 4.6. It is a matrix whose product with its transpose is the identity matrix. Thus A is orthogonal if $AA' = I$, in which case $A^{-1} = A'$ and $A'A = I$. Several properties of orthogonal matrices are worth noting.

i. The inner product of any row (column) of an orthogonal matrix with itself is 1, and that with any other row (column) is zero. This follows immediately from the product $AA' = I$.

ii. A product of orthogonal matrices is itself orthogonal: if A and B are orthogonal $AA' = BB' = I$, and the product $(AB)(AB)'$ readily simplifies to I.

iii. The determinant of every orthogonal matrix is either $+1$ or -1. Taking the determinant of both sides of the equation $AA' = I$ yields this result.

iv. If λ is a latent root of an orthogonal matrix then so is $1/\lambda$. This is so because the characteristic equation $|A - \lambda I| = 0$ is equivalent to

$$|I - \lambda A'| = 0 \qquad \text{for} \quad AA' = I.$$

Hence

$$\left|\frac{1}{\lambda}I - A'\right| = 0$$

and so

$$\left|A - \frac{1}{\lambda}I\right| = 0;$$

i.e. $1/\lambda$ is a latent root of A. A consequence of this result is that all orthogonal matrices of odd order have at least one latent root equal to $+1$ or -1, the remaining roots occurring in pairs, λ and $1/\lambda$.

v. If S is a matrix such that $S' = -S$, then $P = (I - S)(I + S)^{-1}$ is orthogonal. The reader should satisfy himself that $P'P = I$.

* Matrices of the form $S' = -S$ are called *skew-symmetric*, or sometimes just *skew*. They are square and have all elements on one side of the diagonal equal to minus their counterparts on the other side, the diagonal elements being zero. If a skew-symmetric matrix is of odd order its determinant is zero and it has no inverse; if it is of even order the determinant is a perfect square and the inverse is skew. Any square matrix M can be expressed uniquely as the sum of a symmetric and a skew-symmetric matrix: $S = \frac{1}{2}(M + M')$ and $T = \frac{1}{2}(M - M')$. S is symmetric and T is skew.

Demonstration and proof of these properties are left as an exercise for the reader.

2. MATRICES HAVING ALL ELEMENTS EQUAL

On many occasions it happens that matrices can be partitioned into submatrices each of which has all its elements the same. On other occasions matrices can be expressed as linear functions of such matrices. In either case it is advantageous to make use of certain properties of these equi-element matrices.

Let $J_{r \times c}$ denote an $r \times c$ matrix having every element equal to 1. Then λJ is a matrix having every element equal to λ. For example,

$$J_{2 \times 3} = \begin{bmatrix} 1 & 1 & 1 \\ 1 & 1 & 1 \end{bmatrix} \quad \text{and} \quad 5J_{2 \times 3} = \begin{bmatrix} 5 & 5 & 5 \\ 5 & 5 & 5 \end{bmatrix}.$$

The following properties are easily demonstrated:

$$J'_{r \times c} = J_{c \times r},$$
$$J_{r \times c}J_{c \times k} = cJ_{r \times k},$$

and

$$J_{r \times c}J_{c \times r} = cJ_r,$$

where J_r is square, of order r, with all elements unity. Properties pertaining to J_r are:

J_r is symmetric, for $J'_r = J_r$;

$$J_r^2 = rJ_r, \text{ and } J_r^k = r^{k-1}J_r;$$
$$|J_r| = 0, \text{ and so } J_r^{-1} \text{ does not exist};$$
$$\text{rank }(J_r) = 1, \text{ and the only non-zero latent root is } r.$$

Pre-multiplication of any matrix A by a J matrix leads to a matrix having every row the same, the elements being the column totals of A;

and post-multiplication by J yields a matrix having every column the same, the elements being the row totals of A. These two results lead to

$$J_{r\times s}A_{s\times t}J_{t\times w} = \left(\sum_{i=1}^{s}\sum_{j=1}^{t} a_{ij}\right)J_{r\times w}$$

Example. For

$$A = \begin{bmatrix} 2 & 5 & -4 \\ 3 & 4 & 6 \end{bmatrix},$$

$$J_{4\times 2}A = \begin{bmatrix} 1 & 1 \\ 1 & 1 \\ 1 & 1 \\ 1 & 1 \end{bmatrix}\begin{bmatrix} 2 & 5 & -4 \\ 3 & 4 & 6 \end{bmatrix} = \begin{bmatrix} 5 & 9 & 2 \\ 5 & 9 & 2 \\ 5 & 9 & 2 \\ 5 & 9 & 2 \end{bmatrix},$$

$$AJ_{3\times 2} = \begin{bmatrix} 2 & 5 & -4 \\ 3 & 4 & 6 \end{bmatrix}\begin{bmatrix} 1 & 1 \\ 1 & 1 \\ 1 & 1 \end{bmatrix} = \begin{bmatrix} 3 & 3 \\ 13 & 13 \end{bmatrix},$$

and

$$J_{4\times 2}AJ_{3\times 2} = 16J_{4\times 2}.$$

A useful extension of these J matrices is a matrix which is a linear function of I_r and J_r, namely,

$$V_r = aI_r + bJ_r,$$

where a and b are non-zero scalars. For example

$$V_3 = \begin{bmatrix} a+b & b & b \\ b & a+b & b \\ b & b & a+b \end{bmatrix}.$$

The reader might satisfy himself that by writing V_r symbolically as

$$V_r = V_r(a, b),$$

the following results are true:

$$V_r' = V_r,$$

$$[V_r(a, b)]^{-1} = V_r\left(\frac{1}{a}, \frac{-b}{a(a + rb)}\right),$$

$$|V_r(a, b)| = a^{r-1}(a + rb),$$

$$V_r(a, b)V_r(x, y) = V_r(ax, ay + bx + rby),$$

and from this last

$$[V_r(a, b)]^2 = V_r(a^2, 2ab + rb^2).$$

By replacing a by $a - \lambda$ in the expression for $|V_r(a, b)|$ it is seen that $V_r(a, b)$ has one latent root equal to $(a + rb)$ and $(r - 1)$ roots equal to a.

Illustration. It is shown in Section 2.9 that the sum of squares

$$SS = \sum_{i=1}^{n} (x_i - \bar{x})^2$$

can be expressed as

$$SS = x'(I - U_n)x$$

where x is a vector of n scalar observations $x_1, x_2, \ldots, x_n$, and U_n is of order n with every element equal to $1/n$. Hence

$$\begin{aligned} I - U_n &= I_n - \frac{1}{n}J_n \\ &= V_n(1, -1/n). \end{aligned}$$

Therefore, from the results just derived,

$$\begin{aligned} |I - U_n| &= 1^{n-1}(1 - n/n) = 0; \\ (I - U_n)^2 &= V_n[1^2, 2(1)(-1/n) + n(-1/n)^2] \\ &= V_n(1, -1/n) \\ &= (I - U_n); \end{aligned}$$

and $(I - U_n)$ has one latent root equal to zero and $(n - 1)$ roots equal to 1. Hence its rank is $n - 1$.

Should x be a vector of random variables each having mean μ and variance σ^2, x_i can be written as $x_i = \mu + e_i$ and so

$$\begin{aligned} SS &= \sum(x_i - \bar{x})^2 = x'(I - U_n)x \\ &= \sum(e_i - \bar{e})^2 = e'(I - U_n)e \end{aligned}$$

where e is the vector of random terms e_i. Now $I - U_n$ has the latent root $\lambda = 1$ with multiplicity $n - 1$; but for $\lambda = 1$, $(I - U_n - \lambda I)$ reduces to $-U_n$, which has rank $1 = n - (n - 1)$. Hence (Section 7.4) $I - U_n$ can be reduced to canonical form. Furthermore, if the distribution of the x_i's is normal, e represents a vector of normally distributed variables having zero mean and variance-covariance matrix $\sigma^2 I$. The illustration of Section 7.8 therefore applies to $SS = e'(I - U_n)e$, and consequently SS is distributed as $\sigma^2\chi^2$ with $n - 1$ degrees of freedom.

3. IDEMPOTENT MATRICES

In Section 6.4 use was made of a matrix H for which $H^2 = H$; such a matrix is said to be *idempotent*. Thus A is idempotent if $A^2 = A$, whence

$A^3 = A$, $A^4 = A$, and so on. Since A^2 exists only when A is square, all idempotent matrices are square.

Example.

If
$$A = \begin{bmatrix} 2 & 4 & 6 \\ 4 & 8 & 12 \\ -3 & -6 & -9 \end{bmatrix}$$

$A^2 = A$ and for any positive integer k, $A^k = A$.

Graybill (1961) gives an extensive discussion of the properties of idempotent matrices, some of which we now summarize. It is assumed in all cases that A is idempotent.

i. The only nonsingular idempotent matrix is the identity matrix. [If A^{-1} exists, multiplying both sides of the equation $A^2 = A$ by A^{-1} leads to $A = I$.]

ii. $I - A$ is idempotent.

$$[(I - A)^2 = I + A^2 - 2A = I + A - 2A = I - A.]$$

iii. If A and B are idempotent AB is also, provided $AB = BA$.

$$[(AB)^2 = ABAB = A^2B^2 \text{ if } BA = AB, \text{ and } A^2B^2 = AB.]$$

iv. If P is orthogonal and A idempotent, $P'AP$ is idempotent.

$$[(P'AP)^2 = P'APP'AP = P'A^2P = P'AP.]$$

v. The latent roots of an idempotent matrix are either 0 or 1. [If λ and u are a latent root and corresponding vector of A, $Au = \lambda u$, and $A^2u = \lambda^2 u$; but because $A^2 = A$, $A^2u = Au$ and hence $\lambda^2 u = \lambda u$, i.e., $(\lambda^2 - \lambda)u = 0$. But u is non-null, and so $\lambda^2 - \lambda = 0$, the only solutions to which are $\lambda = 0$ or 1.]

vi. The number of latent roots of A equal to 1 is the same as the rank of A. [If $r(A) = r$ and D is the diagonal matrix of the latent roots of A, $r(D) = r(A) = r$. Therefore, since the only non-zero elements in D are 1's there must be r of them.]

vii. The trace of an idempotent matrix equals its rank. [Trace (A) equals the sum of the latent roots, which by (vi) above must be r.]

viii. A general form for an idempotent matrix is $A = X(YX)^{-1}Y$ provided that $(YX)^{-1}$ exists.

$$[A^2 = X(YX)^{-1}YX(YX)^{-1}Y = X(YX)^{-1}Y = A.]$$

A general symmetric form is $A = X(X'X)^{-1}X'$ provided $X'X$ is non-singular.

The properties just listed apply to any idempotent matrix. Applications of symmetric idempotent matrices to problems in mathematical statistics are discussed in some detail in Graybill (1961). In this context idempotency plays an important part in establishing the distributional properties of quadratic forms. We cite two theorems of Graybill's as illustration; they are but two from a lengthy array of theorems that he considers.

Illustrations.

Theorem 1. (Theorem 4.19 of Graybill, 1961.) If x is a vector of random variables having zero means and variance-covariance matrix $\sigma^2 I$, the expected value of $x'Ax$, where A is idempotent of rank r, is $r\sigma^2$.

Proof. From the general expression for a quadratic form given in Section 2.9

$$\begin{aligned} E(x'Ax) &= E\left(\sum_i x_i^2 a_{ii} + \sum_{i \neq j}\sum x_i x_j a_{ij}\right) \\ &= \sum_i a_{ii}(Ex_i^2) + \sum_{i \neq j}\sum a_{ij}E(x_i x_j). \end{aligned}$$

And since the x's have zero means, are independent and have variance σ^2 this leads to

$$\begin{aligned} E(x'Ax) &= \sum_i a_{ii}\sigma^2 + \sum_{i \neq j}\sum a_{ij}(0) \\ &= \sigma^2 \text{ trace}(A) \\ &= \sigma^2 \textstyle\sum(\text{latent roots of } A) \end{aligned}$$

and with A being idempotent of rank r this gives

$$E(x'Ax) = r\sigma^2.$$

Notice that no knowledge of the form of the distribution of the x's is required, only that their means be zero, that they be independent and have uniform variance.

A simple use of Theorem 1 is in the sum of squares considered at the end of the previous section:

$$SS = \sum_{i=1}^{n} (x_i - \bar{x})^2 = x'(I - U_n)x = e'(I - U_n)e.$$

The vector e is now the x vector of Theorem 1 and therefore, because $I - U_n$ is idempotent of rank $n - 1$,

$$\begin{aligned} E(SS) &= E[e'(I - U_n)e] \\ &= \sigma^2 \text{ rank } (I - U_n) \\ &= (n - 1)\sigma^2. \end{aligned}$$

Theorem 2. (Theorem 4.6 of Graybill, 1961.) If x is a vector of random variables independently and normally distributed with zero means and unit variance, the quadratic form $x'Ax$ is distributed as χ^2 with r degrees of freedom, provided that A is an idempotent symmetric matrix of rank r.

Proof. It is shown at the end of Section 7.8 that $x'Ax$, where A is symmetric, can be expressed as a sum $\sum_{i=1}^{r} \lambda_i y_i^2$ where the $\lambda_i y_i^2$ are independent random variables distributed as $\lambda_i \sigma^2 \chi_1^2$, every λ_i being positive. Under the conditions of the theorem the x's have unit variance so therefore $\sigma^2 = 1$, and A is idempotent so therefore $\lambda_i = 1$ [property (v) just given]. Hence $\sum \lambda_i y_i^2$ is a sum of independent variables each distributed as χ_1^2 and so their sum has the χ_r^2 distribution.

The sum of squares SS provides a simple illustration of Theorem 2. If, in addition to having zero mean and variance matrix $\sigma^2 I$, the vector e represents a vector of normally distributed variables, then Theorem 2 applies and we conclude that SS is distributed as χ^2 with $n - 1$ degrees of freedom.

At first sight these procedures may appear somewhat clumsy for deriving results that are well known. But the methods and notation involved apply universally to any such problem that can be expressed in matrix format. It is for this reason that matrix algebra is being used on an increasingly wider scale for investigating distributional properties of both linear functions and quadratic forms that arise in analysis of variance and other statistical techniques.

4. NILPOTENT MATRICES

A matrix one of whose powers is a null matrix is called a nilpotent matrix. Generally $A^2 = 0$ is sufficient definition.

Example.

$$A = \begin{bmatrix} 1 & 2 & 5 \\ 2 & 4 & 10 \\ -1 & -2 & 5 \end{bmatrix}, \qquad A^2 = \begin{bmatrix} 0 & 0 & 0 \\ 0 & 0 & 0 \\ 0 & 0 & 0 \end{bmatrix}.$$

Idempotent matrices satisfy $A^k = A$ and nilpotent matrices satisfy $A^k = 0$. Matrices for which $A^k = I$ also exist. Indeed, one has already been considered (Section 7.2):

$$A = \begin{bmatrix} 0 & 0 & 6 \\ \frac{1}{2} & 0 & 0 \\ 0 & \frac{1}{3} & 0 \end{bmatrix},$$

with $A^3 = I$. Perhaps the name "uni-potent" might be appropriate for such matrices. The partitioned matrix

$$A = \begin{bmatrix} I & B \\ 0 & -I \end{bmatrix}$$

is always of this form. No matter what matrix is used for B, $A^2 = I$ and $A = A^{-1}$.

5. A VECTOR OF DIFFERENTIAL OPERATORS

This section and the next make simple use of the differential calculus, and should be omitted by readers not familiar with this branch of mathematics.

Sometimes a function of several variables has to be differentiated with respect to each of the variables concerned. In such cases it is often convenient to write the derivatives as a column vector. Since the most common usage of this situation for biologists is likely to involve only linear and quadratic functions of variables we confine ourselves to these. For example, if

$$\lambda = 3x_1 + 4x_2 + 9x_3$$

where λ, x_1, x_2 and x_3 are scalars, we write the derivatives of λ with respect to x_1, x_2 and x_3 as

$$\frac{\partial \lambda}{\partial x} = \begin{bmatrix} \dfrac{\partial \lambda}{\partial x_1} \\ \dfrac{\partial \lambda}{\partial x_2} \\ \dfrac{\partial \lambda}{\partial x_3} \end{bmatrix} = \begin{bmatrix} \dfrac{\partial}{\partial x_1} \\ \dfrac{\partial}{\partial x_2} \\ \dfrac{\partial}{\partial x_3} \end{bmatrix} \lambda = \begin{bmatrix} 3 \\ 4 \\ 9 \end{bmatrix}.$$

The second vector shows how the symbol $\dfrac{\partial}{\partial x}$ can represent a whole vector of differential operators. Noting that

$$\lambda = 3x_1 + 4x_2 + 9x_3$$

$$= [3 \quad 4 \quad 9]\begin{bmatrix} x_1 \\ x_2 \\ x_3 \end{bmatrix}$$

$$= a'x,$$

so defining vectors a and x, we therefore see that with $\dfrac{\partial}{\partial x}$ representing a column of operators the differential of the scalar $a'x = x'a$ is

$$\frac{\partial}{\partial x}(a'x) = \frac{\partial}{\partial x}(x'a) = a. \tag{1}$$

Suppose this principle is applied in turn to every element of a vector $y = Ax$ expressed as

$$\begin{bmatrix} y_1 \\ y_2 \\ \cdot \\ \cdot \\ \cdot \\ y_n \end{bmatrix} = \begin{bmatrix} a'_1x \\ a'_2x \\ \cdot \\ \cdot \\ \cdot \\ a'_nx \end{bmatrix}, \tag{2}$$

where the y_i's are the elements of y and $a'_1, a'_2, \ldots, a'_n$ are the rows of the matrix A. The vector of operators $\dfrac{\partial}{\partial x}$ is now to be used on the elements of y. For each element it yields a column vector, $\dfrac{\partial y_i}{\partial x}$ for $i = 1, 2, \ldots, n$. By writing the n values of this alongside one another, as a matrix, the differential $\dfrac{\partial y}{\partial x}$ of the vector $y = Ax$ is found to be the matrix A'. Thus

$$\frac{\partial y}{\partial x} = \begin{bmatrix} \dfrac{\partial y_1}{\partial x} & \dfrac{\partial y_2}{\partial x} & \cdots & \dfrac{\partial y_n}{\partial x} \end{bmatrix} \tag{3}$$

$$= \begin{bmatrix} \dfrac{\partial a'_1x}{\partial x} & \dfrac{\partial a'_2x}{\partial x} & \cdots & \dfrac{\partial a'_nx}{\partial x} \end{bmatrix}, \quad \text{from (2)}$$

$$= [a_1 \quad a_2 \quad \ldots \quad a_n], \quad \text{from (1)}$$
$$= A',$$

from the definition of the a'_i as rows of A. Hence

$$\frac{\partial}{\partial x}(Ax) = A'.$$

Similarly, if $\alpha_1, \alpha_2, \ldots, \alpha_n$ are the columns of A

$$x'A = [x'\alpha_1 \quad x'\alpha_2 \quad \ldots \quad x'\alpha_n]$$

and

$$\frac{\partial}{\partial x}(x'A) = A.$$

Note from (3) that the complete expression for $\partial y/\partial x$ in the case of A being 3×3, for example, is

$$\frac{\partial y}{\partial x} = \frac{\partial}{\partial x}(Ax) = \begin{bmatrix} \frac{\partial y_1}{\partial x_1} & \frac{\partial y_2}{\partial x_1} & \frac{\partial y_3}{\partial x_1} \\ \frac{\partial y_1}{\partial x_2} & \frac{\partial y_2}{\partial x_2} & \frac{\partial y_3}{\partial x_2} \\ \frac{\partial y_1}{\partial x_3} & \frac{\partial y_2}{\partial x_3} & \frac{\partial y_3}{\partial x_3} \end{bmatrix} = A'. \tag{4}$$

Example. For

$$y = Ax = \begin{bmatrix} 2 & 6 & -1 \\ 3 & -2 & 4 \\ 3 & 4 & 7 \end{bmatrix} \begin{bmatrix} x_1 \\ x_2 \\ x_3 \end{bmatrix},$$

i.e.,

$$\begin{bmatrix} y_1 \\ y_2 \\ y_3 \end{bmatrix} = \begin{bmatrix} 2x_1 + 6x_2 - x_3 \\ 3x_1 - 2x_2 + 4x_3 \\ 3x_1 + 4x_2 + 7x_3 \end{bmatrix},$$

the value of $\partial y/\partial x$ given by (4) is

$$\frac{\partial y}{\partial x} = \begin{bmatrix} 2 & 3 & 3 \\ 6 & -2 & 4 \\ -1 & 4 & 7 \end{bmatrix} = A'.$$

Consider the quadratic form of order n:

$$\begin{aligned} y &= x'Ax \\ &= \sum_i \sum_j a_{ij} x_i x_j, \end{aligned}$$

the summations being for values of i and j from 1 through n. Its differential with respect to the vector $\dfrac{\partial}{\partial x}$ is

$$\frac{\partial y}{\partial x} = \begin{bmatrix} \frac{\partial}{\partial x_1} \\ \frac{\partial}{\partial x_2} \\ \frac{\partial}{\partial x_3} \\ \cdot \\ \cdot \\ \cdot \\ \frac{\partial}{\partial x_n} \end{bmatrix} \sum_i \sum_j a_{ij} x_i x_j$$

$$= \begin{bmatrix} \sum_j a_{1j} x_j & + & \sum_i a_{i1} x_i \\ \sum_j a_{2j} x_j & + & \sum_i a_{i2} x_i \\ & \cdot & \\ & \cdot & \\ & \cdot & \\ \sum_j a_{nj} x_j & + & \sum_i a_{in} x_i \end{bmatrix}$$

$$= \begin{bmatrix} a'_1 x \\ a'_2 x \\ \cdot \\ \cdot \\ \cdot \\ a'_n x \end{bmatrix} + \begin{bmatrix} \alpha'_1 x \\ \alpha'_2 x \\ \cdot \\ \cdot \\ \cdot \\ \alpha'_n x \end{bmatrix}$$

where a'_i and α_i are the ith row and column respectively of A. Hence the two vectors are Ax and $A'x$ and so

$$\frac{\partial}{\partial x}(x'Ax) = Ax + A'x.$$

For symmetric A, $A = A'$ and

$$\frac{\partial}{\partial x}(x'Ax) = 2Ax. \tag{5}$$

Example. For

$$
\begin{aligned}
y &= x'Ax \\
&= \begin{bmatrix} x_1 & x_2 & x_3 \end{bmatrix} \begin{bmatrix} 1 & 3 & 5 \\ 3 & 4 & 7 \\ 5 & 7 & 9 \end{bmatrix} \begin{bmatrix} x_1 \\ x_2 \\ x_3 \end{bmatrix} \\
&= x_1^2 + 6x_1x_2 + 10x_1x_3 + 4x_2^2 + 14x_2x_3 + 9x_3^2,
\end{aligned}
$$

$$
\frac{\partial y}{\partial x} = \begin{bmatrix} \dfrac{\partial y}{\partial x_1} \\ \dfrac{\partial y}{\partial x_2} \\ \dfrac{\partial y}{\partial x_3} \end{bmatrix} = \begin{bmatrix} 2x_1 + 6x_2 + 10x_3 \\ 6x_1 + 8x_2 + 14x_3 \\ 10x_1 + 14x_2 + 18x_3 \end{bmatrix} = 2Ax.
$$

Illustration. The method of least squares in statistics involves minimizing the sum of squares of the elements of the vector e in the equation

$$y = Xb + e. \tag{6}$$

If $e = \{e_i\}$ for $i = 1, 2, \ldots, n$, the sum of squares is

$$
\begin{aligned}
S &= \sum_{i=1}^{n} e_i^2 \\
&= e'e \\
&= (y - Xb)'(y - Xb) \\
&= y'y - b'X'y - y'Xb + b'X'Xb
\end{aligned}
$$

and because $y'Xb$ is a scalar it equals its transpose, $b'Xy$, and hence

$$S = y'y - 2b'X'y + b'X'Xb.$$

Minimizing this with respect to elements of the vector b involves equating to zero the expression $\partial S/\partial b$ where $\partial/\partial b$ is a vector of differential operators of the nature just discussed. Using results (1) and (5) gives

$$\frac{\partial S}{\partial b} = -2X'y + 2X'Xb,$$

and equating this to a null vector leads to

$$b = (X'X)^{-1}X'y, \tag{7}$$

provided that $(X'X)^{-1}$ exists.

Illustration. The selection index method of genetics as proposed by Hazel (1943) can be formulated as follows. Let x be a vector of observations (phenotypes) on certain traits in an animal, and b a vector of coefficients by which these observations are combined into an index $I = b'x$. Corresponding to the traits represented in x let g be a vector of (unknown) genetic values and let a be a vector of the relative economic values of these traits. Then the coefficients represented by b are determined by maximizing, for values of b, the correlation between $I = b'x$ and the combined genetic value $H = a'g$. This correlation is

$$r_{IH} = \frac{\text{cov}(IH)}{\sigma_I \sigma_H} = \frac{b'Ga}{\sqrt{b'Pb}\sqrt{a'Ga}},$$

where P is the matrix of phenotypic variances and covariances, namely the variance-covariance matrix appropriate to the vector of phenotypic values x; and G is the matrix of genetic variances and covariances, the variance-covariance matrix corresponding to the genetic values g. Maximizing this with respect to the elements in b is achieved by maximizing $b'Ga/\sqrt{b'Pb}$. Equating to zero the derivative of this with respect to b, using results (1) and (5) to carry out the differentiation, P and G being symmetric, gives

$$(\sqrt{b'Pb})Ga = \tfrac{1}{2}(b'Pb)^{-\frac{1}{2}}2Pb(b'Ga);$$

i.e.

$$(b'Ga)Pb = (b'Pb)Ga.$$

Apart from a scalar, all solutions to this equation are proportional to solutions to $Pb = Ga$. The required value of b is therefore taken as the solution of the latter equation:

$$b = P^{-1}Ga. \tag{8}$$

The corresponding (maximized) value of r_{IH} is then

$$r_{IH} = \sqrt{\frac{b'Ga}{a'Ga}}.$$

Estimates of b and r_{IH} are obtained by using a predetermined value of a and estimates of the phenotypic and genetic variances and covariances involved in P and G.

Bernard et al. (1954) used these methods in a study of swine selection. Two of the traits considered were litter size and individual pig weight, to which were attached relative economic values of 10:1; that is to say, an increase of one pig in a sow's litter is ten times as valuable economically as an increase of one pound weight in a single pig. The index derived for combining these traits was

$$I = 0.95x_1 + 0.10x_2,$$

based on the equation (8),

$$b = P^{-1}Ga$$

$$= \begin{bmatrix} 3.69 & 0 \\ 0 & 521 \end{bmatrix}^{-1} \begin{bmatrix} 0.35 & 0 \\ 0 & 54 \end{bmatrix} \begin{bmatrix} 10 \\ 1 \end{bmatrix},$$

the values for P and G having been obtained from appropriate analyses of data.

6. JACOBIANS

An example of the matrix

$$J = \left\{ \frac{\partial y_i}{\partial x_j} \right\} \quad \text{for} \quad i, j = 1, 2, \ldots, n$$

is shown in equation (4). It is known as a *Jacobian*. It applies to any situation where variables denoted by y's are functions of other variables x, and as such is sometimes referred to as the Jacobian of the functions y_i with respect to the variables x_j. When the relationship between the x's and y's is linear and can be expressed as $y = Tx$ then equation (4) clearly shows that $J = T'$.

The Jacobian finds widespread use in the transformation of variables in the integral calculus. For an integral involving the differentials $dy_1, dy_2, \ldots, dy_n$, a change of variables is made from $y_1, y_2, \ldots, y_n$ to $x_1, x_2, \ldots, x_n$ by substituting for the y's in terms of the x's and by replacing the differentials by $|J| dx_1, dx_2, \ldots, dx_n$ where J is the Jacobian of the y's with respect to the x's, and $|J|$ is its determinant.

* Another matrix of partial differentials is the Hessian. If y is a function of n variables $x_1, x_2, \ldots, x_n$, the matrix of second-order partial derivatives of y with respect to the x's is called a Hessian:

$$H = \left\{ \frac{\partial^2 y}{\partial x_i \partial x_j} \right\}, \quad \text{for} \quad i, j = 1, 2, \ldots, n.$$

Illustration. One of the properties of the principle of maximum likelihood in statistics is that, under certain conditions (see, for example, Mood and Graybill, 1963, Section 10.8), the variance-covariance matrix of a set of simultaneously estimated parameters is the inverse of minus the expected value of the Hessian of the likelihood with respect to those parameters.

7. INVERSION BY PARTITIONING

Obtaining the inverse of a matrix can often be simplified by partitioning it into four sub-matrices in a manner such that those at the top left and bottom right are square; i.e.

$$M = \begin{bmatrix} A & B \\ C & D \end{bmatrix} \tag{9}$$

where A and D are square. If the corresponding partitioned form of M^{-1} is

$$M^{-1} = \begin{bmatrix} X & Y \\ Z & W \end{bmatrix}, \tag{10}$$

post-multiplying (9) by (10) gives

$$\begin{bmatrix} A & B \\ C & D \end{bmatrix}\begin{bmatrix} X & Y \\ Z & W \end{bmatrix} = I = \begin{bmatrix} I & 0 \\ 0 & I \end{bmatrix}.$$

Hence

$$AX + BZ = I, \tag{11}$$
$$AY + BW = 0, \tag{12}$$

and

$$CX + DZ = 0 \tag{13}$$
$$CY + DW = I. \tag{14}$$

For the moment it is supposed that both A and D are nonsingular. Then from (13)

$$Z = -D^{-1}CX, \tag{15}$$

which substituted in (11) leads to

$$X = (A - BD^{-1}C)^{-1}. \tag{16}$$

Similarly from (12)

$$Y = -A^{-1}BW \tag{17}$$

and so in (14)

$$W = (D - CA^{-1}B)^{-1}. \tag{18}$$

These results demand inverting four matrices. Expressions more convenient from a computational point of view can, however, be derived. First note that the identity $M^{-1}M = I$ also holds, so that from (9) and (10)

$$\begin{bmatrix} X & Y \\ Z & W \end{bmatrix}\begin{bmatrix} A & B \\ C & D \end{bmatrix} = \begin{bmatrix} I & 0 \\ 0 & I \end{bmatrix}$$

and hence

$$XA + YC = I, \tag{19}$$
$$XB + YD = 0, \tag{20}$$
$$ZA + WC = 0, \tag{21}$$

and

$$ZB + WD = I. \tag{22}$$

From (20)

$$Y = -XBD^{-1}$$

and from (22)

$$W = D^{-1} - ZBD^{-1}.$$

Using X and Z from (16) and (15) we have expressions that require inverting only two matrices:

$$\begin{aligned} X &= (A - BD^{-1}C)^{-1} \\ Y &= -XBD^{-1} \\ Z &= -D^{-1}CX \\ W &= D^{-1} - ZBD^{-1}. \end{aligned}$$

A step-by-step procedure for obtaining these expressions is as follows:

Procedure a.

i. D^{-1}	vi. BD^{-1}
ii. $D^{-1}C$	[vii.] $Y = -XBD^{-1}$
iii. $BD^{-1}C$	[viii.] $Z = -D^{-1}CX$
iv. $A - BD^{-1}C$	ix. ZBD^{-1}
[v.] $X = (A - BD^{-1}C)^{-1}$	[x.] $W = D^{-1} - ZBD^{-1}$

The steps enclosed in square brackets are those which yield the elements of M^{-1}. Note that this procedure, evolved from equations (15), (16), (20) and (22), requires the existence of D^{-1} but not A^{-1}. In a similar manner equations (17), (18), (19) and (21) can be used to derive an alternative procedure that requires the existence of A^{-1} but not D^{-1}:

Procedure b.

i. A^{-1}	vi. CA^{-1}
ii. $A^{-1}B$	[vii.] $Z = -WCA^{-1}$
iii. $CA^{-1}B$	[viii.] $Y = -A^{-1}BW$
iv. $D - CA^{-1}B$	ix. YCA^{-1}
[v.] $W = (D - CA^{-1}B)^{-1}$	[x.] $X = A^{-1} - YCA^{-1}$

The choice of which of these procedures shall be used in any particular case will depend on any properties pertaining to M and its sub-matrices that may simplify the calculations. For example, if

$$M = \begin{bmatrix} A & B \\ C & D \end{bmatrix} = \left[\begin{array}{cc|ccc} 1 & 2 & 1 & 2 & 3 \\ 3 & 4 & 9 & 4 & 1 \\ \hline 6 & 3 & 7 & 0 & 0 \\ 1 & 0 & 0 & 8 & 0 \\ 4 & 2 & 0 & 0 & 9 \end{array}\right]$$

procedure **a** is preferable because it involves D^{-1} and $(A - BD^{-1}C)^{-1}$, which are relatively easy to calculate since D is diagonal and A is only

2×2. Procedure **b** would require inverting A, which is easy enough, but $(D - CA^{-1}B)$ has also to be inverted—a non-diagonal 3×3. The difference in effort between the two procedures is small in this example, but it would not be were D of order 30 say, and diagonal, and A only 4×4 say. The nature of M should therefore be carefully studied before embarking on the inversion.

These procedures simplify a little when M is symmetric, $M = M'$, for then $A = A'$, $C = B'$, $D = D'$, and $X = X'$, $Z = Y'$ and $W = W'$. That is, $M = M' = \begin{bmatrix} A & B \\ B' & D \end{bmatrix}$ and $M^{-1} = (M^{-1})' = \begin{bmatrix} X & Y \\ Y' & W \end{bmatrix}$. The steps for calculating M^{-1} then resolve themselves into being as follows.

Procedure aa.

i. D^{-1}	[v.] $X = (A - BD^{-1}B')^{-1}$
ii. $BD^{-1} = (D^{-1}B')'$	[vi.] $Y = -XBD^{-1}$
iii. $BD^{-1}B'$	vii. $Y'BD^{-1}$
iv. $A - BD^{-1}B'$	[viii.] $W = D^{-1} - Y'BD^{-1}$.

Compared to procedure **a** there are two less steps, resulting from the symmetry of M. In a similar manner procedure **b** becomes

Procedure bb.

i. A^{-1}	[v.] $W = (D - B'A^{-1}B)^{-1}$
ii. $B'A^{-1} = (A^{-1}B)'$	[vi.] $Y = -A^{-1}BW$
iii. $B'A^{-1}B$	vii. $YB'A^{-1}$
iv. $D - B'A^{-1}B$	[viii.] $X = A^{-1} - YB'A^{-1}$.

It is an interesting exercise to show algebraically that the sub-matrices of M^{-1} as given by procedures **a** and **b** are equivalent, as are those of procedures **aa** and **bb**.

8. MATRIX FUNCTIONS

Turner et al. (1963), reporting on a medical study involving concentrations of an injected substance in the bloodstream, urine and kidneys, utilize a matrix K whose elements relate to the transport of the material between various compartments of the body. In doing so they use a matrix exponential function e^K. Defining such a function in the same way that e^λ for scalar λ is e raised to the λ power would be pointless, because raising something to the Kth power for K being a matrix has no meaning. On the other hand, invoking the power series

$$e^\lambda = 1 + \lambda + \tfrac{1}{2}\lambda^2 + \lambda^3/(3!) + \dots$$

and adapting it to the case of K gives

$$e^K = I + K + K^2/2 + K^3/(3!) + \cdots$$
$$= I + \sum_{r=1}^{\infty} K^r/(r!)$$

which is meaningful for K. Immediately one sees, of course, that K must be square in order for this definition to hold and also, that e^K is a matrix of the same order as K, a power series in K, in fact. Turner et al. (1963) give definitions of other functions in the same manner.

Another use of matrix functions of this style is given by Pielou (1964) in a study of the patch-and-gap patterns of vegetation in a forest. There we find the relationship

$$R = \log_e P \tag{23}$$

where R and P are probability transition matrices. To obtain R directly a power series might again be invoked from scalar algebra,

$$\log_e x = \log[1 - (1 - x)] = -\sum_{i=1}^{\infty} (1 - x)^{2i-1}/(2i - 1),$$

but as this holds true only for limited values of x it is preferable to take (23) as being equivalent to

$$P = e^R$$
$$= I + \sum_{k=1}^{\infty} R^k/(k!),$$

as already discussed. This can be used to establish a relationship between the elements of P and R:

$$p_{ij} = \delta_{ij} + \sum_{k=1}^{\infty} r_{ij}^{(k)}/(k!)$$

where $P = \{p_{ij}\}$, $\delta_{ij} = 1$ if $i = j$ and 0 otherwise, and $r_{ij}^{(k)}$ is the ijth element of R^k. In the case of Pielou's matrices, which are of order 2, R is such that this result reduces to a relatively simple form, as indicated in Exercise 13.

9. DIRECT SUMS

The direct sum of two matrices A and B is defined as

$$A \oplus B = \begin{bmatrix} A & 0 \\ 0 & B \end{bmatrix}$$

where the zeros are null matrices of appropriate order. The definition applies whether or not A and B are of the same order; for example

$$[1 \quad 2 \quad 3] \oplus \begin{bmatrix} 6 & 7 \\ 8 & 9 \end{bmatrix} = \begin{bmatrix} 1 & 2 & 3 & 0 & 0 \\ 0 & 0 & 0 & 6 & 7 \\ 0 & 0 & 0 & 8 & 9 \end{bmatrix}.$$

The transpose of a direct sum is the direct sum of the transposes. The rank of a direct sum is the sum of the ranks, as is evident from the definition of rank (Section 5.6). It is clear from the definition of a direct sum that $A \oplus (-A) \neq 0$ unless A is null. Furthermore,

$$(A \oplus B) + (C \oplus D) = (A + C) \oplus (B + D)$$

only if the necessary conditions of conformability for addition are met. Similarly

$$(A \oplus B)(C \oplus D) = AC \oplus BD$$

provided that conformability for multiplication is satisfied. Should A and B be nonsingular, the preceding result leads to

$$(A \oplus B)^{-1} = A^{-1} \oplus B^{-1}.$$

The direct sum $A \oplus B$ is square either if both A and B are square or if the number of rows in the two matrices equals the number of columns. The determinant of $A \oplus B$ equals $|A||B|$ if both A and B are square, but otherwise it is zero (as may be seen from considering the Laplace expansion of $|A \oplus B|$). When both A and B are square the latent roots of their direct sum are the latent roots of A and B, as is evident from the characteristic equation of $A \oplus B$. If u and v are latent vectors of A and B respectively, then $\begin{bmatrix} u \\ 0 \end{bmatrix}$ and $\begin{bmatrix} 0 \\ v \end{bmatrix}$ are latent vectors of $A \oplus B$. However, when A and B are not square but $A \oplus B$ is, it appears that the only conclusion to be drawn about the latent roots of $A \oplus B$ is that at least d of them will be zero where d is the difference between the number of rows and columns in A (or B). The latent vectors will be found as for any matrix.

Illustration. The variance-covariance matrix of observations on n_i litter mates in a nutrition study of young animals, pigs for example, is

$$A_i = \sigma_w^2 I_{n_i} + \sigma_L^2 J_{n_i}$$

where J_{n_i} is a J matrix as defined in Section 2. Variation due to litters is denoted by σ_L^2 and variation due to animal differences within litters by σ_w^2. For observations on k litters the variance-covariance matrix is

$$V = A_1 \oplus A_2 \oplus \ldots \oplus A_k.$$

Certain analyses of this situation (Searle, 1956, for example) require $|V|$ and V^{-1}, which are readily obtained from the results just discussed (see Exercise 14).

10. DIRECT PRODUCTS

The direct product of two matrices $A_{p\times q}$ and $B_{m\times n}$ is defined as

$$A_{p\times q} * B_{m\times n} = \begin{bmatrix} a_{11}B & \dots & a_{1q}B \\ \cdot & & \cdot \\ \cdot & & \cdot \\ \cdot & & \cdot \\ a_{p1}B & \dots & a_{pq}B \end{bmatrix}. \tag{24}$$

It is clear from this that the direct product can be partitioned into as many sub-matrices as there are elements of A, each sub-matrix being B multiplied by an element of A. The elements of the direct product consist of all possible products of an element of A multiplied by an element of B, and its order is $pm \times qn$. For example,

$$[1 \quad 2 \quad 3] * \begin{bmatrix} 6 & 7 \\ 8 & 9 \end{bmatrix} = \begin{bmatrix} 6 & 7 & 12 & 14 & 18 & 21 \\ 8 & 9 & 16 & 18 & 24 & 27 \end{bmatrix}.$$

The transpose of a direct product is the direct product of the transposes—as is evident from transposing equation (24). Note, however, that $(A * B)' = A' * B'$, in contrast to $(AB)' = B'A'$. As expressed in (24), $A * B$ consists of p sets of m rows each. If the rank of B is b there will be b independent rows in each of these sets. But if the rank of A is a, all rows of the last $p - a$ sets will be linear combinations of all rows of the first a sets. Therefore only the b rows in each of the first a sets are independent of each other, and so the number of independent rows in $A * B$ is ab. Hence the rank of $A * B$ is ab, the product of the rank of A and the rank of B.

Direct products are frequently called Kronecker products [Cornish (1957) and Vartak (1955), for example]. Although MacDuffee (1933) cites many references from the late nineteenth and early twentieth centuries in an extensive discussion of direct products, nowhere does he mention Kronecker in this regard. One might speculate, therefore, as to how his name has become associated with direct products. Some writers call $A * B$ the *right direct product* to distinguish it from $B * A$ which is then called the *left direct product*. On some occasions $B * A$ will be found defined as the right-hand side of equation (24), but the definition given there, of $A * B$, is the one usually employed, and is the definition used in this book for direct product.

* It is apparent from (24) that for $A_{p\times q} = \{a_{ij}\}$ and $B_{m\times n} = \{b_{rs}\}$ the elements of both $A * B$ and $B * A$ consist of all possible products $a_{ij}b_{rs}$. In fact, $B * A$ is simply $A * B$ with the rows and columns each in a different order. Thus

$$B * A = P(A * B)Q \tag{25}$$

where P and Q are each a product of E- type elementary operators (Section 5.7).

For any values of i, j, r and s the element $a_{ij}b_{rs}$ is located in $A * B$ in the rth row and sth column of the i, jth sub-matrix of the partitioned form shown in (24). It is therefore in row $[m(i-1) + r]$ and column $[n(j-1) + s]$ of $A * B$. In $B * A$, however, $b_{rs}a_{ij}$ is in row i and column j of the r,sth sub-matrix of the form analogous to (24). It is therefore in row $[p(r-1) + i]$ and column $[q(s-1) + j]$ of $B * A$. Consequently the interchanging of rows and columns implied in (25), to obtain $B * A$ from $A * B$, can be specified as follows. The $[i + (j-1)p]$th row of P is the $[(i-1)m + j]$th row of I_{pm}, for $i = 1, 2, \ldots, p$ and $j = 1, 2, \ldots, m$, and the $[i + (j-1)q]$th column of Q is the $[(i-1)n + j]$th column of I_{qn}, for $i = 1, 2, \ldots, q$ and $j = 1, 2, \ldots, n$.

The following results are of interest.

i. For x and y being vectors:

$$x' * y = yx' = y * x'.$$

ii. For D being a diagonal matrix of order k:

$$D * A = \mathrm{d}_1 A \oplus d_2 A \oplus \cdots \oplus d_k A,$$

but $$A_{p\times q} * D_k = (A * I_k)[D \oplus D \oplus \cdots \oplus D, \text{ for } q \text{ terms}].$$

iii. For λ being a scalar:

$$\lambda * A = \lambda A = A * \lambda = A\lambda.$$

iv. When using partitioned matrices,

$$[A_1 \quad A_2] * B = [A_1 * B \quad A_2 * B],$$

but $$A * [B_1 \quad B_2] \neq [A * B_1 \quad A * B_2].$$

v. Provided that conformability conditions for regular matrix multiplication are satisfied

$$(A * B)(X * Y) = AX * BY.$$

vi. For A and B both square and nonsingular

$$(A * B)^{-1} = A^{-1} * B^{-1}.$$

These results are readily verified with simple numerical examples. It is recommended that the reader satisfy himself of their validity.

In contemplating the determinant of $A * B$ we need be concerned with only the situations in which both A and B are square, because when they are not, consideration of rank reveals that $A * B$ will not be of full rank, and its determinant is therefore zero. If A and B are square, of order p and m respectively, then from result (v) just given

$$A_p * B_m = (A_p * I_m)(I_p * B_m)$$

and hence

$$|A_p * B_m| = |A_p * I_m||I_p * B_m|. \tag{26}$$

Now, because of (25), either $|A * B| = |B * A|$ or $|A * B| = -|B * A|$. The latter cannot be true because, if it were, putting $B = A$ would give $|A * A| = -|A * A|$. Therefore

$$|A * B| = |B * A|,$$

and applying this to (26) gives

$$|A_p * B_m| = |I_m * A_p||I_p * B_m|.$$

But from result (ii)

$$|I_m * A_p| = |A|^m,$$

so that

$$|A_p * B_m| = |A|^m|B|^p.$$

The characteristic equation of $A * B$ exists only when $A * B$ is square. If A and B are both square with latent vectors u and v corresponding to latent roots λ and μ respectively, then $Au = \lambda u$ and $Bv = \mu v$. Therefore

$$Au * Bv = \lambda u * \mu v$$

and so

$$(A * B)(u * v) = \lambda\mu(u * v).$$

Hence the latent roots of $A * B$ are the products of the latent roots of A with those of B, and the corresponding latent vectors are the direct products of the latent vectors of A and B. This result means that if $U^{-1}AU = D_a$ and $V^{-1}BV = D_b$ are the canonical forms under similarity, then the canonical reduction of $A * B$ is

$$(U^{-1} * V^{-1})(A * B)(U * V) = D_a * D_b.$$

When $A * B$ is square but A and B are not, the only conclusion to be drawn about latent roots of $A * B$ is that $r(A) \cdot r(B)$ of them will be non-zero.

Illustration. Hill (1964) has recently utilized direct products in a problem concerned with n generations of random mating starting with the progeny obtained from crossing two autotetraploid plants which both have genotype *AAaa*. Normally the original plants would produce gametes *AA*, *Aa* and *aa* in the proportion 1:4:1. But suppose the proportion is $u:1-2u:u$ where, for example, u might take the value $(1-\alpha)/6$, for α being a measure of "diploidization" of the plants: $\alpha = 0$ is the case of autotetraploids with chromosome segregation and $\alpha = 1$ is the diploid case with all gametes being Aa. The question now is, what are the genotypic frequencies in the population after n generations of random mating? Let u_i be the vector of gametic frequencies and f_i the vector of genotype frequencies in the ith generation of random mating, where $u_0 = v$ is the vector of gametic frequencies in the initial plants. Then

$$u_0 = v = \begin{bmatrix} u \\ 1 - 2u \\ u \end{bmatrix}$$

and $f_{i+1} = u_i * u_i$ for $i = 0, 1, 2, \ldots, n$. Furthermore, the relationship between u_i and f_i at any generation is

$$u_i = Bf_i \text{ where } B = \begin{bmatrix} 1 & \frac{1}{2} & u & \frac{1}{2} & u & 0 & u & 0 & 0 \\ 0 & \frac{1}{2} & 1 - 2u & \frac{1}{2} & 1 - 2u & \frac{1}{2} & 1 - 2u & \frac{1}{2} & 0 \\ 0 & 0 & u & 0 & u & \frac{1}{2} & u & \frac{1}{2} & 1 \end{bmatrix}.$$

Thus

$$\begin{aligned} f_i &= u_{i-1} * u_{i-1} \\ &= Bf_{i-1} * Bf_{i-1} \\ &= (B * B)(f_{i-1} * f_{i-1}) \\ &= (B * B)[(B * B)(f_{i-2} * f_{i-2}) * (B * B)(f_{i-2} * f_{i-2})] \\ &= (B * B)[(B * B) * (B * B)][(f_{i-2} * f_{i-2}) * (f_{i-2} * f_{i-2})]. \end{aligned}$$

By making repeated use of the general result $(P * Q)(X * Y) = (PX * QY)$, and of the associative law for direct products, it can be proved by induction that

$$f_i = B^{*2}B^{*4}B^{*8} \ldots B^{*2^{i-1}}v^{*2^i}$$

where the notation B^{*n} means $B * B * B * \ldots$ for n B's. With B being 3×3^2 and v being 3×1, the products involved in f_i are, of course, conformable.

Illustration. The statistical methods of analyzing factorial experiments are given in many texts (for example, Federer, 1955; Snedecor, 1956; and Steel and Torrie, 1960). We will show here how estimating contrasts from observed means can be derived through the use of direct products of matrices and vectors. The procedure is outlined in terms of a simple, specific example.

First, a system of notation must be established. Suppose the experiment is one carried out by a horticulturist involving two fertilizer treatments on each of three varieties of plant. The treatments will be denoted generally by the letter A and the varieties by B, the observed treatment means being

a_1 and a_2, the variety means being b_1, b_2 and b_3. With analogous notation the complete table of means is represented as follows.

TABLE 1. TABLE OF MEANS

Treatments (A)	Varieties (B) 1	2	3	Treatment Mean
1	ab_{11}	ab_{12}	ab_{13}	a_1
2	ab_{21}	ab_{22}	ab_{23}	a_2
Variety mean	b_1	b_2	b_3	Grand mean

Furthermore, let I, a scalar, represent the overall population mean, let A_1 be the linear effect among the treatments, B_1 the linear effect among the varieties and B_2 the quadratic effect among the varieties. These are, in many instances, the effects of interest to be estimated.

Suppose, first of all, that there were only one variety of plant. The estimates of I and A_1 would then be

$$\begin{bmatrix} \hat{I} \\ \hat{A}_1 \end{bmatrix} = \tfrac{1}{2}\begin{bmatrix} 1 & 1 \\ -1 & 1 \end{bmatrix}\begin{bmatrix} a_1 \\ a_2 \end{bmatrix}$$

where $\hat{I}$, for example, indicates the estimate of I. Likewise, had there been no fertilizer treatments the estimates of I, B_1 and B_2 would be

$$\begin{bmatrix} \hat{I} \\ \hat{B}_1 \\ \hat{B}_2 \end{bmatrix} = \tfrac{1}{3}\begin{bmatrix} 1 & 1 & 1 \\ -1 & 0 & 1 \\ -1 & 2 & -1 \end{bmatrix}\begin{bmatrix} b_1 \\ b_2 \\ b_3 \end{bmatrix}.$$

Taking treatments and varieties into account simultaneously, however, the estimates of I, A_1, B_1, B_2 and the interactions A_1B_1 and A_1B_2 are given by the direct product of the preceding equations.

$$\begin{bmatrix} \hat{I} \\ \hat{B}_1 \\ \hat{B}_2 \end{bmatrix} * \begin{bmatrix} \hat{I} \\ \hat{A}_1 \end{bmatrix} = \tfrac{1}{3}\begin{bmatrix} 1 & 1 & 1 \\ -1 & 0 & 1 \\ -1 & 2 & -1 \end{bmatrix}\begin{bmatrix} b_1 \\ b_2 \\ b_3 \end{bmatrix} * \tfrac{1}{2}\begin{bmatrix} 1 & 1 \\ -1 & 1 \end{bmatrix}\begin{bmatrix} a_1 \\ a_2 \end{bmatrix}$$

$$= \tfrac{1}{6}\left\{\begin{bmatrix} 1 & 1 & 1 \\ -1 & 0 & 1 \\ -1 & 2 & -1 \end{bmatrix} * \begin{bmatrix} 1 & 1 \\ -1 & 1 \end{bmatrix}\right\}\left\{\begin{bmatrix} b_1 \\ b_2 \\ b_3 \end{bmatrix} * \begin{bmatrix} a_1 \\ a_2 \end{bmatrix}\right\}.$$

In expanding the direct products of vectors in this equation the following convention is to be adopted: the product a_1b_1, for example, is to be taken as ab_{11}; and likewise $\hat{A}_1\hat{B}_2$ is $\widehat{AB}_{12}$. Although a_1b_1 as a product is the product of two means, a_1 and b_1, it is to be interpreted as the mean ab_{11}. With this convention the preceding equation expands to

$$\begin{bmatrix} \hat{I} \\ \hat{A}_1 \\ \hat{B}_1 \\ \widehat{AB}_{11} \\ \hat{B}_2 \\ \widehat{AB}_{12} \end{bmatrix} = \frac{1}{6} \begin{bmatrix} 1 & 1 & 1 & 1 & 1 & 1 \\ -1 & 1 & -1 & 1 & -1 & 1 \\ -1 & -1 & 0 & 0 & 1 & 1 \\ 1 & -1 & 0 & 0 & -1 & 1 \\ -1 & -1 & 2 & 2 & -1 & -1 \\ 1 & -1 & -2 & 2 & 1 & -1 \end{bmatrix} \begin{bmatrix} ab_{11} \\ ab_{21} \\ ab_{12} \\ ab_{22} \\ ab_{13} \\ ab_{23} \end{bmatrix}.$$

The general result is clear: if $\hat{\mu}_\beta$ is the vector of estimated effects for the β factor, e.g. $\hat{\mu}_\beta = \begin{bmatrix} \hat{I} \\ \hat{B}_1 \\ \hat{B}_2 \end{bmatrix}$; if m_β is the corresponding vector of observed means, e.g. $m_\beta = \begin{bmatrix} b_1 \\ b_2 \\ b_3 \end{bmatrix}$; and if A_β is the matrix such that $\hat{\mu}_\beta = A_\beta m_\beta$ when all other factors are ignored, then the estimates for a series of factors α, β, γ, δ say, are given by

$$\hat{\mu}_\delta * \hat{\mu}_\gamma * \hat{\mu}_\beta * \hat{\mu}_\alpha = (A_\delta * A_\gamma * A_\beta * A_\alpha)(m_\delta * m_\gamma * m_\beta * m_\alpha).$$

The convention relative to the elements of direct products of vectors applies here too.

Rules for establishing equations such as these have been suggested by several authors, for example, Yates (1937) and Fisher (1951), and are summarized by Federer (1955). The generality of the expression is appealing. It is easily remembered, is universally true no matter how many factors are involved, it eliminates the need for remembering lengthy detailed results and it is easily derived from the general direct product result. Extensions of these ideas are to be found in many places, for example, Robson (1959), Kurkjian and Zelen (1962) and Federer and Zelen (1964).

11. EXERCISES

1. Evaluate:

$$\begin{vmatrix} 0 & -1 & 7 \\ 1 & 0 & 3 \\ -7 & -3 & 0 \end{vmatrix}, \quad \begin{vmatrix} 0 & -2 & 4 & 6 \\ 2 & 0 & 3 & -7 \\ -4 & -3 & 0 & 9 \\ -6 & 7 & -9 & 0 \end{vmatrix} \quad \text{and} \quad \begin{vmatrix} 0 & 16 & -1 & 2 \\ -16 & 0 & 7 & 0 \\ 1 & -7 & 0 & 6 \\ -2 & 0 & -6 & 0 \end{vmatrix}.$$

2. For

$$A = \begin{bmatrix} 0 & -a & b & c \\ a & 0 & -d & e \\ -b & d & 0 & -f \\ c & -e & f & 0 \end{bmatrix}$$

explain why

$$|I + A| = 1 + (a^2 + b^2 + c^2 + d^2 + e^2 + f^2) + |A|.$$

Evaluate $|A|$.

3. Derive orthogonal matrices from

$$\begin{bmatrix} 0 & -1 & 1 \\ 1 & 0 & 2 \\ -1 & -2 & 0 \end{bmatrix} \quad \text{and} \quad \begin{bmatrix} 0 & 1 & -2 \\ -1 & 0 & 5 \\ 2 & -5 & 0 \end{bmatrix},$$

evaluate their determinants, show that their product is orthogonal, find their latent roots and reduce each to canonical form.

4. Derive a symmetric idempotent matrix from $X = \begin{bmatrix} 1 & 2 \\ 0 & -1 \\ -1 & 0 \end{bmatrix}$; find its rank and latent roots. Denoting it by M show that $P'MP$ is idempotent where P is one of the orthogonal matrices derived in Exercise 3.

5. For the J matrices defined in Section 2 show that

(a) $J_{1 \times m} J_{m \times 1} = m$,

(b) $J_{m \times 1} J_{1 \times n} = J_{m \times n}$,

(c) $J_{1 \times m} J_m J_{m \times 1} = m^2$,

(d) $J_m(aI_m + bJ_m) = (a + mb)J_m$.

6. By partitioning $M = \begin{bmatrix} -5 & 3 & 5 & 1 & 0 \\ -4 & 8 & 10 & 0 & -1 \\ 2 & 13 & 11 & 2 & 1 \\ 0 & 1 & 1 & 3 & 2 \\ 3 & 1 & 0 & 7 & 5 \end{bmatrix}$, find its inverse.

7. Show that with respect to x_1 and x_2, the Jacobian of

$$y_1 = 6x_1^2x_2 + 2x_1x_2 + x_2^2 \quad \text{and} \quad y_2 = 2x_1^3 + x_1^2 + 2x_1x_2$$

is the same as the Hessian of

$$y = 2x_1^3x_2 + x_1^2x_2 + x_1x_2^2.$$

8. Express the quadratic form

$$7x_1^2 + 4x_1x_2 - 5x_2^2 - 6x_2x_3 + 3x_3^2 + 6x_1x_3$$

in the form $x'Ax$ where $A = A'$, and show that

$$\frac{\partial}{\partial x}(x'Ax) = 2Ax.$$

9. (*a*) Show that $X(YX)^{-1}Y$ is idempotent and is the identity matrix when X and Y are diagonal.

(*b*) Show that a symmetric orthogonal matrix equals its inverse. Construct such a matrix, of order 2.

(*c*) If $W = R^{-1} - R^{-1}Z(Z'R^{-1}Z + D^{-1})^{-1}Z'R^{-1}$ show that $(R+ZDZ')$ is the inverse of W.

(*d*) If $D = X'(XV^{-1}X')^{-1}XV^{-1}$ show that where V is symmetric

i. $D' = V^{-1}DV$,

ii. D is idempotent,

iii. $(I - D')V^{-1}(I - D) = (I - D')V^{-1} = V^{-1}(I - D)$.

(*e*) Prove that the trace of $X(YX)^{-1}Y$ equals the number of rows in Y.

(*f*) Inverting a partitioned matrix is discussed in Section 7. Show that the expressions for X, Y, Z and W as given in procedure **a** are equivalent to those of procedure **b**.

10. Show that

(*a*) $$\begin{bmatrix} I & P \\ Q & I \end{bmatrix}^{-1} = \begin{bmatrix} (I-PQ)^{-1} & -(I-PQ)^{-1}P \\ -Q(I-PQ)^{-1} & I + Q(I-PQ)^{-1}P \end{bmatrix},$$

(*b*) $$\begin{bmatrix} 0 & P \\ Q & I \end{bmatrix}^{-1} = \begin{bmatrix} -(PQ)^{-1} & (PQ)^{-1}P \\ Q(PQ)^{-1} & I - Q(PQ)^{-1}P \end{bmatrix},$$

(*c*) $$\begin{bmatrix} 0 & P \\ Q & R \end{bmatrix}^{-1} = \begin{bmatrix} -(PR^{-1}Q)^{-1} & (PR^{-1}Q)^{-1}PR^{-1} \\ R^{-1}Q(PR^{-1}Q)^{-1} & R^{-1} - R^{-1}Q(PR^{-1}Q)^{-1}PR^{-1} \end{bmatrix},$$

(*d*) $$\begin{bmatrix} S & P \\ 0 & R \end{bmatrix}^{-1} = \begin{bmatrix} S^{-1} & -S^{-1}PR^{-1} \\ 0 & R^{-1} \end{bmatrix}.$$

11. If K is a square matrix of order n, what values of K make e^K equal to the following?

(*a*) eI

(*b*) $I - K + eK$

(*c*) I

(*d*) $e^{\lambda}I$

(*e*) $I + K$

(*f*) $I + K(e^n - 1)/n$

12. With $A = V_r(a, b)$ as defined in Section 4, show that

$$A^n = a^n I_r + \frac{1}{r}[(a + rb)^n - a^n]J_r$$

$$= V_r\left\{a^n, \frac{1}{r}[(a + rb)^n - a^n]\right\}$$

and

$$e^A = V_r\left\{e^a, \frac{1}{r}(e^{a+rb} - e^a)\right\}.$$

13. With
$$R = \begin{bmatrix} -a & a \\ b & -b \end{bmatrix}$$
show that
$$R^n = [-(a+b)]^{n-1}R;$$
and for
$$P = \begin{bmatrix} p_1 & 1-p_1 \\ p_2 & 1-p_2 \end{bmatrix}$$
show that the functional relationship
$$R = \log_e P$$
implies
$$a = \frac{-(1-p_1)\log_e(p_1-p_2)}{1-p_1+p_2}$$
and
$$a + b = -\log_e(p_1 - p_2).$$
These are the matrices used by Pielou (1964).

14. Show that for the matrix V defined at the bottom of page 214
$$|V| = \sigma_w^{2(n-k)} \prod_{i=1}^{k} (\sigma_w^2 + n_i\sigma_L^2)$$
and
$$V^{-1} = B_1 \oplus B_2 \oplus \ldots \oplus B_k$$
where
$$B_i = (1/\sigma_w^2)[I_{n_i} - \sigma_L^2/(\sigma_w^2 + n_i\sigma_L^2)J_{n_i}].$$
The matrix V, its determinant and inverse are used by Searle (1956) in studying maximum likelihood procedures for estimating σ_w^2 and σ_L^2.

15. Find the latent roots of $A = \begin{bmatrix} 4 & -2 \\ 5 & -3 \end{bmatrix}$ and of $B = \begin{bmatrix} 2 & -1 \\ 2 & 5 \end{bmatrix}$ and show that their products are the latent roots of $A * B$.

16. For vectors x and y, matrices A and B, and scalars μ and λ, show that

(*a*) $(x' * B)(A * y) = Byx'A$

(*b*) $(\lambda * y)(x' * \mu) = \mu\lambda yx'$

(*c*) $(B * B)(x * x) = (B * Bx)x \neq B(x * Bx)$

(*d*) $(A \oplus B)^{-1}$ exists only if A^{-1} and B^{-1} exist.

17. For matrices A, B and C show that
$$A * (B \oplus C) \neq (A * B) \oplus (A * C)$$
but
$$(A \oplus B) * C = (A * C) \oplus (B * C).$$

REFERENCES

Bernard, C. S., A. B. Chapman and R. H. Grummer (1954). Selection of pigs under farm conditions: kind and amount practiced and a recommended selection index. *J. Anim. Sci.*, **13**, 389–404.

Cornish, E. A. (1957). An application of the Kronecker product of matrices in multiple regression. *Biometrics*, **13**, 19–27.

Federer, W. T. (1955). *Experimental Design.* Macmillan, New York.

Federer, W. T. and M. Zelen (1964). Application of the calculus for factorial arrangements. III. Analysis of factorials with unequal numbers of observations. (Unpublished.) Biometrics Unit, Cornell University and The National Cancer Institute.

Fisher, R. A. (1951). *The Design of Experiments*. Sixth Edition, Oliver and Boyd, Edinburgh.

Graybill, Franklin A. (1961). *An Introduction to Linear Statistical Models*. Vol. I, McGraw-Hill, New York.

Hazel, L. N. (1943). The genetic basis for constructing selection indexes. *Genetics*, **28,** 476–490.

Hill, R. R. (1964). *Personal Communication*.

Kurkjian, B. and M. Zelen (1962). A calculus for factorial arrangements. *Ann. Math. Stat.*, **33,** 600–619.

MacDuffee, C. C. (1933). *The Theory of Matrices*. Springer, Berlin. (Also, Chelsea Publishing Company, New York, 1946, 1956.)

Mood, Alexander M. and Franklin A. Graybill (1963). *Introduction to the Theory of Statistics*. Second Edition, McGraw-Hill, New York.

Pielou, E. C. (1964). The spatial pattern of two-phase patchworks of vegetation. *Biometrics*, **20,** 156–167.

Robson, D. S. (1959). A simple method for constructing orthogonal polynomials when the independent variable is unequally spaced. *Biometrics*, **15,** 187–191.

Searle, S. R. (1956). Matrix methods in components of variance and covariance analysis. *Ann. Math. Stat.*, **27,** 737–748.

Snedecor, G. W. (1956). *Statistical Methods*. Fifth Edition. Iowa State College Press, Ames, Iowa.

Steel, R. G. D. and J. H. Torrie (1960). *Principles and Procedures of Statistics*. McGraw-Hill, New York.

Turner, M. E., R. J. Monroe and L. D. Homer (1963). Generalized kinetic regression analysis: hypergeometric kinetics. *Biometrics*, **19,** 406–428.

Vartak, M. N. (1955). On an application of Kronecker product of matrices to statistical designs. *Ann. Math. Stat.*, **26,** 420–438.

Yates, F. (1937). The design and analysis of factorial experiments. *Imp. Bur. Soil. Sci. Tech. Comm.*, **35,** 1–95.

CHAPTER 9

THE MATRIX ALGEBRA OF REGRESSION ANALYSIS

The intention of this book is to help the biologist attain an understanding of matrix algebra that it might assist him with mathematical problems that arise in biology. Many of these problems are statistical in nature. We therefore conclude with two chapters devoted to topics in statistics that are likely to interest biologists: regression is discussed in this chapter and linear models in Chapter 10. Both topics are ideally suited to being described in matrix terminology and gain considerable clarity therefrom. The matrix descriptions also lead to clarification of the necessary computing procedures.

A certain acquaintance with statistics is assumed, but it is hoped that the material given is of a nature to promote interest both in the statistical methods themselves and in the advantages of using matrix algebra to describe them. Such references as Snedecor (1956), Kendall and Stuart (1958), Steel and Torrie (1960) and Mood and Graybill (1963) will give the reader the relevant statistical details, while texts like Kempthorne (1952) and Graybill (1961) will give him more extensive matrix descriptions than are presented here.

1. GENERAL DESCRIPTION

Illustration. The feed consumed by a dairy cow is utilized for three main purposes: to produce milk, to maintain her body weight and to provide for gains in body weight. In studying the feed intake of dairy cows Stone et al. (1960) therefore considered daily intake as a linear combination of the daily weight of milk produced by a cow, of her body weight and her daily weight change. With measurements of these four variables on some 175 cows they estimated the linear combination as

daily weight of forage dry matter intake
= 5.92 + 0.097 (daily weight of 4% fat-corrected milk yield)
+ 0.012 (body weight)
+ 0.95 (daily weight gain),

all weights being measured in pounds avoirdupois.

The technique of regression analysis is just this: of studying situations in which one variable measured on an object is thought to be some function of one or more other variables measured on the same object; and from a series of such measurements estimating the form of this function. To be specific, suppose that for each of n objects we have observations on four variables y, x_1, x_2 and x_3. Suppose further, that for given values of x_1, x_2 and x_3 the average value of y in the whole population of objects (of which we are assuming we have a random sample) is of the form

$$b_1x_1 + b_2x_2 + b_3x_3$$

where b_1, b_2 and b_3 are some (unknown) constants. Using $E(y)$ to denote this average value of y (the "expected" value of y) we write

$$E(y) = b_1x_1 + b_2x_2 + b_3x_3.$$

The procedure of regression analysis is simply that of using the n sets of observations to obtain estimates of the b's in this equation. More often we have the equation

$$E(y) = a + b_1x_1 + b_2x_2 + b_3x_3;$$

this is considered in Section 8.

The variable denoted by y is usually called the *dependent variable* (because it depends on the x's) and the x variables are correspondingly referred to as the *independent variables.*

The relationship envisaged in the previous equations is linear in the x's, and the analysis we are to describe, referred to so far as regression analysis, is more correctly called *linear regression*, or (for more than one x variable) *multiple linear regression* analysis. Sometimes, however, the underlying relationship envisaged between variables may not be a linear one, although it is the most common and is the only one considered here. Situations that are not linear can frequently be handled as if they were; for example, in another study on dairy cow feed intake Wallace (1956) uses $W^{0 \cdot 73}$ to represent weight, where W is the cow's actual weight. The relationship between feed intake (digestible organic matter in this case) and weight is therefore not linear, but between feed intake and the function $W^{0 \cdot 73}$ it is.

Likewise the equation $y = b_1x + b_2x^2 + b_3x^3$ is not linear in x but is linear in the three variables, x, x^2 and x^3. By these means non-linear relationships between variables can often be reduced to linear ones and studied analytically using linear regression. It is therefore a technique that has wide application.

The first use of matrices is to represent the n sets of observations in matrix form: y will be the vector of the y observations y_i for $i = 1, 2, \ldots, n$,

$$y = \begin{bmatrix} y_1 \\ y_2 \\ \cdot \\ \cdot \\ \cdot \\ y_n \end{bmatrix}, \tag{1}$$

and X will be the matrix of the n observations on each x variable: x_{ij} for $i = 1, 2, \ldots, n$ and $j = 1, 2$ and 3:

$$X = \begin{bmatrix} x_{11} & x_{12} & x_{13} \\ x_{21} & x_{22} & x_{23} \\ x_{31} & x_{32} & x_{33} \\ \cdot & \cdot & \cdot \\ \cdot & \cdot & \cdot \\ \cdot & \cdot & \cdot \\ x_{n1} & x_{n2} & x_{n3} \end{bmatrix}. \tag{2}$$

If

$$e = \begin{bmatrix} e_1 \\ e_2 \\ \cdot \\ \cdot \\ \cdot \\ e_n \end{bmatrix} \tag{3}$$

is the vector of terms e_i that represent the deviation of the observed y_i from its expected value

$$E(y_i) = b_1x_{i1} + b_2x_{i2} + b_3x_{i3}, \tag{4}$$

we have

$$y_i = b_1x_{i1} + b_2x_{i2} + b_3x_{i3} + e_i. \tag{5}$$

Equation (5) occurs for every set of observations y_i, x_{i1}, x_{i2} and x_{i3}. Using the matrix and vectors defined in (1), (2) and (3) these equations can be summarized as

$$y = Xb + e \tag{6}$$

where b is the vector of b_i's to be estimated,

$$b = \begin{bmatrix} b_1 \\ b_2 \\ b_3 \end{bmatrix}.$$

Equation (6) is known as the equation of the mathematical model. The problem is to obtain an estimate of b using the observations arrayed in X and y. Derivation of the estimate is now considered, and its properties are discussed following Section 4 which deals with characteristics of the e terms in (6).

2. ESTIMATION

The customary estimation procedure is to apply the method of "Least Squares" to equation (6). It results in choosing as the estimator of b those values of b_1, b_2 and b_3 which minimize the expression $\sum_{i=1}^{n} [y_i - E(y_i)]^2$. From (4) and (5) this sum of squares is

$$\begin{aligned} \sum_{i=1}^{n} [y_i - E(y_i)]^2 &= \sum_{i=1}^{n} [y_i - (b_1 x_{i1} + b_2 x_{i2} + b_3 x_{i3})]^2 \\ &= \sum_{i=1}^{n} e_i^2 \\ &= e'e. \end{aligned}$$

This is the expression considered in the first illustration of Section 8.5; equation (6) there is identical to equation (6) of this chapter and consequently the estimator of b, which shall be denoted by $\hat{b}$, is as derived earlier, namely from

$$X'X\hat{b} = X'y, \tag{7}$$

comes

$$\hat{b} = (X'X)^{-1}X'y. \tag{8}$$

This is the general solution[1]; it holds provided $(X'X)^{-1}$ exists. In practically all instances of regression analysis this inverse does exist and for the moment we will not be concerned with situations when it does not. These are considered later.

[1] A notation frequently employed in this situation is β for the vector of parameters and b for its estimator. However, the scheme being adopted here is b for the parameters and $\hat{b}$ (read as "b-hat") for the estimators.

Now $X'X$ is square, and from the nature of X is the matrix of sums of squares and cross-products of the x observations: in this instance, the 3×3 matrix

$$X'X = \begin{bmatrix} \sum x_{i1}^2 & \sum x_{i1}x_{i2} & \sum x_{i1}x_{i3} \\ \sum x_{i1}x_{i2} & \sum x_{i2}^2 & \sum x_{i2}x_{i3} \\ \sum x_{i1}x_{i3} & \sum x_{i2}x_{i3} & \sum x_{i3}^2 \end{bmatrix}. \tag{9}$$

$X'y$ is likewise the vector of the sums of cross-products of the x observations with the y observations; on this occasion of order 3×1:

$$X'y = \begin{bmatrix} \sum x_{i1}y_i \\ \sum x_{i2}y_i \\ \sum x_{i3}y_i \end{bmatrix}. \tag{10}$$

In all instances summation is over the n sets of observations: i.e. for $i = 1, 2, \ldots, n$.

Visualizing (9) and (10) substituted in (8) we see that $\hat{b}$ is the vector of sums of products of the x's and y's pre-multiplied by the inverse of the matrix of sums of squares and products of the x's.

Example. Suppose we have the following 5 sets of observations:

i	y_i	x_{i1}	x_{i2}	x_{i3}
1	24	1	1	3
2	26	1	0	1
3	20	2	1	1
4	20	3	2	1
5	23	1	2	0

The b's are estimated as

$$\hat{b} = \begin{bmatrix} \hat{b}_1 \\ \hat{b}_2 \\ \hat{b}_3 \end{bmatrix} = (X'X)^{-1}X'y$$

where

$$X = \begin{bmatrix} 1 & 1 & 3 \\ 1 & 0 & 1 \\ 2 & 1 & 1 \\ 3 & 2 & 1 \\ 1 & 2 & 0 \end{bmatrix} \quad \text{and} \quad y = \begin{bmatrix} 24 \\ 26 \\ 20 \\ 20 \\ 23 \end{bmatrix}.$$

Hence

$$(X'X)^{-1} = \begin{bmatrix} 16 & 11 & 9 \\ 11 & 10 & 6 \\ 9 & 6 & 12 \end{bmatrix}^{-1} = \tfrac{1}{90}\begin{bmatrix} 28 & -26 & -8 \\ -26 & 37 & 1 \\ -8 & 1 & 13 \end{bmatrix},$$

$$X'y = \begin{bmatrix} 173 \\ 130 \\ 138 \end{bmatrix},$$

and

$$\hat{b} = \tfrac{1}{90}\begin{bmatrix} 28 & -26 & -8 \\ -26 & 37 & 1 \\ -8 & 1 & 13 \end{bmatrix}\begin{bmatrix} 173 \\ 130 \\ 138 \end{bmatrix} = \begin{bmatrix} 4 \\ 5 \\ 6 \end{bmatrix}.$$

3. THE CASE OF k x-VARIABLES

The procedure for fitting more than three x-variables, k of them say, is a direct extension of the preceding section. Analogous to (5) the model is

$$y_i = b_1 x_{i1} + b_2 x_{i2} + \cdots + b_k x_{ik} + e_i. \tag{11}$$

The vector equation for the n sets of observations is still as in (6),

$$y = Xb + e,$$

where y and e are as defined in (1) and (3); X, similar to (2), is now of order $n \times k$:

$$X = \begin{bmatrix} x_{11} & x_{12} & \cdots & x_{1k} \\ x_{21} & x_{22} & \cdots & x_{2k} \\ \cdot & & & \cdot \\ \cdot & & & \cdot \\ \cdot & & & \cdot \\ x_{n1} & x_{n2} & \cdots & x_{nk} \end{bmatrix}. \tag{12}$$

Equations for estimating

$$b = \begin{bmatrix} b_1 \\ b_2 \\ \cdot \\ \cdot \\ \cdot \\ b_k \end{bmatrix} \tag{13}$$

are exactly as before, namely,

$$X'X\hat{b} = X'y \tag{14}$$

for which the solution is

$$\hat{b} = (X'X)^{-1}X'y \tag{15}$$

just as in (7) and (8). The matrix $X'X$ and the vector $X'y$ are exactly as in (9) and (10) only for k x variables instead of 3:

$$X'X = \begin{bmatrix} \sum x_{i1}^2 & \sum x_{i1}x_{i2} & \cdots & \sum x_{i1}x_{ik} \\ \sum x_{i1}x_{i2} & \sum x_{i2}^2 & \cdots & \sum x_{i2}x_{ik} \\ \cdot & \cdot & & \cdot \\ \cdot & \cdot & & \cdot \\ \cdot & \cdot & & \cdot \\ \sum x_{i1}x_{ik} & \sum x_{i2}x_{ik} & \cdots & \sum x_{ik}^2 \end{bmatrix}, \tag{16}$$

and

$$X'y = \begin{bmatrix} \sum x_{i1}y_i \\ \sum x_{i2}y_i \\ \cdot \\ \cdot \\ \cdot \\ \sum x_{ik}y_i \end{bmatrix}. \tag{17}$$

Basing the analysis on equation (11) implies that when the x_i's are zero the corresponding expected value of y_i is zero also. This is sometimes referred to as the *no-intercept model.* Frequently it is inappropriate and, as indicated at the outset, the more usual formulation is

$$y_i = a + b_1x_{i1} + b_2x_{i2} + \cdots + b_kx_{ik} + e_i$$

in which the expected value of y_i for zero x_i's is a. We return to this in Section 8, after considering the properties of $\hat{b}$ given by (15).

4. THE MATHEMATICAL MODEL

We have just seen that when the equation of the model is

$$y = Xb + e$$

b is estimated as

$$\hat{b} = (X'X)^{-1}X'y. \tag{15}$$

Statistical properties of $\hat{b}$ arise from assumptions made about the elements of e: they are customarily assumed to be a random sample from a distribution having zero mean and variance σ^2. Thus each e_i has zero mean and variance σ^2 and is uncorrelated with every other e_i. Hence, using E to denote expected value, $E(e) = 0$,

$$E(y) = Xb + E(e) = Xb,$$

and

$$y - E(y) = e.$$

Furthermore, the variance-covariance matrix of the e_i terms is, from Section 2.10,

$$V(e) = E[e - E(e)][e' - E(e')] = E(ee') = \sigma^2 I.$$

5. UNBIASEDNESS AND VARIANCES

The estimator $\hat{b}$ is unbiased. For,

$$E(\hat{b}) = E[(X'X)^{-1}X'y] = (X'X)^{-1}X'Xb = b.$$

Similarly, the variance-covariance matrix of $\hat{b}$ is $(X'X)^{-1}\sigma^2$:

$$\begin{aligned} \text{var}(\hat{b}) &= E[\hat{b} - E(\hat{b})][\hat{b}' - E(\hat{b}')] \\ &= (X'X)^{-1}X'E[y - E(y)][y' - E(y')]X(X'X)^{-1}, \text{ from (15)} \\ &= (X'X)^{-1}X'E(ee')X(X'X)^{-1}, \\ &= (X'X)^{-1}X'IX(X'X)^{-1}\sigma^2, \\ &= (X'X)^{-1}\sigma^2. \end{aligned}$$

The inverse matrix used for obtaining $\hat{b}$ therefore also determines the variances and covariances of the elements of $\hat{b}$.

6. PREDICTED y VALUES

The estimator $\hat{b}$ can be used to obtain predicted y values corresponding to the observed y's, namely

$$\hat{y}_i = \hat{b}_1 x_{i1} + \hat{b}_2 x_{i2} + \cdots + \hat{b}_k x_{ik},$$

and in keeping with the vector of observations y is the vector of predicted values $\hat{y} = \{\hat{y}_i\}$ for $i = 1, 2, \ldots, n$, namely

$$\hat{y} = X\hat{b} = X(X'X)^{-1}X'y.$$

The variances of these predicted values are given in the variance-covariance matrix of $\hat{y}$:

$$\text{var}(\hat{y}) = X \,\text{var}(\hat{b})X' = X(X'X)^{-1}X'\sigma^2.$$

7. ESTIMATING THE ERROR VARIANCE

The sum of squares of the deviations of the observed y's from their predicted values is known as the error sum of squares:

$$SSE = \sum_{i=1}^{n} (y_i - \hat{y}_i)^2.$$

It can be expressed as

$$SSE = (y' - \hat{y}')(y - \hat{y}),$$

and on substituting the $\hat{y}$ from the previous section it becomes

$$\begin{aligned} SSE &= y'[I - X(X'X)^{-1}X'][I - X(X'X)^{-1}X']y \\ &= y'[I - X(X'X)^{-1}X']y \\ &= (b'X' + e')[I - X(X'X)^{-1}X'](Xb + e), \end{aligned}$$

which simplifies to

$$SSE = e'[I - X(X'X)^{-1}X']e.$$

Because $E(e) = 0$ and $\text{var}\,(e) = \sigma^2 I$, and because $[I - X(X'X)^{-1}X']$ is idempotent with rank $n - k$, Theorem 1 of Section 8.3 applies and the expected value of SSE is

$$E(SSE) = (n - k)\sigma^2.$$

Hence

$$\hat{\sigma}^2 = \frac{SSE}{n - k}$$

is an unbiased estimator of σ^2.

8. DEVIATIONS FROM MEANS

a. Estimators

Consider the intercept model mentioned at the end of Section 3:

$$y_i = a + b_1 x_{i1} + b_2 x_{i2} + \cdots + b_k x_{ik} + e_i. \tag{18}$$

This is the model for fitting k x-variables when the expected value of y_i for zero x_i's is a, a value sometimes referred to as the *intercept*. Although in this model the number of x-variables is k, the estimation procedure can be developed as if it were the one already discussed (having no intercept) but with $k + 1$ such variables.

First note that (18) is identical to

$$\begin{aligned} y_i = a + b_1\bar{x}_1 + b_2\bar{x}_2 + \cdots + b_k\bar{x}_k \\ + b_1(x_{i1} - \bar{x}_1) + b_2(x_{i2} - \bar{x}_2) + \cdots + b_k(x_{ik} - \bar{x}_k) + e_i, \end{aligned}$$

where $\bar{x}_1, \bar{x}_2, \ldots, \bar{x}_k$ are means of the x observations, for example, $\bar{x}_1 = \left(\sum_{i=1}^{n} x_{i1}\right)\Big/ n$. Let us now write

$$b_0 = a + b_1\bar{x}_1 + b_2\bar{x}_2 + \cdots + b_k\bar{x}_k \tag{19}$$

and

$$\begin{aligned} d_{i1} &= x_{i1} - \bar{x}_1 \\ d_{i2} &= x_{i2} - \bar{x}_2 \\ &\;\;\vdots \\ d_{ik} &= x_{ik} - \bar{x}_k, \end{aligned} \tag{20}$$

the d's being deviations of the individual observations from their respective means. The expression for y_i then becomes

$$y_i = b_0 + b_1 d_{i1} + b_2 d_{i2} + \ldots + b_k d_{ik} + e_i.$$

Suppose we now introduce a dummy variable x_0 whose every value x_{i0} is unity. In this way

$$y_i = b_0 x_{i0} + b_1 d_{i1} + b_2 d_{i2} + \cdots + b_k d_{ik} + e_i$$

with b_0 as defined in (19) and $x_{i0} = 1$ for $i = 1, 2, \ldots, n$.

This form of y_i is exactly the same as that of equation (11), except instead of involving k variables it involves $k + 1$, namely x_0 and the k d's. Corresponding to b defined in (13) we now have

$$b^* = \begin{bmatrix} b_0 \\ b_1 \\ b_2 \\ \cdot \\ \cdot \\ \cdot \\ b_k \end{bmatrix} = \begin{bmatrix} b_0 \\ b \end{bmatrix},$$

and analogous to (6) the model is $y = Xb^* + e$, where y and e are the same vectors as before and, corresponding to (12), X is

$$X = \begin{bmatrix} x_{10} & d_{11} & d_{12} & \dots & d_{1k} \\ x_{20} & d_{21} & d_{22} & \dots & d_{2k} \\ \cdot & \cdot & \cdot & & \cdot \\ \cdot & \cdot & \cdot & & \cdot \\ \cdot & \cdot & \cdot & & \cdot \\ x_{n0} & d_{n1} & d_{n2} & \dots & d_{nk} \end{bmatrix}. \tag{21}$$

Therefore, just as in (14), the estimator of b^* is derived from

$$X'X\hat{b}^* = X'y. \tag{22}$$

Substituting (21) into (22) and using

$$\hat{b}^* = \begin{bmatrix} \hat{b}_0 \\ \hat{b} \end{bmatrix}$$

it will be found that, because of the nature of the dummy variable x_0 and the deviations defined in (20), equation (22) becomes

$$\begin{bmatrix} n & 0 \\ 0 & S \end{bmatrix} \begin{bmatrix} \hat{b}_0 \\ \hat{b} \end{bmatrix} = \begin{bmatrix} n\bar{y} \\ u \end{bmatrix},$$

where

$$S = \begin{bmatrix} \sum d_{i1}^2 & \sum d_{i1}d_{i2} & \dots & \sum d_{i1}d_{ik} \\ \sum d_{i1}d_{i2} & \sum d_{i2}^2 & \dots & \sum d_{i2}d_{ik} \\ \cdot & \cdot & & \cdot \\ \cdot & \cdot & & \cdot \\ \cdot & \cdot & & \cdot \\ \sum d_{i1}d_{ik} & \sum d_{i2}d_{ik} & \dots & \sum d_{ik}^2 \end{bmatrix}. \tag{23}$$

If, similar to (20), we define d_{i0} as

$$d_{i0} = y_i - \bar{y},$$

for the deviations of the y observations from their mean, $\bar{y} = \left(\sum_{i=1}^{n} y_i\right)\Big/n$, then the vector u just used is

$$u = \begin{bmatrix} \sum d_{i1}d_{i0} \\ \sum d_{i2}d_{i0} \\ \cdot \\ \cdot \\ \cdot \\ \sum d_{ik}d_{i0} \end{bmatrix}, \tag{24}$$

all summations being for $i = 1, 2, \ldots, n$. With these values the estimates $\hat{b}_0$ and $\hat{b}$ are then given by

$$\begin{bmatrix} \hat{b}_0 \\ \hat{b} \end{bmatrix} = \begin{bmatrix} n & 0 \\ 0 & S \end{bmatrix}^{-1} \begin{bmatrix} n\bar{y} \\ u \end{bmatrix} = \begin{bmatrix} 1/n & 0 \\ 0 & S^{-1} \end{bmatrix} \begin{bmatrix} n\bar{y} \\ u \end{bmatrix}$$

and so

$$\hat{b} = S^{-1}u \tag{25}$$

and

$$\hat{b}_0 = \bar{y}.$$

From the latter, based on (19), we get

$$\hat{a} = \bar{y} - \hat{b}_1\bar{x}_1 - \hat{b}_2\bar{x}_2 - \cdots - \hat{b}_k\bar{x}_k. \tag{26}$$

Equations (25) and (26) give the estimators of the parameters b_1, $b_2, \ldots, b_k$ and a used in the equation of the intercept model (18). And in (25) we see that the form of $\hat{b}$ is exactly like that in (15) except that S and u are used instead of $X'X$ and $X'y$. Furthermore, from (23) and (24) it is at once apparent that S and u are exactly the same form as $X'X$ and $X'y$ given in (16) and (17) respectively, except that they are in terms of deviations of the observations from their means whereas $X'X$ and $X'y$ are in terms of the observations themselves. But with this difference the expression for $\hat{b}$ given in (25) has the same form as that given in (15). Note further, that because of the definitions given in (21), the elements of S and u are the familiar corrected sums of squares and products of the x's and y's: for example,

$$\sum d_{i1}^2 = \sum(x_{i1} - \bar{x}_1)^2 = \sum x_{i1}^2 - n\bar{x}_1^2,$$

$$\sum d_{i1}d_{i2} = \sum(x_{i1} - \bar{x}_1)(x_{i2} - \bar{x}_2) = \sum x_{i1}x_{i2} - n\bar{x}_1\bar{x}_2, \tag{27}$$

and

$$\sum d_{i1}d_{i0} = \sum(x_{i1} - \bar{x}_1)(y_i - \bar{y}) = \sum x_{i1}y_i - n\bar{x}_1\bar{y}.$$

Thus $\hat{b} = S^{-1}u$ is the vector of corrected sums of products of the x's with the y's, pre-multiplied by the inverse of the matrix of corrected sums of squares and products of the x's.

Computing $\hat{b}$ from (25) and $\hat{a}$ from (26) is advantageous for several reasons. First, the elements of S, when divided by $n - 1$, give variance and covariance estimates of the x's; and likewise dividing the elements of u by $n - 1$ gives covariance estimates of the x's with y. Correlations are therefore readily obtained as a by-product. Secondly, the elements of S and u, being corrected sums of squares and cross-products, are smaller in magnitude than those of $X'X$ and $X'y$, and this is sometimes an advantage in computing.

Example. Let us estimate the regression

$$E(y) = a + b_1x_1 + b_2x_2$$

from the six sets of observations:

i	y_i	x_{i1}	x_{i2}
1	10	1	0
2	17	4	6
3	13	2	4
4	14	2	3
5	12	1	1
6	15	3	5

The basic calculations are as follows:

$$\begin{aligned} \textstyle\sum y_i &= 81, & \textstyle\sum x_{i1} &= 13, & \textstyle\sum x_{i2} &= 19,\\ \textstyle\sum y_i^2 &= 1623, & \textstyle\sum x_{i1}^2 &= 35, & \textstyle\sum x_{i2}^2 &= 87,\\ \textstyle\sum y_i x_{i1} &= 189, & \textstyle\sum x_{i1}x_{i2} &= 54, & \textstyle\sum y_i x_{i2} &= 283. \end{aligned}$$

From (23) and (27)

$$S = \begin{bmatrix} 35 - 13^2/6 & 54 - 13(19)/6 \\ 54 - 13(19)/6 & 87 - 19^2/6 \end{bmatrix} = \begin{bmatrix} 41/6 & 77/6 \\ 77/6 & 161/6 \end{bmatrix}$$

and hence

$$S^{-1} = \tfrac{1}{112}\begin{bmatrix} 161 & -77 \\ -77 & 41 \end{bmatrix};$$

and from (24) and (27)

$$u = \begin{bmatrix} 189 - 13(81)/6 \\ 283 - 19(81)/6 \end{bmatrix} = \begin{bmatrix} 81/6 \\ 159/6 \end{bmatrix}.$$

Therefore in (25)

$$\hat{b} = \begin{bmatrix} \hat{b}_1 \\ \hat{b}_2 \end{bmatrix} = \tfrac{1}{112}\begin{bmatrix} 161 & -77 \\ -77 & 41 \end{bmatrix}\begin{bmatrix} 81/6 \\ 159/6 \end{bmatrix} = \begin{bmatrix} 133/112 \\ 47/112 \end{bmatrix}.$$

Using this in (26) gives

$$\hat{a} = 81/6 - [133/112)(13/6) + (47/112)(19/6)] = 1075/112.$$

This example is a simple case of estimating a regression using just two x-variables and six sets of observations. But the principles are no different for situations considerably larger. No matter how many variables there are, nor how numerous the observations, S and u are calculated as in (23) and (24), $\hat{b}$ is obtained from (25) and $\hat{a}$ from (26). The technique of fitting a regression is therefore reduced to a series of simple matrix expressions, whose calculation on any occasion can readily be achieved on modern computing equipment. And this, of course, is just what is being done. Computer programs for fitting regressions now abound, and it is probably safe to say that almost all computer establishments have programs available for carrying out the necessary calculations. As can be appreciated, these calculations will be tedious on a desk calculator should the number of variables, k, be at all large, for then S^{-1} demands inverting a matrix of order k. On high-speed computers, however, this may take only a minute or two, even for k as large as 100. Understanding the mechanics of regression in terms of matrices therefore seems very advantageous; for by their succinctness they enable one small series of equations, (21) through (27), to be used for all situations, and these same equations are highly suitable for speedy processing on modern computers. The same equations are also well suited to obtaining the variances discussed in Sections 5, 6 and 7.

b. Variances

From Section 5 the variance of $\hat{b}^*$ is

$$\text{var}(\hat{b}^*) = (X'X)^{-1}\sigma^2$$

with X as in (21), and hence

$$\begin{bmatrix} \text{var}(\hat{b}_0) & \text{cov}(\hat{b}_0, \hat{b}') \\ \text{cov}(\hat{b}_0, \hat{b}) & \text{var}(\hat{b}) \end{bmatrix} = \begin{bmatrix} 1/n & 0 \\ 0 & S^{-1} \end{bmatrix}\sigma^2.$$

Thus

$$\text{var}(\hat{b}) = S^{-1}\sigma^2,$$
$$\text{var}(\hat{b}_0) = \sigma^2/n$$

and

$$\text{cov}(\hat{b}_0, \hat{b}_i) = 0, \qquad \text{for} \quad i = 1, 2, \ldots, n.$$

Recalling that $\hat{b}_0 = \bar{y}$ we therefore find from (26) that for the vector of means of the x-observations

$$m' = [\bar{x}_1 \quad \bar{x}_2 \quad \ldots \quad \bar{x}_k],$$

we have

$$\hat{a} = \hat{b}_0 - \hat{b}'m$$

and hence, by the results just derived,

$$\text{var}(\hat{a}) = \text{var}(\hat{b}_0) + \text{var}(\hat{b}'m)$$
$$= (1/n + m'S^{-1}m)\sigma^2$$

and

$$\text{cov}(\hat{a}, \hat{b}) = -S^{-1}m\sigma^2.$$

Thus the preferred computing procedure for obtaining $\hat{b}$ also yields $\text{var}(\hat{b})$ very easily, with $\text{var}(\hat{a})$ and $\text{cov}(\hat{a}, \hat{b})$ being derived from using m, the vector of means of the observed x's.

Example (*continued*). From previous results

$$\text{var}(\hat{b}) = \begin{bmatrix} \text{var}(\hat{b}_1) & \text{cov}(\hat{b}_1, \hat{b}_2) \\ \text{cov}(\hat{b}_1, \hat{b}_2) & \text{var}(\hat{b}_2) \end{bmatrix} = S^{-1}\sigma^2 = \tfrac{1}{112}\begin{bmatrix} 161 & -77 \\ -77 & 41 \end{bmatrix}\sigma^2,$$

and with $m' = (13/6 \quad 19/6)$, the variance of $\hat{a}$ is

$$\text{var}(\hat{a}) = \left\{\frac{1}{6} + (13/6 \quad 19/6)\frac{1}{112}\begin{bmatrix} 161 & -77 \\ -77 & 41 \end{bmatrix}\begin{bmatrix} 13/6 \\ 19/6 \end{bmatrix}\right\}\sigma^2$$

$$= \left\{\frac{1}{6} + \frac{13^2(161) - 2(13)(19)(77) + 19^2(41)}{36(112)}\right\}\sigma^2 = \frac{129\sigma^2}{112},$$

and

$$\text{cov}(\hat{a}, \hat{b}) = \begin{bmatrix} \text{cov}(\hat{a}, \hat{b}_1) \\ \text{cov}(\hat{a}, \hat{b}_2) \end{bmatrix} = -S^{-1}m\sigma^2 = \begin{bmatrix} -105/112 \\ 37/112 \end{bmatrix}\sigma^2.$$

The reader might confirm for himself that the equations $X'X\hat{b}^* = X'y$ in this example are

$$\begin{bmatrix} 6 & 13 & 19 \\ 13 & 35 & 54 \\ 19 & 54 & 87 \end{bmatrix}\begin{bmatrix} \hat{a} \\ \hat{b}_1 \\ \hat{b}_2 \end{bmatrix} = \begin{bmatrix} 81 \\ 189 \\ 283 \end{bmatrix}$$

and that the variance of $\hat{b}^*$ is

$$\text{var}(\hat{b}^*) = \begin{bmatrix} 6 & 13 & 19 \\ 13 & 35 & 54 \\ 19 & 54 & 87 \end{bmatrix}^{-1}\sigma^2,$$

namely,

$$\begin{bmatrix} \text{var}(\hat{a}) & \text{cov}(\hat{a}, \hat{b}_1) & \text{cov}(\hat{a}, \hat{b}_2) \\ \text{cov}(\hat{a}, \hat{b}_1) & \text{var}(\hat{b}_1) & \text{cov}(\hat{b}_1, \hat{b}_2) \\ \text{cov}(\hat{a}, \hat{b}_2) & \text{cov}(\hat{b}_1, \hat{b}_2) & \text{var}(\hat{b}_2) \end{bmatrix} = \tfrac{1}{112}\begin{bmatrix} 129 & -105 & 37 \\ -105 & 161 & -77 \\ 37 & -77 & 41 \end{bmatrix}\sigma^2$$

as previously obtained.

c. Error variance

In the no-intercept model of equation (11) we had the vector of predicted y's as $\hat{y} = X\hat{b}$, and $y - \hat{y}$ thus equalled $y - X\hat{b}$. In terms of the intercept model, equation (18), we now have

$$\hat{y}_i = \hat{a} + \hat{b}_1 x_{i1} + \cdots + \hat{b}_k x_{ik}$$

and on substituting for $\hat{a}$ from (26) this is

$$\hat{y}_i = \bar{y} + \hat{b}_1 d_{i1} + \cdots + \hat{b}_k d_{ik}$$

Using D for the matrix of deviations $d_{ij} = x_{ij} - \bar{x}_i$ and c for the vector of deviations $d_{i0} = y_i - \bar{y}$, it may be noted that $D'D = S$ and $D'c = u$. The vector of differences $y - \hat{y}$ is accordingly

$$\begin{aligned} y - \hat{y} &= c - D\hat{b} \\ &= c - DS^{-1}u \\ &= c - DS^{-1}D'c \\ &= (I - DS^{-1}D')c. \end{aligned}$$

Hence the error sum of squares is

$$\begin{aligned} SSE &= (y' - \hat{y}')(y - \hat{y}) \\ &= c'(I - DS^{-1}D')(I - DS^{-1}D')c \\ &= c'(I - DS^{-1}D')c \\ &= c'c - u'S^{-1}u \\ &= c'c - \hat{b}'u. \end{aligned}$$

This is a convenient form for computing SSE: by the definition of c, $c'c$ is the corrected sum of squares of the y observations; $\hat{b}$ is the vector of estimated regression coefficients corresponding to the k x-variables; and u is the vector of right-hand sides of the equations from which $\hat{b}$ is obtained, namely $S\hat{b} = u$.

The unbiased estimator of σ^2 is derived from SSE exactly as at the end of Section 7, only in this context we recall that the intercept model being considered here is equivalent to the no-intercept model with $k + 1$ variables (x_0 and the k d's) rather than k. Hence

$$E(SSE) = [n - (k + 1)]\sigma^2,$$

and so

$$\hat{\sigma}^2 = \frac{SSE}{n - k - 1} = \frac{c'c - \hat{b}'u}{n - k - 1}$$

is an unbiased estimator of the error variance σ^2.

Example (*continued*).

$$\begin{aligned} SSE &= c'c - \hat{b}'u \\ &= \sum y_i^2 - n\bar{y}^2 - \hat{b}'u \\ &= 1123 - 81^2/6 - [133/112 \quad 47/112]\begin{bmatrix} 81/6 \\ 159/6 \end{bmatrix} = 263/112, \end{aligned}$$

and

$$\hat{\sigma}^2 = \frac{263/112}{6 - 2 - 1} = 263/336.$$

d. Variances of predicted *y*'s

The matrix counterpart of

$$\begin{aligned}\hat{y}_i &= \hat{a} + \hat{b}_1 x_{i1} + \cdots + \hat{b}_k x_{ik} \\ &= \bar{y} + \hat{b}_1 d_{i1} + \cdots + \hat{b}_k d_{ik}\end{aligned}$$

is

$$\hat{y} = \bar{y} J_{n \times 1} + D\hat{b}$$

where $J_{n \times 1}$ is a matrix of n rows and 1 column, i.e. a column vector, with all elements equal to unity (Section 8.2). Thus

$$\begin{aligned}\text{var}(\hat{y}) &= (\sigma^2/n) J_{n \times n} + D \,\text{var}(\hat{b}) D' \\ &= (\sigma^2/n) J_{n \times n} + D S^{-1} D' \sigma^2.\end{aligned}$$

9. ANALYSIS OF VARIANCE

The sum of squares of the observed *y*'s about their mean is

$$\sum_{i=1}^{n} (y_i - \bar{y})^2 = SST$$

say, and the sum of squares of the deviations of the observed *y*'s from their predicted values is $\sum_{i=1}^{n} (y_i - \hat{y}_i)^2 = SSE$. The difference between these two, $SSR = SST - SSE$, represents that portion of SST attributable to having fitted the regression, and is usually referred to as the sum of squares due to regression. Splitting SST into two portions in this fashion is the underlying process of the analysis of variance technique and can be summarized in the familiar analysis of variance table:

TABLE 1. ANALYSIS OF VARIANCE

Source of Variation	Degrees of Freedom	Sums of Squares
Regression on k variables	k	$SSR = SST - SSE$
Error	$n - k - 1$	$SSE = \Sigma(y_i - \hat{y}_i)^2$
Total	$n - 1$	$SST = \Sigma y_i^2 - n\bar{y}^2$

We have just seen that $SSE = c'c - \hat{b}'u$ and by definition of c

$$SST = \sum y_i^2 - n\bar{y}^2 = c'c.$$

Therefore

$$SSR = SST - SSE = \hat{b}'u,$$

which, incidentally, can also be written as

$$SSR = u'S^{-1}u. \tag{28}$$

The easiest procedure for computing the analysis of variance table after obtaining the estimator $\hat{b}$ is, therefore, to calculate $SST = \sum y_i^2 - n\bar{y}^2$ (the customary corrected sum of squares of the y's) and then $SSR = \hat{b}'u$, with SSE being the difference. In this way the analysis of variance table is obtained as in Table 2.

TABLE 2. ANALYSIS OF VARIANCE

Source of Variation	Degrees of Freedom	Sums of Squares
Regression	k	$SSR = \hat{b}'u$
Error	$n - k - 1$	$SSE = SST - SSR$
Total	$n - 1$	$SST = \Sigma y_i^2 - n\bar{y}^2$

Example (*continued*).

$$SST = 1123 - 81^2/6 = 177/6$$

$$SSR = (133/112 \quad 47/112)\begin{bmatrix} 81/6 \\ 159/6 \end{bmatrix} = 3041/112.$$

Hence

$SSE = 177/6 - 3041/112 = 263/112$, as before, and the analysis of variance is that shown in Table 3.

TABLE 3. ANALYSIS OF VARIANCE

Source of Variation	d.f.	Sums of Squares
Regression	2	3041/112
Error	3	263/112
Total	5	177/6

10. MULTIPLE CORRELATION

A measure of the goodness of fit of the regression is the multiple correlation coefficient, estimated as the product-moment correlation between the predicted $\hat{y}_i$'s and the observed y_i's. Denoted by R it can be calculated as

$$R^2 = \frac{[\sum(\hat{y}_i - \hat{\bar{y}})(y_i - \bar{y})]^2}{\sum(\hat{y}_i - \hat{\bar{y}})^2 \sum(y_i - \bar{y})^2}$$

$$= \frac{(\hat{c}'c)^2}{(\hat{c}'\hat{c})(c'c)}$$

where $\hat{c}$ is the vector of deviations of the predicted y's from their mean,

$$\hat{c} = \{\hat{d}_{i0}\}$$
$$= \{\hat{y}_i - \hat{\bar{y}}\} \quad \text{for} \quad i = 1, 2, \ldots, n.$$

As shown earlier,

$$\hat{y}_i = \bar{y} + \hat{b}_1 d_{i1} + \hat{b}_2 d_{i2} + \ldots + \hat{b}_k d_{ik},$$

so that the mean of the $\hat{y}_i$'s, namely $\hat{\bar{y}}$, is $\bar{y}$. Therefore

$$\hat{c} = \{\hat{y}_i - \bar{y}\} = D\hat{b}.$$

Thus the terms in R^2 are

$$\hat{c}'\hat{c} = \hat{b}'D'D\hat{b} = \hat{b}'SS^{-1}u = \hat{b}'u = SSR$$

and

$$\hat{c}'c = \hat{b}'D'c = \hat{b}'u = SSR,$$

which gives

$$R^2 = \frac{(SSR)^2}{SSR(SST)} = \frac{SSR}{SST}.$$

Hence the square of the multiple regression coefficient is directly obtainable from the analysis of variance table as the ratio of the regression sum of squares to the total sum of squares.

Example (*continued*). As seen in the analysis of variance table

$$SSR = 3041/112 \quad \text{and} \quad SST = 177/6.$$

Therefore $R^2 = 3041(6)/112(177)$ giving $R = 0.96$.

11. TESTS OF SIGNIFICANCE

Up to now the only properties implied for the e_i terms of the model have been that they are independently distributed with zero means and variance σ^2. If, in addition, it is assumed that they are normally distributed a significance test can be made of the goodness of fit of the regression. For then SSE/σ^2 has a χ^2 distribution with $n - k - 1$ degrees of freedom, and, under the null hypothesis (i.e., on the presumption that all the b_i's are zero), SSR/σ^2 also has a χ^2 distribution, with k degrees of freedom, independently of SSE/σ^2. Hence

$$F = \frac{SSR}{k} \bigg/ \frac{SSE}{n - k - 1} = \frac{(n - k - 1)SSR}{k(SST - SSR)}$$

has an F-distribution with k and $n - k - 1$ degrees of freedom, and computed values of this ratio may be compared with tabulated values to test the significance of the regression on $x_1, x_2, \ldots, x_k$. Note from the expression for R^2 that

$$F = \frac{(n - k - 1)R^2}{k(1 - R^2)}.$$

Significance tests are also available for testing the significance of fitting just *some* of the x's from among $x_1, x_2, \ldots, x_k$.

Illustration. Young et al. (1962), in predicting specific gravity of young girls from weight and skin-fold measurements, consider skin-fold measurements taken at twelve different places on the body. Regressions are then obtained of specific gravity on weight and the twelve skin folds, on weight and just some of the skin-fold measurements, and on skin-fold measurements alone. Comparing the regressions leads to significance tests to see if, for example, weight contributed significantly to the regression, or if certain of the skin-fold measurements contributed significantly. This procedure is a standard part of multiple regression analysis.

Suppose we are dealing with five x variables, x_1, x_2, x_3, x_4 and x_5. First of all, a regression could be fitted to all five of them, obtaining SSR_5 as the sum of squares due to regression and using it in the manner of Table 2 to derive the analysis of variance:

TABLE 4. ANALYSIS OF VARIANCE

Source of Variation	Degrees of Freedom	Sums of Squares
Regression on x_1, x_2, x_3, x_4 and x_5	5	SSR_5
Error	$n - 6$	$SST - SSR_5$
Total	$n - 1$	SST

Suppose now that the five x variates fall naturally into two groups; for example, in a study of crop growth x_1, x_2 and x_3 might be seedling measurements at the time of transplanting (e.g., age, leaf development and height of transplant) and x_4 and x_5 might relate to climate (e.g., temperature and rainfall). We might then wish to ascertain whether or not the two climatic variables make a significant contribution to the regression. If they do not, the regression would be based solely on the seedling measurements x_1, x_2 and x_3, and for just these three variables a regression could be estimated and the corresponding analysis of variance table obtained.

TABLE 5. ANALYSIS OF VARIANCE

Source of Variation	Degrees of Freedom	Sums of Squares
Regression on x_1, x_2 and x_3	3	SSR_3
Error	$n - 4$	$SST - SSR_3$
Total	$n - 1$	SST

Now SSR_5 of Table 4 will always be larger than SSR_3 of Table 5, but the point of interest is "How much larger?"—or more particularly, is SSR_5 significantly larger than SSR_3? If it is, then since the difference between them is a consequence of fitting x_4 and x_5 as well as x_1, x_2 and x_3, we would say that x_4 and x_5 do contribute significantly to the regression involving all five variables. The test is set up by splitting SSR_5 into two parts, one being SSR_3 of Table 5 with three degrees of freedom and the other being $SSR_5 - SSR_3$ with two degrees of freedom. Thus Table 4 becomes

TABLE 6. ANALYSIS OF VARIANCE

Source of Variation	Degrees of Freedom	Sums of Squares
Regression on x_1, x_2 and x_3	3	SSR_3
Regression on x_4 and x_5 over and above that on x_1, x_2 and x_3	2	$SSR_5 - SSR_3$
Error	$n - 6$	$SST - SSR_5$
Total	$n - 1$	SST

Again the sums of squares are statistically independent and, under the null hypothesis that x_4 and x_5 contribute nothing to the regression, the second and third of them (when divided by σ^2) are distributed as χ^2.

Hence an F-test can be made, comparing

$$F = \frac{(n - 6)(SSR_5 - SSR_3)}{2(SST - SSR_5)}$$

with tabulated values of the F-distribution with 2 and $(n - 6)$ degrees of freedom. This provides a test of the significance of the contribution of x_4 and x_5 to the regression involving all five variables. If R_5 is the multiple correlation estimate when fitting all five variables and R_3 when fitting only three of them, the computed F can be written as

$$F = \frac{(n - 6)(R_5^2 - R_3^2)}{2(1 - R_5^2)}.$$

Generalizations and extensions of this procedure are clearly evident. For testing whether m variables from among $k + m$ variables contribute to the regression the F-value to be calculated is

$$F = \frac{(n - k - m - 1)(SSR_{k+m} - SSR_k)}{m(SST - SSR_{k+m})}$$

having m and $n - k - m - 1$ degrees of freedom. A useful computing procedure for obtaining the difference $SSR_{k+m} - SSR_k$ can be derived as follows. Suppose the equations for estimating all $k + m$ regression coefficients are

$$\begin{bmatrix} S_k & S_{km} \\ S_{mk} & S_m \end{bmatrix} \begin{bmatrix} \tilde{b}_k \\ \tilde{b}_m \end{bmatrix} = \begin{bmatrix} u_k \\ u_m \end{bmatrix}, \tag{29}$$

where $\tilde{b}_k$ and $\tilde{b}_m$ denote the estimators of b_k and b_m when estimated simultaneously—as distinct from $\hat{b}_k = S_k^{-1}u_k$ and $\hat{b}_m = S_m^{-1}u_m$ when estimating them separately. In this formulation S_k (and S_m) are the matrices of corrected sums of squares and products of the k (and the m) x-variables respectively, and S_{km} is the matrix of corrected sums of products between the two sets, with $S_{mk} = S'_{km}$. Then, from (28),

$$SSR_{k+m} = [u'_k \quad u'_m] \begin{bmatrix} S_k & S_{km} \\ S_{mk} & S_m \end{bmatrix}^{-1} \begin{bmatrix} u_k \\ u_m \end{bmatrix}$$

and if

$$\begin{bmatrix} S_k & S_{km} \\ S_{mk} & S_m \end{bmatrix}^{-1} = \begin{bmatrix} T_k & T_{km} \\ T_{mk} & T_m \end{bmatrix}, \tag{30}$$

$$SSR_{k+m} = u'_k T_k u_k + 2u'_k T_{km} u_m + u'_m T_m u_m. \tag{31}$$

Also,

$$SSR_k = u'_k S_k^{-1} u_k. \tag{32}$$

Now from (30) it can be shown (Section 8.7) that

$$\begin{aligned} T_k &= S_k^{-1} - S_k^{-1} S_{km} T_{mk}, \\ T_{km} &= -S_k^{-1} S_{km} T_m \end{aligned}$$

and

$$T_m = (S_m - S_{mk}S_k^{-1}S_{km})^{-1}.$$

If these results are used in (31) and (32) following substitution for u_k and u_m from (29), the difference between (31) and (32) reduces to

$$SSR_{k+m} - SSR_k = \tilde{b}_m' T_m^{-1} \tilde{b}_m. \tag{33}$$

As set out in (31), SSR_{k+m} requires inverting a matrix of order $k + m$; and SSR_k of (32) demands inversion of a matrix of order k. But from (33) this can be derived by inverting T_m, a matrix of order m. Hence for large k and small m, (33) is preferable to (32) as a computing procedure for obtaining SSR_k.

12. FITTING VARIABLES ONE AT A TIME

In the work by Young et al. mentioned earlier a series of estimated regressions are reported, of the form

$$y = 1.06 - .0006x_1,$$
$$y = 1.09 - .0004x_1 - .0006x_2,$$
$$y = 1.08 - .0003x_1 - .0003x_2 - .0003x_3,$$

and so on. To obtain results such as these by the methods already described would require inverting matrices of orders 1, 2, 3 and so on. In each case, however, the matrix to be inverted would be the same as the immediately previous one with but a single row and column added. Thus if $S_k\hat{b}_k = u_k$ is the set of equations to be solved for fitting k variables $x_1, x_2, \ldots, x_k$, the S matrix for fitting those same k plus one more, x_{k+1}, will be

$$S_{k+1} = \begin{bmatrix} S_k & v_k \\ v_k' & \lambda \end{bmatrix}$$

where v_k corresponds to S_{km} of (29) and is the vector of corrected sums of products of x_{k+1} with $x_1, \ldots, x_k$, and λ corresponds to S_m of (29) and is the corrected sum of squares of x_{k+1}. Inverting S_{k+1} in its partitioned form (Section 8.7) gives

$$S_{k+1}^{-1} = \begin{bmatrix} S_k^{-1} & 0 \\ 0 & 0 \end{bmatrix} + \frac{1}{\lambda - v_k'S_k^{-1}v_k}\begin{bmatrix} -S_k^{-1}v_k \\ 1 \end{bmatrix}[-v_k'S_k^{-1} \quad 1]. \tag{34}$$

From this we can obtain the regression estimates and regression sum of squares for fitting $k + 1$ variables in terms of those for fitting k variables.

We have already defined S_k and S_{k+1}, matrices of (corrected) sums of squares and products of k and $k + 1$ x-variables. Corresponding to these let u_k and u_{k+1} be the vectors of corrected sums of products of the k and the $k+1$ x-variables with y; let $\hat{b}_k$ and $\hat{b}_{k+1}$ be the respective vectors of estimated regression coefficients, and SSR_k and SSR_{k+1} the corresponding sums of squares due to regression. Then

$$\hat{b}_k = S_k^{-1}u_k \quad \text{and} \quad SSR_k = u_k'S_k^{-1}u_k,$$

and

$$u_{k+1} = \begin{bmatrix} u_k \\ \mu \end{bmatrix} = \begin{bmatrix} u_k \\ \Sigma d_{i,k+1}d_{i0} \end{bmatrix} \tag{35}$$

where μ is the corrected sum of products of the x_{k+1} observations with the y observations, $\Sigma d_{i,k+1}\, d_{i0}$. Furthermore,

$$\hat{b}_{k+1} = S_{k+1}^{-1}u_{k+1},$$

and substituting (34) and (35) into this leads to

$$\hat{b}_{k+1} = \begin{bmatrix} \hat{b}_k \\ 0 \end{bmatrix} + \left\{\frac{\mu - \hat{b}_k'v_k}{\lambda - v_k'S_k^{-1}v_k}\right\}\begin{bmatrix} -S_k^{-1}v_k \\ 1 \end{bmatrix}. \tag{36}$$

Thus $\hat{b}_{k+1}$ can be obtained from $\hat{b}_k$ by calculating a vector and a scalar [the contents of the curly brackets in (36)] and adding their product to $\begin{bmatrix} \hat{b}_k \\ 0 \end{bmatrix}$. Note here the expression $S_k^{-1}v_k$ that arises both in (34) and (36); by the definition of S_k and v_k it is the vector of regression estimates of x_{k+1} on $x_1, x_2, \ldots, x_k$ and $v_k'S_k^{-1}v_k$ is the corresponding regression sum of squares.

The regression sum of squares SSR_{k+1} corresponding to (36) is now easily derived: for,

$$SSR_{k+1} = u_{k+1}'S_{k+1}^{-1}u_{k+1} = u_{k+1}'\hat{b}_{k+1},$$

and substituting (35) and (36) into this gives

$$SSR_{k+1} = SSR_k + \frac{(\mu - \hat{b}_k'v_k)^2}{\lambda - v_k'S_k^{-1}v_k} \tag{37}$$

which is the form that (33) reduces to when $m = 1$. Immediately we see that SSR_{k+1} exceeds SSR_k, that is to say, adding another variable to the regression increases the sum of squares due to regression. And this is always true, although it is not necessarily a significant increase, of course. It follows from this too that SSR_{k+m} always exceeds SSR_k, as indicated earlier.

The real value of (36) and (37) is, however, that $\hat{b}_{k+1}$ and SSR_{k+1} can be obtained without inverting a matrix, and SSR_{k+1} can be calculated without even calculating the regression coefficients in $\hat{b}_{k+1}$. Thus, before obtaining $\hat{b}_{k+1}$ the significance of including x_{k+1} in the regression can be tested and a decision made as to whether or not to include it. This is very convenient computationally.

Example.
Suppose

$$S_2 = \begin{bmatrix} 45 & 18 \\ 18 & 32 \end{bmatrix} \quad \text{and} \quad u_2 = \begin{bmatrix} 144 \\ 132 \end{bmatrix}.$$

Then

$$\hat{b}_2 = S_2^{-1}u_2 = \tfrac{1}{1116}\begin{bmatrix} 32 & -18 \\ -18 & 45 \end{bmatrix}\begin{bmatrix} 144 \\ 132 \end{bmatrix} = \begin{bmatrix} 2 \\ 3 \end{bmatrix}.$$

Suppose further that

$$S_3 = \begin{bmatrix} S_2 & v_2 \\ v_2' & \lambda \end{bmatrix} = \begin{bmatrix} 45 & 18 & 90 \\ 18 & 32 & 36 \\ 90 & 36 & 204 \end{bmatrix},$$

so defining $v_2 = \begin{bmatrix} 90 \\ 36 \end{bmatrix}$ and $\lambda = 204$; and that $u_3 = \begin{bmatrix} u_2 \\ \mu \end{bmatrix} = \begin{bmatrix} 144 \\ 132 \\ 300 \end{bmatrix}$

with $\mu = 300$. Hence in (36)

$$\tilde{b}_3 = \begin{bmatrix} 2 \\ 3 \\ 0 \end{bmatrix}$$

$$+ \frac{300 - [2 \;\; 3]\begin{bmatrix} 90 \\ 36 \end{bmatrix}}{204 - [90 \;\; 36]\tfrac{1}{1116}\begin{bmatrix} 32 & -18 \\ -18 & 45 \end{bmatrix}\begin{bmatrix} 90 \\ 36 \end{bmatrix}} \begin{bmatrix} -\tfrac{1}{1116}\begin{bmatrix} 32 & -18 \\ -18 & 45 \end{bmatrix}\begin{bmatrix} 90 \\ 36 \end{bmatrix} \\ 1 \end{bmatrix}$$

$$= \begin{bmatrix} 2 \\ 3 \\ 0 \end{bmatrix} + \frac{300 - 288}{204 - 180}\begin{bmatrix} -2 \\ 0 \\ 1 \end{bmatrix} = \begin{bmatrix} 1 \\ 3 \\ \tfrac{1}{2} \end{bmatrix}. \tag{38}$$

It will be found that this is the value of $S_3^{-1}u_3$ obtained directly. From (37) and the values in (38)

$$SSR_3 = SSR_2 + 12^2/24 = SSR_2 + 6.$$

13. INVERTING THE MATRIX OF ALL VARIABLES

Another useful computing technique is based on inverting not S but

$$\begin{bmatrix} S & u \\ u' & SST \end{bmatrix},$$

the matrix of corrected sums of squares and products of the x's and the y's. By (34) the inverse of this matrix is

$$\begin{bmatrix} S & u \\ u' & SST \end{bmatrix}^{-1} = \begin{bmatrix} S^{-1} & 0 \\ 0 & 0 \end{bmatrix} + \frac{1}{SST - u'S^{-1}u} \begin{bmatrix} -S^{-1}u \\ 1 \end{bmatrix} [-u'S^{-1} \quad 1]$$

$$= \begin{bmatrix} S^{-1} & 0 \\ 0 & 0 \end{bmatrix} + (1/SSE) \begin{bmatrix} -\hat{b} \\ 1 \end{bmatrix} [-\hat{b}' \quad 1]$$

$$= \begin{bmatrix} S^{-1} + \hat{b}\hat{b}'/SSE & -\hat{b}/SSE \\ -\hat{b}'/SSE & 1/SSE \end{bmatrix}.$$

Thus by writing

$$\begin{bmatrix} S & u \\ u' & SST \end{bmatrix}^{-1} = \begin{bmatrix} H & g \\ g' & \lambda \end{bmatrix}$$

we obtain the terms of the regression analysis directly:

$$SSE = 1/\lambda,$$
$$\hat{b} = -(1/\lambda)g,$$
$$S^{-1} = H - (1/\lambda)gg'.$$

The analysis of variance is accordingly as given in Table 7.

TABLE 7. ANALYSIS OF VARIANCE

Source of Variation	Degrees of Freedom	Sums of Squares
Regression	k	$SST - 1/\lambda$
Error	$n - k - 1$	$1/\lambda$
Total	$n - 1$	SST

Example (*continued*). In the example introduced in Section 8

$$S = \begin{bmatrix} 41/6 & 77/6 \\ 77/6 & 161/6 \end{bmatrix}, \quad u = \begin{bmatrix} 81/6 \\ 159/6 \end{bmatrix} \quad \text{and} \quad SST = 177/6.$$

Thus

$$\begin{bmatrix} S & u \\ u' & SST \end{bmatrix}^{-1} = \begin{bmatrix} H & g \\ g' & \lambda \end{bmatrix}$$

is

$$\begin{bmatrix} 41/6 & 77/6 & 81/6 \\ 77/6 & 161/6 & 159/6 \\ 81/6 & 159/6 & 177/6 \end{bmatrix}^{-1} = \tfrac{1}{263} \begin{bmatrix} 536 & -125 & -133 \\ -125 & 116 & -47 \\ -133 & -47 & 112 \end{bmatrix}.$$

Therefore

$$SSE = 1/\lambda = 263/112$$

and

$$\hat{b} = -(1/\lambda)g = \begin{bmatrix} 133/112 \\ 47/112 \end{bmatrix}, \text{ as before.}$$

The matrix inverted in this procedure is the matrix of corrected sums of squares and products of all variables involved in the analysis, the y variable and all the x's. Should, therefore, we wish to estimate the regression of y on just some of the x's this procedure can be used directly, applying it to the matrix formed by taking from the above just the rows and columns appropriate to y and the required x's.

Example (*continued*). To fit the regression of y on x_1 we find that

$$\begin{bmatrix} 41/6 & 81/6 \\ 81/6 & 177/6 \end{bmatrix}^{-1} = \frac{1}{116}\begin{bmatrix} 177 & -81 \\ -81 & 41 \end{bmatrix},$$

and so the error sum of squares is 116/41 and the regression coefficient is 81/41.

An additional use for this technique arises in situations where we wish to use one of the x's as the y variable (independent variable) and obtain the regression of it on the other x's and/or the variable initially observed as the y variable. This can be done by re-ordering the rows and columns of the matrix so that the last row and column relate to the variable being used as the independent variable.

Example (*continued*). To obtain the regression of x_2 on x_1 and y we re-order the rows and columns of the matrix to correspond with x_1, y and x_2 and derive

$$\begin{bmatrix} 41/6 & 81/6 & 77/6 \\ 81/6 & 177/6 & 159/6 \\ 77/6 & 159/6 & 161/6 \end{bmatrix}^{-1} = \frac{1}{263}\begin{bmatrix} 536 & -133 & -125 \\ -133 & 112 & -47 \\ -125 & -47 & 116 \end{bmatrix}.$$

Thus the error sum of squares is 263/116, and the regression coefficients are

$$-(263/116)\begin{bmatrix} -125/263 \\ -47/263 \end{bmatrix} = \begin{bmatrix} 125/116 \\ 47/116 \end{bmatrix}.$$

The estimated regression is therefore

$$\hat{x}_2 = \bar{x}_2 + (125/116)(x_1 - \bar{x}_1) + (47/116)(y - \bar{y})$$

and the associated analysis of variance is that shown in Table 8.

TABLE 8. ANALYSIS OF VARIANCE FOR FITTING X_2 ON X_1 AND Y

Term	d.f.	Sums of Squares
Regression	2	17098/696
Error	3	263/116
Total	5	161/6

Computing procedures based on the results of this and the preceding section may not seem particularly useful from the simple examples used here, but the generality of the results is ideally suited to the demands of high-speed computers and the procedures are used extensively in programming regression analysis on modern computers (see Ralston and Wilf, 1959, for example).

14. SUMMARY OF CALCULATIONS

The two previous sections contain a variety of devices that can be used for computing the items of a regression analysis. The more general expressions developed earlier in the chapter are now summarized and listed in terms of computing formulae suitable for estimating the regression equation

$$y = a + b_1x_1 + b_2x_2 + \cdots + b_kx_k$$

from n sets of observations, denoting the vector of b's by b.

S = matrix of corrected sums of squares and products of the x's.
u = vector of corrected sums of products of the x's with the y's.
m = vector of means of the x's.
$\bar{y}$ = mean of the y's.
SST = corrected sum of squares of the y's.

$\hat{b} = S^{-1}u$: estimated regression coefficients.

$\hat{a} = \bar{y} - \hat{b}'m$: estimate of the "constant" term.

$SSR = \hat{b}'u = u'S^{-1}u$: regression sum of squares.

$SSE = SST - SSR$: error sum of squares.

$\hat{\sigma}^2 = SSE/(n - k - 1)$: estimated error variance.

$R = \sqrt{SSR/SST}$: multiple correlation coefficient.

$F_{k,n-k-1} = \dfrac{(n - k - 1)R^2}{k(1 - R^2)}$: F-value for F-test.

$\text{var}(\hat{b}) = S^{-1}\sigma^2$: variance matrix of $\hat{b}$.

$\text{var}(\hat{a}) = (1/n + m'S^{-1}m)\sigma^2$: variance of $\hat{a}$.

$\text{cov}(\hat{a}, \hat{b}) = -S^{-1}m\sigma^2$: covariance of $\hat{a}$ with $\hat{b}$.

REFERENCES

Graybill, Franklin A. (1961). *An Introduction to Linear Statistical Models.* Vol. I, McGraw-Hill, New York.

Kempthorne, Oscar (1952). *Design and Analysis of Experiments.* Wiley, New York.

Kendall, M. G., and A. Stuart (1958). *The Advanced Theory of Statistics.* Vol. I, Griffin, London.

Mood, Alexander M., and Franklin A. Graybill (1963). *Introduction to the Theory of Statistics.* Second Edition, McGraw-Hill, New York.

Ralston, Anthony, and Herbert S. Wilf (Editors) (1960). *Mathematical Methods for Digital Computers.* Wiley, New York.

Snedecor, G. W. (1956). *Statistical Methods.* Fifth Edition. Iowa State College Press, Ames, Iowa.

Steel, R. G. D. and James H. Torrie (1960). *Principles and Procedures of Statistics.* McGraw-Hill, New York.

Stone, J. B., G. W. Trimberger, C. R. Henderson, J. T. Reid, K. L. Turk, and J. K. Loosli (1960). Forage intake and efficiency of feed utilization in dairy cattle. *J. Dairy Sci.*, **43**, 1275–1281.

Wallace, L. R. (1956). The intake and utilisation of pasture by grazing dairy cattle. *Proceedings of the Seventh International Grassland Congress*, 134–145.

Young, Charlotte M., M. Elizabeth Kerr Martin, Rosalinda Tensuan, and Joan Blondin (1962). Predicting specific gravity and body fatness in young women. *J. Am. Dietet. Assoc.*, **40**, 102–107.

CHAPTER 10

SOME MATRIX ALGEBRA OF LINEAR STATISTICAL MODELS

The general subject of linear statistical models is a large one and is dealt with in numerous texts, for example, Kempthorne (1952), Anderson (1958), Scheffée (1959), Graybill (1961) and Mood and Graybill (1963). It is a subject of great importance to biologists, indeed to anyone using either analysis of variance techniques, or the method of least squares for estimating what might generally be termed "treatment effects". It is also a subject that is exceedingly amenable to presentation in terms of matrices, and they in turn bring considerable clarity to it. Because of the subject's vastness just its fringe is touched here, in a discussion that is far less complete than that on regression in the preceding chapter. Great reliance is placed on results given there, however, and considerable use is made of the generalized inverse of a matrix discussed in Chapter 6.

For those familiar with linear models it can be said that the main objective of this chapter is to present a simple and general method of deriving estimable functions and their unique, least squares (minimum variance, linear, unbiased) estimators. The method is based entirely on the properties of generalized inverses and is applicable to any situation whatever, whether the data are balanced or unbalanced.

1. GENERAL DESCRIPTION

A simple illustration serves to introduce the general idea of linear models and the matrix notation applicable to them.

Illustration. Federer (1955) reports an analysis of rubber-producing plants called guayule, for which the plant weights were available for 54 plants of three different kinds, 27 of them normal, 15 off-types and 12 aberrants. We will consider just 6 plants for purposes of illustration, 3 normals, 2 off-types and 1 aberrant:

TABLE 1. WEIGHTS OF SIX PLANTS ACCORDING TO TYPE

	Normal	Off-Type	Aberrant
	101	84	32
	105	88	
	94		
Totals	300	172	32

For the entries in this table let y_{ij} denote the weight of the jth plant of the ith class, i taking values 1, 2, 3 for normal, off-type and aberrant respectively, and $j = 1, 2, \ldots, n_i$, where n_i is the number of observations in the ith class. The problem is to estimate the effect of class on weight of plant. To do this we assume that the observation y_{ij} is the sum of three parts

$$y_{ij} = \mu + \alpha_i + e_{ij}, \tag{1}$$

where μ represents the population mean of the weight of plant, α_i is the effect of class i on weight, and e_{ij} is a random error term peculiar to the observation y_{ij}. As in regression it is assumed that the e_{ij}'s are independently distributed with zero mean, i.e., $E(e_{ij}) = 0$. Then $E(y_{ij}) = \mu + \alpha_i$. It is also assumed that each e_{ij} has the same variance, σ^2, so that the variance-covariance matrix of the vector of e-terms is $\sigma^2 I$. This is the model specified, in this case the model of the general *1-way classification*; it is clearly a linear model since it is based on the assumption that y_{ij} consists of the simple sum of its three component parts μ, α_i and e_{ij}.

The problem is to estimate μ and the α_i terms, and also σ^2, the variance of the error terms. We will find that not all of the terms μ and α_i can be estimated satisfactorily, only certain linear functions of them can. At first thought this may seem to be a matter for concern; sometimes it is, but on many occasions it is not because the number of linear functions that can be estimated is large and frequently includes those in which we are interested; e.g., differences between the effects, such as $\alpha_1 - \alpha_2$ for example. Sometimes, however, functions of interest cannot be estimated, because of a paucity of data. On all occasions though, a method is needed for ascertaining which functions can be estimated and which cannot. This is provided in what follows.

To develop the method of estimation we write down the six observations in terms of equation (1) of the model:

$$\begin{aligned}
101 &= y_{11} = \mu + \alpha_1 && + e_{11}\\
105 &= y_{12} = \mu + \alpha_1 && + e_{12}\\
94 &= y_{13} = \mu + \alpha_1 && + e_{13}\\
84 &= y_{21} = \mu \quad + \alpha_2 && + e_{21}\\
88 &= y_{22} = \mu \quad + \alpha_2 && + e_{22}\\
30 &= y_{31} = \mu \quad\quad + \alpha_3 && + e_{31}
\end{aligned}$$

These equations are easily written in matrix form

$$\begin{bmatrix}101\\105\\94\\84\\88\\30\end{bmatrix} = \begin{bmatrix}y_{11}\\y_{12}\\y_{13}\\y_{21}\\y_{22}\\y_{31}\end{bmatrix} = \begin{bmatrix}1&1&0&0\\1&1&0&0\\1&1&0&0\\1&0&1&0\\1&0&1&0\\1&0&0&1\end{bmatrix}\begin{bmatrix}\mu\\\alpha_1\\\alpha_2\\\alpha_3\end{bmatrix} + \begin{bmatrix}e_{11}\\e_{12}\\e_{13}\\e_{21}\\e_{22}\\e_{31}\end{bmatrix}$$

as

$$y = Xb + e \tag{2}$$

where y is the matrix observations, X is the matrix of 0's and 1's, b is the vector of parameters to be estimated,

$$b = \begin{bmatrix}\mu\\\alpha_1\\\alpha_2\\\alpha_3\end{bmatrix},$$

and e is the vector of error terms.

Note that equation (2) is exactly the same form as (6) of Chapter 9. In both cases the vectors y and e are of observations and errors, respectively, and in both cases b is a vector of parameters whose estimates we seek. Also, the properties of e are the same: $E(e) = 0$, so that $E(y) = Xb$, and $E(ee') = \sigma^2 I$. The only difference is in the form of X: in regression it is a matrix of observations on x-variables, whereas in (2) its elements are either 0 or 1, depending on the absence or presence of particular terms of the model in each y_{ij} observation. But, with respect to applying the principle of least squares for estimating the elements of b, there is no difference at all between equation (2) of this chapter and equation (6) of Chapter 9. Consequently we can go directly to equation (14) of Chapter 9 to obtain the least squares estimator of b as $\hat{b}$ defined by

$$X'X\hat{b} = X'y. \tag{3}$$

In regression analysis the solution $\hat{b}$ was obtained from this as

$$\hat{b} = (X'X)^{-1}X'y,$$

$X'X$ being nonsingular and therefore having an inverse. Now, however, $X'X$ has no inverse and we have to use the methods of generalized inverses to obtain a solution.

2. THE NORMAL EQUATIONS

The equations represented by (3) are usually referred to as the *normal equations.* Before discussing their solution let us look briefly at the form they take. First, the vector of parameters, b: it is the vector of all the elements of the model, in this case the elements μ, α_1, α_2 and α_3. And this is so in general; for example, if data can be arranged in rows and columns according to two different classifications the vector b will have as its elements the term μ, the terms representing the row effects, those representing the column effects and those representing the interaction effects between rows and columns. For r rows and c columns it can therefore have as many as $1 + r + c + rc$ elements.

Second, the matrix X: it is often called the *design matrix* because the location of the 0's and 1's throughout its elements represents the occurrence of terms of the model among the observations—and hence of the classifications in which the observations lie. This is particularly evident if one writes X as a 2-way table with the parameters as headings to the columns and the observations as headings to the rows:

TABLE 2. THE MATRIX X AS A 2-WAY TABLE

Observations	Parameters of Model			
	μ	α_1	α_2	α_3
y_{11}	1	1	0	0
y_{12}	1	1	0	0
y_{13}	1	1	0	0
y_{21}	1	0	1	0
y_{22}	1	0	1	0
y_{31}	1	0	0	1

The matrix $X'X$ is easily obtained: it is obviously square and symmetric, and its elements are the inner products of the columns of X with each other. Hence in this instance

$$X'X = \begin{bmatrix} 6 & 3 & 2 & 1 \\ 3 & 3 & 0 & 0 \\ 2 & 0 & 2 & 0 \\ 1 & 0 & 0 & 1 \end{bmatrix}.$$

Finally, the vector $X'y$; it has as elements the inner products of the columns of X with the vector y, and since the only non-zero elements of X are ones the elements of $X'y$ are certain sums of elements of y. In fact they are the sums corresponding to the elements of the vector b. Thus $X'y$ in the example is

$$X'y = \begin{bmatrix} 1 & 1 & 1 & 1 & 1 & 1 \\ 1 & 1 & 1 & 0 & 0 & 0 \\ 0 & 0 & 0 & 1 & 1 & 0 \\ 0 & 0 & 0 & 0 & 0 & 1 \end{bmatrix} \begin{bmatrix} y_{11} \\ y_{12} \\ y_{13} \\ y_{21} \\ y_{22} \\ y_{31} \end{bmatrix}$$

$$= \begin{bmatrix} y_{11} + y_{12} + y_{13} + y_{21} + y_{22} + y_{31} \\ y_{11} + y_{12} + y_{13} \\ y_{21} + y_{22} \\ y_{31} \end{bmatrix} = \begin{bmatrix} y_{..} \\ y_{1.} \\ y_{2.} \\ y_{3.} \end{bmatrix} = \begin{bmatrix} 504 \\ 300 \\ 172 \\ 32 \end{bmatrix}.$$

This is the general nature of $X'y$ in linear models—a vector of totals of the y observations.

Let us now write the normal equations in full: that is to say, $X'X\hat{b} = X'y$ is

$$\begin{bmatrix} 6 & 3 & 2 & 1 \\ 3 & 3 & 0 & 0 \\ 2 & 0 & 2 & 0 \\ 1 & 0 & 0 & 1 \end{bmatrix} \begin{bmatrix} \hat{\mu} \\ \hat{\alpha}_1 \\ \hat{\alpha}_2 \\ \hat{\alpha}_3 \end{bmatrix} = \begin{bmatrix} y_{..} \\ y_{1.} \\ y_{2.} \\ y_{3.} \end{bmatrix} = \begin{bmatrix} 504 \\ 300 \\ 172 \\ 32 \end{bmatrix}. \tag{4}$$

It is instructive to retain the algebraic form of $X'y$ as well as the arithmetic form. By so doing we see that if $X'X$ is written in a 2-way table in the same way that X has been represented earlier, its row headings will be the totals in $X'y$ and its column headings the parameters. Indeed the elements of $X'X$ are the numbers of times that a parameter of the model occurs in a total; for example, μ occurs six times in $y_{..}$ and α_1 occurs three times; likewise α_2 does not occur at all in $y_{1.}$; and so on. Another way of looking at $X'X$ is that its elements are the coefficients of the parameters of the model in the expected values of the totals in $X'y$. In this sense we might write the normal equations as $\widehat{E(X'y)} = X'y$; however, the easiest way of deriving $X'y$ and $X'X$ other than carrying out the matrix products explicitly is to form $X'y$ as the vector of all class and subclass totals of the observations (including the grand total), and to form $X'X$ as the matrix of the number of times that each parameter arises in each total that occurs in $X'y$.

3. SOLVING THE NORMAL EQUATIONS

We seek a solution for $\hat{b}$ to the equations $X'X\hat{b} = X'y$. But notice in the example given in (4) that $X'X$ is singular (its first row equals the sum of its last three rows). This is usually the case; in analysing a linear model of the general nature being discussed here $X'X$ is almost invariably singular. Accordingly, its rank, r say, is less than its order n, and $\hat{b}$ cannot be obtained as $(X'X)^{-1}X'y$. We go at once to the relevant sections of Chapter 6 to find a solution by means of a generalized inverse of $X'X$.

Let G be a generalized inverse of $X'X$. Since $X'X$ is symmetric its canonical reduction under equivalence will be of the form $PX'XP' = \Delta$ (Section 5.8). Hence its generalized inverse as derived in Section 6.4 will be $G = P'\Delta^{-}P$, which is symmetric. And for $H = GX'X$ we also have $HG = GX'XG = G$ and $X'XH = X'XGX'X = X'X$. These results are used extensively in what follows. They are fully discussed in Chapter 6.

Using the definitions of G and H just given, the solution to the normal equations comes immediately from Theorem 4 of Section 6.5*a* as

$$\hat{b} = GX'y + (H - I)z, \tag{5}$$

where z is arbitrary. This means there is no unique solution $\hat{b}$; by the nature of (3), with $X'X$ being singular, this is to be expected; it comes as no surprise to those familiar with linear statistical models.

This lack of uniqueness in the solutions of the normal equations is discussed in many places, usually in terms of making use of the fact that for equations like (3) the addition of what is often called a "convenient restraint" or "obvious restriction" (such as $\alpha_1 + \alpha_2 + \alpha_3 = 0$) does lead to a single solution. But indeed this is convenient only for very particular cases of (3), and for the general case there *are* no "convenient" or "obvious" restrictions leading to unique solutions. Although a great deal has been written about obtaining solutions through adding extra equations ("conditions", "restraints" or "constraints") of this nature—and almost any text on experimental design refers to them—we avoid them here by considering the solution, given in (5), obtained from using the generalized inverse. Properties of $\hat{b}$ derived in this way we now pursue.

4. EXPECTED VALUES AND VARIANCES

It is clear from (5) that the expected value of $\hat{b}$ is

$$\begin{aligned} E(\hat{b}) &= GX'E(y) + (H - I)E(z) \\ &= GX'Xb + (H - I)z \\ &= Hb + (H - I)z, \end{aligned} \tag{6}$$

which is obviously different from b, even if z is taken as a null vector, i.e. $\hat{b}$ is not an unbiased estimator of b. We return to this in Section 6.

Since z is an arbitrary constant it does not affect the variance of $\hat{b}$. Therefore in using (5) and (6) to obtain var($\hat{b}$), z can be treated as a null vector to give

$$\begin{aligned} \text{var}(\hat{b}) &= E[\hat{b} - E(\hat{b})][\hat{b}' - E(\hat{b}')] \\ &= E(GX'y - Hb)(y'XG' - b'H') \\ &= GX'E(y - Xb)(y' - b'X')XG' \\ &= GX'E(ee')XG \\ &= GX'XG\sigma^2 \\ &= G\sigma^2. \end{aligned}$$

This result is exactly analogous to that pertaining to the regression situation when $X'X$ was nonsingular. In that case (Section 9.5) the variance-covariance matrix of $\hat{b}$ was based on the inverse of $X'X$—now it is based on a generalized inverse.

5. ESTIMATING THE ERROR VARIANCE

Corresponding to the vector of observations y, a vector of predicted values $\hat{y} = X\hat{b}$ can be obtained from $\hat{b}$, just as was done in the case of regression analysis (Section 9.6). Following this we can compute the sum of squares of the deviations of the observed y's from their predicted values. This is usually referred to as the residual, or error sum of squares, SSE, and in regression analysis (Section 9.7) it was expressed as

$$SSE = y'y - \hat{b}'X'y,$$

although more easily computed as $c'c - \hat{b}'u$. We now investigate the situation in the case of the linear model. As before,

$$\begin{aligned} SSE &= \sum_i \sum_j (y_{ij} - \hat{y}_{ij})^2 \\ &= (y' - \hat{y}')(y - \hat{y}). \end{aligned}$$

But now,

$$\hat{y} = X\hat{b} = XGX'y + X(H - I)z.$$

Furthermore,

$$\begin{aligned}[X(H - I)]'X(H - I) &= (H' - I)X'X(H - I) \\ &= H'X'XH - X'XH - H'X'X + X'X \\ &= X'XGX'XGX'X - X'XGX'X \\ &\qquad - X'XGX'X + X'X \\ &= X'X - X'X - X'X + X'X \\ &= 0;\end{aligned}$$

i.e., $X(H - I)$ is null. Hence $XH = X$ (see Exercise 14*d* of Chapter 2) and consequently $\hat{y} = XGX'y$. Therefore

$$SSE = y'(I - XGX')(I - XGX')y$$

which, on expansion, becomes

$$SSE = y'(I - XGX')y.$$

Although G is not unique XGX' is (see Exercise 8, Chapter 6), and hence SSE has the same, unique value no matter what generalized inverse of $X'X$ is used. Furthermore, because $X(H - I)$ is null, SSE is equivalent to

$$\begin{aligned}SSE &= y'y - y'X[GX'y + (H - I)z] \\ &= y'y - y'X\hat{b} \\ &= y'y - \hat{b}'X'y.\end{aligned} \tag{7}$$

Hence the error sum of squares is the total uncorrected sum of squares $y'y$ after subtracting from it the sum of products of the elements in $\hat{b}$ each multiplied by the corresponding right-hand side of the equation $X'X\hat{b} = X'y$. This is exactly the same result as in regression analysis; however, in regression $\hat{b}$ was unique, whereas here $\hat{b}$ represents just one of many solutions to the normal equations $X'X\hat{b} = X'y$. And yet (7) still remains true, that for *any* solution $\hat{b}$ the error sum of squares is $y'y - \hat{b}'X'y$. This is also the best form for computing, there being no counterpart to the $c'c - \hat{b}'u$ used in regression.

We now use one of Graybill's two theorems discussed in Section 8.3 to find the expected value of SSE. As just shown, $SSE = y'(I - XGX')y$ and, by substituting $y = Xb + e$, this becomes

$$SSE = (b'X' + e')(I - XGX')(Xb + e),$$

and using $XGX'X = X$ it reduces to

$$SSE = e'(I - XGX')e.$$

Then, just as in Section 9.7, $E(e) = 0$ and $\text{var}(e) = \sigma^2 I$, and $I - XGX'$ is idempotent with rank $n - r$, r being the rank of X, $X'X$ and hence of G also. Therefore, by Theorem 1 of Section 8.3,

$$E(SSE) = (n - r)\sigma^2$$

and so an unbiased estimator of σ^2 is

$$\hat{\sigma}^2 = \frac{SSE}{n - r}.$$

6. ESTIMABLE FUNCTIONS

We saw in equation (6) that $\hat{b}$ is not an unbiased estimator of b. But from it can be derived unique, unbaised estimators of certain linear combinations of the elements of b. This is achieved by using another result obtained in Chapter 6: that certain linear combinations of the elements of the solution $\hat{b}$ have a unique value no matter what solution for $\hat{b}$ is obtained from (5). These combinations (Theorem 7, Section 6.5*d*) are $q'\hat{b}$ where q' is such that $q'H = q'$. Note, though, that *any* q' of the form $q' = w'H$, no matter what the vector w' is, satisfies this condition because $q'H$ then equals $w'H^2 = w'H = q'$, since H is idempotent. Thus for any arbitrary vector w', $w'H\hat{b}$ is unique. Consider this expression in terms of the solution (5). We have $q' = w'H$ and hence

$$\begin{aligned} q'\hat{b} &= w'H\hat{b} \\ &= w'H[GX'y + (H - I)z] \\ &= w'HGX'y \\ &= w'GX'y \end{aligned}$$

and this is $w'\hat{b}$ when $\hat{b}$ is taken as $GX'y$. Therefore

$$\begin{aligned} E(q'\hat{b}) &= w'GX'E(y) \\ &= w'GX'Xb \\ &= q'b. \end{aligned}$$

The consequence of this is that any linear function of the estimators that is unique is unbiased. This is just the property of estimability: linear functions of the elements of $\hat{b}$ (parameter estimates) that are unique are the unique, unbiased estimators of the same linear functions of the elements of b (parameters)—and these are the only linear functions of the parameters that can be so estimated (and linear combinations of them, of course); such functions are called *estimable functions*. Hence if q' is such that $q'\hat{b}$ is unique, then $q'\hat{b}$ is the unique unbiased estimator of $q'b$, and $q'b$ is called an *estimable function*.

The variance of the estimator of such a function is

$$\begin{aligned}\text{var}(q'\hat{b}) &= E(q'\hat{b} - q'b)(\hat{b}'q - b'q)\\ &= q'Gq\sigma^2\\ &= w'Gw\sigma^2\end{aligned}$$

because $H = GX'X$ and G is symmetric. This is again a simple result, and is analogous to that pertaining when $X'X$ is nonsingular. Now $\hat{b}$ has been obtained by the method of least squares. Therefore, of all linear unbiased estimators of $q'b$, we can show that $q'\hat{b}$ is the one having minimum variance; i.e., there is no linear unbiased estimator of $q'b$ having variance smaller than $w'Gw\sigma^2$. The following proof of this is based on Rao (1962).

******Proof.*

Suppose $t'y$ is an unbiased estimator of $q'b$ different from $q'\hat{b} = w'GX'y$. Then $E(t'y) = t'Xb = q'b$ is true for all b so that $t'X = q'$. Therefore

$$\begin{aligned}\text{cov}(q'\hat{b}, t'y) &= E(q'\hat{b} - q'b)(y't - b'q) = E(q'GX'e)(e't)\\ &= q'GX't\sigma^2 = q'Gq\sigma^2.\end{aligned}$$

Hence

$$\begin{aligned}\text{var}(q'\hat{b} - t'y) &= \text{var}(q'\hat{b}) + \text{var}(t'y) - 2\,\text{cov}(q'\hat{b}, t'y)\\ &= q'Gq\sigma^2 + \text{var}(t'y) - 2q'Gq\sigma^2\\ &= \text{var}(t'y) - \text{var}(q'\hat{b}).\end{aligned}$$

But $\text{var}(q'\hat{b} - t'y)$ is positive and so therefore the variance of $t'y$ exceeds that of $q'\hat{b}$. Hence $q'\hat{b}$ has a smaller variance than any other unbiased estimator of $q'b$.

The covariance between estimators of two estimable functions $q_1'\hat{b} = w_1'H\hat{b}$ and $q_2'\hat{b} = w_2'H\hat{b}$ can also be obtained. It is

$$\begin{aligned}\text{cov}(q_1'\hat{b}, q_2'\hat{b}) &= q_1'E(\hat{b} - b)(\hat{b}' - b')q_2 = q_1'Gq_2\sigma^2\\ &= w_1'Gw_2\sigma^2.\end{aligned}$$

Summary. For any arbitrary vector w', an estimable function is

$$q'b = w'Hb, \tag{8}$$

whose unique, unbiased, minimum variance estimator is

$$q'\hat{b} = w'GX'y. \tag{9}$$

The variance of the estimator is

$$\text{var}(q'\hat{b}) = w'Gw\sigma^2 \tag{10}$$

where σ^2 is estimated as

$$\hat{\sigma}^2 = (y'y - y'XGX'y)/(n - r), \tag{11}$$

n being the number of observations and r the rank of X. Since the rank of H is also r there are only r linearly independent vectors $q' = w'H$, and hence only r linearly independent estimable functions. Any linear combination of them is, of course, also an estimable function. The covariance between any two estimable functions involving arbitrary vectors w_1' and w_2' is $w_1'Gw_2\sigma^2$.

7. EXAMPLES

The general results just given are now used in three simple examples—that considered at the beginning of this chapter and two illustrations of the 2-way classification, one with one observation in every cell and one with unequal numbers of observations in the cells. All three are hypothetical examples, designed in part to keep the arithmetic simple.

a. One-way classification

In the example of Section 1 equation (4) shows normal equations $X'Xb = X'y$, where

$$X'X = \begin{bmatrix} 6 & 3 & 2 & 1 \\ 3 & 3 & 0 & 0 \\ 2 & 0 & 2 & 0 \\ 1 & 0 & 0 & 1 \end{bmatrix} \quad \text{and} \quad X'y = \begin{bmatrix} y_{..} \\ y_{1.} \\ y_{2.} \\ y_{3.} \end{bmatrix} = \begin{bmatrix} 504 \\ 300 \\ 172 \\ 32 \end{bmatrix}.$$

A generalized inverse of $X'X$ is

$$G = \begin{bmatrix} 0 & 0 & 0 & 0 \\ 0 & 1/3 & 0 & 0 \\ 0 & 0 & 1/2 & 0 \\ 0 & 0 & 0 & 1 \end{bmatrix} \quad \text{for which } H = GX'X = \begin{bmatrix} 0 & 0 & 0 & 0 \\ 1 & 1 & 0 & 0 \\ 1 & 0 & 1 & 0 \\ 1 & 0 & 0 & 1 \end{bmatrix}.$$

Hence for the arbitrary vector $w' = [w_0 \quad w_1 \quad w_2 \quad w_3]$ and the vector of parameters $b' = [\mu \quad \alpha_1 \quad \alpha_2 \quad \alpha_3]$ estimable functions given by (8) are

$$q'b = w'Hb = (w_1 + w_2 + w_3)\mu + w_1\alpha_1 + w_2\alpha_2 + w_3\alpha_3 \qquad (12)$$

for which estimators are, from (9),

$$q'\hat{b} = w'GX'y = w_1\bar{y}_{1.} + w_2\bar{y}_{2.} + w_3\bar{y}_{3.}. \qquad (13)$$

These equations illustrate an important aspect of the general results (8) and (9) from which they come, namely that they apply for *any* vector w'; i.e. (12) and (13) hold true for any values that we care to give to w_1, w_2 and w_3. There are two consequences of this: first, by giving specific values to the w's we can obtain specific estimable functions from (12); for example, for $w_1 = 1 = w_2$, and $w_3 = 0$ we see that $2\mu + \alpha_1 + \alpha_2$ is estimable. Second, we can find out whether particular functions of the parameters that interest us are estimable—by seeing if w's can be found such that (12) reduces to the function of interest; e.g. is $\alpha_1 - \alpha_2$ estimable? Yes, because with $w_1 = 1$, $w_2 = -1$ and $w_3 = 0$, (12) reduces to $\alpha_1 - \alpha_2$. This, of course, is tantamount to seeing whether, for a particular value of $q'b$ that interest us, q' satisfies $q'H = q'$. Thus, for $q'b = \alpha_1 - \alpha_2$ in the example, $q' = [0 \quad 1 \quad 0 \quad -1]$ and we see that

$$q'H = [0 \quad 1 \quad 0 \quad -1]\begin{bmatrix} 0 & 0 & 0 & 0 \\ 1 & 1 & 0 & 0 \\ 1 & 0 & 1 & 0 \\ 1 & 0 & 0 & 1 \end{bmatrix} = [0 \quad 1 \quad 0 \quad -1] = q'$$

and so $\alpha_1 - \alpha_2$ is estimable. Similarly we find that μ is not estimable, for the q' appropriate to $q'b \equiv \mu$ is $q' = [1 \quad 0 \quad 0 \quad 0]$, for which $q'H \neq q'$. Thus it is impossible to choose values for the w's that reduce (12) to be just μ. On all occasions the estimator of (12) is given by the corresponding value of (13). There is, of course, an unlimited number of sets of values that can be given to the w's and hence an unlimited number of estimable functions—but only $n - r = 3$ of them can be linearly independent. Examples are shown in Table 3.

TABLE 3. EXAMPLES OF ESTIMABLE FUNCTIONS AND THEIR ESTIMATORS (IN THE 1-WAY CLASSIFICATION)

Example	Values of w's			Estimable Function	Estimator
	w_1	w_2	w_3	$(w_1 + w_2 + w_3)\mu + w_1\alpha_1 + w_2\alpha_2 + w_3\alpha_3$	$w_1\bar{y}_{1\cdot} + w_2\bar{y}_{2\cdot} + w_3\bar{y}_{3\cdot} = 100w_1 + 86w_2 + 32w_3$
1	1	-1	0	$\alpha_1 - \alpha_2$	$\bar{y}_{1\cdot} - \bar{y}_{2\cdot} = 14$
2	0	1	-1	$\alpha_2 - \alpha_3$	$\bar{y}_{2\cdot} - \bar{y}_{3\cdot} = 54$
3	1/3	1/3	1/3	$\mu + \frac{1}{3}(\alpha_1 + \alpha_2 + \alpha_3)$	$(\bar{y}_{1\cdot} + \bar{y}_{2\cdot} + \bar{y}_{3\cdot})/3 = 72\frac{2}{3}$

It can be shown that the sets of values given to the w's in Table 3 are linearly independent and therefore so are the corresponding estimable functions.

The last line of the table shows why it is a "convenient restraint" to have $\alpha_1 + \alpha_2 + \alpha_3 = 0$; but it is certainly not necessary in order to ascertain what functions of the parameters are estimable nor to obtain the estimators of them. In this instance it merely provides an estimator of μ, and in one particular restricted situation only, namely that in which we care to define the α's such that they sum to zero. This may be a very appropriate form of definition in many situations, but it is useful in the estimation process only if it enters into an estimable function as just exemplified. Otherwise it is of little help.

The variance of the estimator of any estimable function derived from (12) is, as given in (10),

$$\text{var}(q'\hat{b}) = w'Gw\sigma^2$$

and from the value of G this is

$$\text{var}(q'\hat{b}) = (w_1^2/3 + w_2^2/2 + w_1^2/1)\sigma^2.$$

An estimate of σ^2 is obtained from (11) as

$$\hat{\sigma}^2 = (y'y - y'XGX'y)/(6 - 3).$$

In general we know that $y'XGX'y = \hat{b}'X'y$ has the same value no matter what solution we use for $\hat{b}$; it is interesting to demonstrate this in this particular instance. Thus from (5),

$$\hat{b} = GX'y + (H - I)z$$

$$= \begin{bmatrix} 0 & 0 & 0 & 0 \\ 0 & 1/3 & 0 & 0 \\ 0 & 0 & 1/2 & 0 \\ 0 & 0 & 0 & 1 \end{bmatrix} \begin{bmatrix} 504 \\ 300 \\ 172 \\ 32 \end{bmatrix} + \begin{bmatrix} -1 & 0 & 0 & 0 \\ 1 & 0 & 0 & 0 \\ 1 & 0 & 0 & 0 \\ 1 & 0 & 0 & 0 \end{bmatrix} \begin{bmatrix} z_1 \\ z_2 \\ z_3 \\ z_4 \end{bmatrix}$$

$$= \begin{bmatrix} -z_1 \\ 100 + z_1 \\ 86 + z_1 \\ 32 + z_1 \end{bmatrix},$$

where z_1 is arbitrary, the first element of the arbitrary vector z. We then have

$$\hat{b}'X'y = [-z_1 \quad 100 + z_1 \quad 86 + z_1 \quad 32 + z_1] \begin{bmatrix} 504 \\ 300 \\ 172 \\ 32 \end{bmatrix}$$

$$= z_1(-504 + 300 + 172 + 32) + 100(300) + 86(172) + 32(32)$$
$$= 45{,}816$$

independent of z_1. This will be the value of $\hat{b}'X'y$ for any solution $\hat{b}$, and so SSE takes the value $(\sum_i \sum_j y_{ij}^2 - 45{,}816)$ for all solutions $\hat{b}$.

b. Two-way classification, no interactions, balanced data

Countless experiments are undertaken each year in agriculture and the plant sciences to investigate the effect on growth and yield of various fertilizer treatments applied to different varieties of a species. Suppose we have data from six plants, representing three varieties being tested in combination with two fertilizer treatments. Although the experiment would not necessarily be conducted by growing the plants in two rows of three plants each, it is convenient to visualize the data as in Table 4.

TABLE 4. YIELDS OF PLANTS

Variety	Treatment 1	Treatment 2	Totals
1	y_{11}	y_{12}	$y_{1\cdot}$
2	y_{21}	y_{22}	$y_{2\cdot}$
3	y_{31}	y_{32}	$y_{3\cdot}$
Totals	$y_{\cdot 1}$	$y_{\cdot 2}$	$y_{\cdot\cdot}$

The entries in the table are such that y_{ij} represents the yield of the plant of variety i that received fertilizer treatment j. If μ is a general mean for plant yield, if α_i represents the effect on yield due to variety i and β_j the effect due to treatment j the equation of the linear model for y_{ij} is

$$y_{ij} = \mu + \alpha_i + \beta_j + e_{ij}, \tag{14}$$

where e_{ij} is a random error term peculiar to y_{ij}. This is the model for the 2-way classification without interaction. It is a straightforward extension of the model given in (1) for the one-way classification. As before, it is assumed that the e's are independent with zero means and variance σ^2.

On writing down each observation in terms of (14) we get

$$\begin{aligned}
y_{11} &= \mu + \alpha_1 &&&& + \beta_1 && + e_{11}\\
y_{12} &= \mu + \alpha_1 &&&&& + \beta_2 & + e_{12}\\
y_{21} &= \mu && + \alpha_2 && + \beta_1 && + e_{21}\\
y_{22} &= \mu && + \alpha_2 &&& + \beta_2 & + e_{22}\\
y_{31} &= \mu &&& + \alpha_3 & + \beta_1 && + e_{31}\\
y_{32} &= \mu &&& + \alpha_3 && + \beta_2 & + e_{32}
\end{aligned}$$

which can easily be written in matrix form as

$$y = \begin{bmatrix} 1 & 1 & 0 & 0 & 1 & 0\\ 1 & 1 & 0 & 0 & 0 & 1\\ 1 & 0 & 1 & 0 & 1 & 0\\ 1 & 0 & 1 & 0 & 0 & 1\\ 1 & 0 & 0 & 1 & 1 & 0\\ 1 & 0 & 0 & 1 & 0 & 1 \end{bmatrix} \begin{bmatrix} \mu\\ \alpha_1\\ \alpha_2\\ \alpha_3\\ \beta_1\\ \beta_2 \end{bmatrix} + e$$

where y and e are the vectors of observations and error terms respectively. With X representing the matrix of 0's and 1's and b the vector of parameters, this equation is the familiar form $y = Xb + e$. Accordingly we find

$$X'y = \begin{bmatrix} y_{..} \\ y_{1.} \\ y_{2.} \\ y_{3.} \\ y_{.1} \\ y_{.2} \end{bmatrix} \quad \text{and} \quad X'X = \begin{bmatrix} 6 & 2 & 2 & 2 & 3 & 3 \\ 2 & 2 & 0 & 0 & 1 & 1 \\ 2 & 0 & 2 & 0 & 1 & 1 \\ 2 & 0 & 0 & 2 & 1 & 1 \\ 3 & 1 & 1 & 1 & 3 & 0 \\ 3 & 1 & 1 & 1 & 0 & 3 \end{bmatrix},$$

for which a suitable generalized inverse is

$$G = \tfrac{1}{6}\begin{bmatrix} 0 & 0 & 0 & 0 & 0 & 0 \\ 0 & 4 & 1 & 1 & -2 & 0 \\ 0 & 1 & 4 & 1 & -2 & 0 \\ 0 & 1 & 1 & 4 & -2 & 0 \\ 0 & -2 & -2 & -2 & 4 & 0 \\ 0 & 0 & 0 & 0 & 0 & 0 \end{bmatrix}$$

with

$$H = \begin{bmatrix} 0 & 0 & 0 & 0 & 0 & 0 \\ 1 & 1 & 0 & 0 & 0 & 1 \\ 1 & 0 & 1 & 0 & 0 & 1 \\ 1 & 0 & 0 & 1 & 0 & 1 \\ 0 & 0 & 0 & 0 & 1 & -1 \\ 0 & 0 & 0 & 0 & 0 & 0 \end{bmatrix}.$$

By direct multiplication it will be seen that $X'XGX'X = X'X$, $GX'XG = G$, $H^2 = H$ and $XH = X$; and from (8) we find that with

$$w' = [w_0 \quad w_1 \quad w_2 \quad w_3 \quad w_4 \quad w_5],$$

estimable functions are

$$\begin{aligned} q'b &= w'Hb \\ &= (w_1 + w_2 + w_3)\mu + w_1\alpha_1 + w_2\alpha_2 + w_3\alpha_3 + w_4\beta_1 \\ &\qquad + (w_1 + w_2 + w_3 - w_4)\beta_2 \end{aligned} \tag{15}$$

and their unique, unbiased estimators are

$$\begin{aligned} q'\hat{b} &= w'GX'y \\ &= w_1(4y_{1.} + y_{2.} + y_{3.} - 2y_{.1})/6 + w_2(y_{1.} + 4y_{2.} + y_{3.} - 2y_{.1})/6 \\ &\quad + w_3(y_{1.} + y_{2.} + 4y_{3.} - 2y_{.1})/6 - w_4(2y_{1.} + 2y_{2.} + 2y_{3.} - 4y_{.1})/6 \end{aligned}$$

As with (12), (15) is an illustration of (8) and may be used to develop estimable functions (by giving specific values to the w's). It can also be used to investigate particular functions to see if they are estimable (by determining if appropriate values can be selected for the w's). For example, $\alpha_1 - \alpha_2$ is clearly estimable, by putting $w_1 = 1$, $w_2 = -1$, and $w_3 = 0 = w_4$ in (15); but $\mu + \alpha_1$ is not estimable. Again there is an infinite number of estimable functions, although any linearly independent set of them will contain only four. This is so because (15) contains only 4 w's—this arising from the nature of H in $w'Hb$, H having the same rank as X, in this case 4. Table 5 shows such a set of linearly independent estimable functions.

TABLE 5. ESTIMATORS OF ESTIMABLE FUNCTIONS (IN THE 2-WAY CLASSIFICATION, NO INTERACTION MODEL, BALANCED DATA)

Example	Values of w's				Estimable Function, $q'b$ Equation (15)	Estimator, $q'\hat{b}$
	w_1	w_2	w_3	w_4		
1	1	−1	0	0	$\alpha_1 - \alpha_2$	$\bar{y}_{1\cdot} - \bar{y}_{2\cdot}$
2	1	0	−1	0	$\alpha_1 - \alpha_3$	$\bar{y}_{1\cdot} - \bar{y}_{3\cdot}$
3	1/3	1/3	1/3	1/2	$\mu + (\alpha_1 + \alpha_2 + \alpha_3)/3 + (\beta_1 + \beta_2)/2$	$\bar{y}_{\cdot\cdot}$
4	0	0	0	1	$\beta_1 - \beta_2$	$\bar{y}_{\cdot 1} - \bar{y}_{\cdot 2}$*

* $\bar{y}_{\cdot 1} - \bar{y}_{\cdot 2} \equiv -(2y_{1\cdot} + 2y_{2\cdot} + 2y_{3\cdot} - 4y_{\cdot 1})/6$.

These are all familiar results; nevertheless it is interesting to demonstrate their derivation from the general equations (8) and (9), showing again that no "convenient restraints" such as $\alpha_1 + \alpha_2 + \alpha_3 = 0$ or $\beta_1 + \beta_2 = 0$ are needed to obtain them.

Once again it is appropriate to reiterate what has been said before in these pages. That although these matrix procedures may at first appear cumbersome when applied to the simple, familiar cases just considered, it is the universality of their application to situations not so familiar that commands our attention. With balanced data (the same number of observations in each of the smallest subclassifications of the data) the customary procedures used for computing parameter estimates are suitable *solely* because the matrix $X'X$ has a simple form in such cases. As a result, the matrix methods that have been described lead to these customary procedures. But when data are not balanced, as is so often the case in biological experiments or with field survey data, the expressions for estimates of functions of parameters are not at all standard or well known, and they depend a great deal on the nature of the data available. The matrix methods are then a most useful means of deriving the estimable functions and their estimators. We illustrate this further in terms of another example.

c. Two-way classification, no interactions, unbalanced data

Suppose in our previous example there had been two plants of variety 1 receiving treatment 1, and none of variety 3 receiving treatment 2. The numbers of plants and total yields could then be summarized in Table 6, analogous and similar to Table 5.

TABLE 6. NUMBERS OF PLANTS AND TOTAL YIELDS

Variety	Numbers of Plants — Treatment 1	2	Total Yield
1	2	1	$y_{1\cdot}$
2	1	1	$y_{2\cdot}$
3	1	0	$y_{3\cdot}$
Total yield	$y_{\cdot 1}$	$y_{\cdot 2}$	$y_{\cdot\cdot}$

The model is essentially the same as before, as are the vectors b and $X'y$. In solving the normal equations $X'X$ is now

$$X'X = \begin{bmatrix} 6 & 3 & 2 & 1 & 4 & 2 \\ 3 & 3 & 0 & 0 & 2 & 1 \\ 2 & 0 & 2 & 0 & 1 & 1 \\ 1 & 0 & 0 & 1 & 1 & 0 \\ 4 & 2 & 1 & 1 & 4 & 0 \\ 2 & 1 & 1 & 0 & 0 & 2 \end{bmatrix}$$

for which a generalized inverse is

$$G = \frac{1}{7}\begin{bmatrix} 0 & 0 & 0 & 0 & 0 & 0 \\ 0 & 5 & 2 & 4 & -4 & 0 \\ 0 & 2 & 5 & 3 & -3 & 0 \\ 0 & 4 & 3 & 13 & -6 & 0 \\ 0 & -4 & -3 & -6 & 6 & 0 \\ 0 & 0 & 0 & 0 & 0 & 0 \end{bmatrix}$$

with

$$H = \begin{bmatrix} 0 & 0 & 0 & 0 & 0 & 0 \\ 1 & 1 & 0 & 0 & 0 & 1 \\ 1 & 0 & 1 & 0 & 0 & 1 \\ 1 & 0 & 0 & 1 & 0 & 1 \\ 0 & 0 & 0 & 0 & 1 & -1 \\ 0 & 0 & 0 & 0 & 0 & 0 \end{bmatrix}.$$

This H is identical to that of the previous example (prompting the suggestion that it may be the same for all 2-way classifications with 3 rows and 2 columns, and no interaction in the model); therefore the estimable functions are the same as given in (15), namely,

$$q'b = (w_1 + w_2 + w_3)\mu + w_1\alpha_1 + w_2\alpha_2 + w_3\alpha_3 + w_4\beta_1 + (w_1 + w_2 + w_3 - w_4)\beta_2.$$

But now the estimators of these functions are

$$q'\hat{b} = w'GX'y$$

$$= w'G \begin{bmatrix} y_{..} \\ y_{1.} \\ y_{2.} \\ y_{3.} \\ y_{.1} \\ y_{.2} \end{bmatrix}$$

$$\begin{aligned} &= w_1(5y_{1.} + 2y_{2.} + 4y_{3.} - 4y_{.1})/7 \\ &\quad + w_2(2y_{1.} + 5y_{2.} + 3y_{3.} - 3y_{.1})/7 \\ &\quad + w_3(4y_{1.} + 3y_{2.} + 13y_{3.} - 6y_{.1})/7 \\ &\quad - w_4(4y_{1.} + 3y_{2.} + 6y_{3.} - 6y_{.1})/7 \end{aligned}$$

We see at once that this expression differs considerably from its counterpart of the previous example, and yet the expression for the estimable functions is the same in both cases. This emphasizes how even a minor difference in the nature of the data available, such as envisaged here, can lead to quite substantial differences in the estimators. This is further highlighted in Table 7 which shows the same estimable functions as Table 6. Their estimators, however, are noticeably different from the well-known expressions developed earlier.

It is interesting to see that in using results (8) through (11) no direct use is made of any particular solution $\hat{b}$ of the normal equations. Certainly the solution $\hat{b} = GX'y$ does enter into (9), $q'\hat{b} = w'GX'y$, but even here it is

TABLE 7. EXAMPLES OF ESTIMABLE FUNCTIONS AND THEIR ESTIMATORS (IN THE 2-WAY CLASSIFICATION, NO INTERACTION MODEL, UNBALANCED DATA)

Example	Values of w's				Estimable function	Estimator
	w_1	w_2	w_3	w_4	$q'b$	$q'\hat{b}$
1	1	−1	0	0	$\alpha_1 - \alpha_2$	$(3y_{1.} - 3y_{2.} + y_{3.} - y_{.1})/7$
2	1	0	−1	0	$\alpha_1 - \alpha_3$	$(y_{1.} - y_{2.} - 9y_{3.} + 2y_{.1})/7$
3	1/3	1/3	1/3	1/2	$\mu + (\alpha_1 + \alpha_2 + \alpha_3)/3 + (\beta_1 + \beta_2)/2$	$(10y_{1.} + 11y_{2.} + 22y_{3.} - 8y_{.1})/42$
4	0	0	0	1	$\beta_1 - \beta_2$	$-(4y_{1.} + 3y_{2.} + 6y_{3.} - 6y_{.1})/7$

of no particular interest in itself. And this is appropriate—for there is an infinity of solutions to the normal equations and no one of them is any more interesting to us than any other. Furthermore, $\hat{b}$ of itself tells nothing about what functions are estimable; these are given by (8) as $q'b = w'Hb$, so as soon as G and H are derived estimable functions can be ascertained from (8) and their estimators from (9). In practice therefore we first find G and H, then look at (8) to see what functions are estimable, and obtain their estimators from (9). Equation (10) gives their variances, using (11) to estimate the residual variance σ^2.

8. ANALYSIS OF VARIANCE

a. Fitting the model

Use has already been made of the error sum of squares

$$SSE = \sum_i \sum_j (y_{ij} - \hat{y}_{ij})^2,$$

best computed as $y'y - \hat{b}'X'y$. Suppose we now subtract this, just as was done in the case of regression, from the total (corrected) sum of squares $SST = \sum_i \sum_j (y_{ij} - \bar{y})^2$, usually computed as $y'y - n\bar{y}^2$. The difference, $SSM = SST - SSE$, can be described as the sum of squares due to fitting the model, analogous to SSR of regression analysis. This division of SST into two parts can be summarized in an analysis of variance table similar to the analyses given in Chapter 9.

TABLE 8. ANALYSIS OF VARIANCE FOR FITTING A LINEAR MODEL

Source of Variation	Degrees of Freedom	Sums of Squares
Fitting the model	$r - 1$	$SSM = SST - SSE$
Error	$n - r$	$SSE = \Sigma\Sigma(y_{ij} - \hat{y}_{ij})^2$
Total	$n - 1$	$SST = y'y - n\bar{y}^2$

There is one apparent difference between this table and its counterpart in regression (Table 1 of Section 9.9). In regression analysis the degrees of freedom associated with SSR, the sum of squares due to fitting the regression, is k, the number of x-variables being fitted. In the above table the degrees of freedom associated with SSM, the sum of squares due to fitting the model, is $r - 1$, one less than the rank of the matrix X in the model

$y = Xb + e$. Actually this is also the case in regression analysis, because there the rank of X is $k + 1$, and k, $= (k + 1) - 1$, is analogous to $r - 1$ just mentioned. In reality, therefore, there is no difference between the two cases.

Computationally, this table is best derived in a manner a little different from the regression case, because now there is no analogue of the computing procedures based on deviations of means. As shown in Section 5,

$$SSE = y'(I - XGX')y$$

and therefore

$$\begin{aligned} SSM &= SST - SSE \\ &= y'y - n\bar{y}^2 - y'(I - XGX')y \\ &= y'XGX'y - n\bar{y}^2 \\ &= \hat{b}'X'y - n\bar{y}^2 \end{aligned}$$

where $\hat{b}$ is any solution $\hat{b} = GX'y$ to the normal equations. The analysis of variance table is therefore computed as shown in Table 9.

TABLE 9. ANALYSIS OF VARIANCE FOR FITTING A LINEAR MODEL

Source of Variation	Degrees of Freedom	Sums of Squares
Fitting the model	$r - 1$	$SSM = \hat{b}'X'y - n\bar{y}^2$
Error	$n - r$	$SSE = SST - SSM$
Total	$n - 1$	$SST = \Sigma\Sigma y_{ij}^2 - n\bar{y}^2$

Example. In the example of Sections 1 and 7*a*

$$\begin{aligned} y' &= [101 \quad 105 \quad 94 \quad 84 \quad 88 \quad 32], \\ (X'y)' &= [504 \quad 300 \quad 172 \quad 32], \end{aligned}$$

and

$$G = \begin{bmatrix} 0 & 0 & 0 & 0 \\ 0 & \frac{1}{3} & 0 & 0 \\ 0 & 0 & \frac{1}{2} & 0 \\ 0 & 0 & 0 & 1 \end{bmatrix}.$$

Thus

$$SSE = y'y - y'XGX'y = 45{,}886 - 45{,}816 = 70$$

and

$$SSM = 45{,}816 - 504^2/6 = 3480$$

so that the analysis of variance is

TABLE 10. ANALYSIS OF VARIANCE

Source of Variation	Degrees of Freedom	Sums of Squares
Model	2	3480
Error	3	70
Total	5	3550

b. Tests of general linear hypotheses

We now consider the calculations involved in testing linear hypotheses, a linear hypothesis being a hypothesis that some linear function of the parameters equals some arbitrary constant. The discussion is confined to linear hypotheses involving estimable functions, since these are the only ones that can be tested. Thus in Section 7 the function $\alpha_1 - \alpha_2$ is estimable and the hypothesis that $\alpha_1 - \alpha_2$ equals some constant, m_0 say, can be tested.

In general, a linear hypothesis that $q'b = m_0$, where m_0 is arbitrary and $q'b$ is estimable, is tested by amending the model $y = XB + e$ to take account of the hypothesis. In this connection the model $y = Xb + e$ is usually referred to as the *full model*, and amendment in terms of the hypothesis yields what is called the *reduced model*, since it represents the full model reduced by the conditions of the hypothesis. Thus if $q'b = m_0$ led to b being changed into b^* the reduced model would be $y = X^*b^* + e$ where X^* is the form of X corresponding to b^*. For example, in Sections 1 and 7a, $b' = [\mu \quad \alpha_1 \quad \alpha_2 \quad \alpha_3]$, and to test the hypothesis that $\alpha_1 = \alpha_2$, i.e., that $\alpha_1 - \alpha_2 = 0$ (which is testable because $\alpha_1 - \alpha_2$ is estimable), $b^{*\prime}$ would be $b^{*\prime} = [\hat{\mu} \quad \alpha_1 \quad \alpha_3]$. The matrix X is

$$X = \begin{bmatrix} 1 & 1 & 0 & 0 \\ 1 & 1 & 0 & 0 \\ 1 & 1 & 0 & 0 \\ 1 & 0 & 1 & 0 \\ 1 & 0 & 1 & 0 \\ 1 & 0 & 0 & 1 \end{bmatrix}, \quad \text{and} \quad X^* \text{ would be } X^* = \begin{bmatrix} 1 & 1 & 0 \\ 1 & 1 & 0 \\ 1 & 1 & 0 \\ 1 & 1 & 0 \\ 1 & 1 & 0 \\ 1 & 0 & 1 \end{bmatrix}.$$

Many hypotheses cannot be represented by a single equation $q'b = m_0$ but require several such equations. These are usually thought of in vector form $Q'b = m$, where the rows of $Q'b$ are linearly independent estimable functions. In this way $Q'H = Q'$, and if Q' has s rows the rank of Q' is s. As an example, were the hypothesis that $\alpha_1 = \alpha_2 = \alpha_3$ to be considered in Section 7a, it could be represented in the form $Q'b = m$ as

$$\begin{bmatrix} 0 & 1 & -1 & 0 \\ 0 & 1 & 0 & -1 \end{bmatrix} \begin{bmatrix} \mu \\ \alpha_1 \\ \alpha_2 \\ \alpha_3 \end{bmatrix} = \begin{bmatrix} 0 \\ 0 \end{bmatrix}.$$

The hypothesis $Q'b = m$ is known as the *general linear hypothesis*. The test of this hypothesis requires obtaining the sum of squares due to fitting the full model and that due to fitting the reduced model. The first of these is, as shown earlier, $SSM = y'XGX'y - n\bar{y}^2$ where G is a generalized inverse of $X'X$, and since this refers to the full model it shall be denoted by

$$SSM \text{ (full model)} = y'XGX'y - n\bar{y}^2.$$

By analogy, the sum of squares for fitting the reduced model $y = X^*b^* + e$ (derived from the full model by applying to it the hypothesis $Q'b = m$) is

$$SSM \text{ (reduced model)} = y'X^*G^*X^{*\prime}y - n\bar{y}^2,$$

where G^* is a generalized inverse of $X^{*\prime}X^*$. The difference between these two expressions is used to obtain

$$F = [SSM \text{ (full model)} - SSM \text{ (reduced model)}]/s\hat{\sigma}^2,$$

where $\hat{\sigma}^2 = SSE/(n - r)$ as in Section 5. Comparing F with tabulated values of the F-distribution for s and $n - r$ degrees of freedom provides the test of the hypothesis.

Clearly, the numerator of F could be calculated as $y'XGX'y - y'X^*G^*X^{*\prime}y$, but to do so would require $X^*G^*X^{*\prime}$ for each and every hypothesis that is to be tested. However, because in both models $SSM + SSE = SST$, the expression for F can also be written as

$$F = [SSE \text{ (reduced model)} - SSE \text{ (full model)}]/s\hat{\sigma}^2$$

and from this a computing formula can be developed which avoids altogether the necessity of obtaining X^*. The formula utilizes $\hat{b}$ as well as Q' and m, the specifications of the hypothesis, and it is relatively easy to compute. It is given in the following theorem.

Theorem. When fitting the linear model $y = Xb + e$, the numerator sum of squares of the F value used for testing the (testable) general linear hypothesis $Q'b = m$, for Q' consisting of s linearly independent rows, is $(Q'\hat{b} - m)'(Q'GQ)^{-1}(Q'\hat{b} - m)$ where $\hat{b} = GX'y$ is a solution to the normal equations $X'Xb = X'y$ and G is a symmetric generalized inverse of $X'X$.

The following lemma is used in proving the theorem.

Lemma. $Q'GQ$ is nonsingular.

Proof of lemma. Because $Q'b = m$ is a testable hypothesis, the rows of $Q'b$ are estimable functions and therefore $Q'H = Q'$ where $H = GX'X$. Hence

$$Q'GQ = Q'HGQ = Q'GX'XGQ = Q'GX'(Q'GX')',$$

so that $r(Q'GQ) = r(Q'GX')$. But $Q' = Q'H = Q'GX'X$; therefore, by the rule for the rank of a product matrix (Section 5.13), $r(Q') = s \leqq r(Q'GX')$, and also

$$r(Q'GX') \leqq r(Q') = s.$$

Hence $r(Q'GX') = s$, and so therefore does the rank of $Q'GQ$. But s is the order of $Q'GQ$. Therefore $Q'GQ$ is nonsingular.

Proof of theorem. Fitting the reduced model is equivalent to fitting the full model $y = Xb + e$ subject to the condition $Q'b = m$. The appropriate normal equations are derived by minimizing $(y - Xb)'(y - Xb) + 2t'(Q'b - m)$ where t' is a vector of Lagrange multipliers. The resulting equations are

$$X'X\tilde{b} + Qt = X'y \tag{16}$$

and

$$Q'\tilde{b} = m. \tag{17}$$

Using G and $\hat{b} = GX'y$, equation (16) can be solved as

$$\tilde{b} = \hat{b} - GQt. \tag{18}$$

Pre-multiplying (18) by Q', substituting from (17) and using the lemma gives

$$t = (Q'GQ)^{-1}(Q'\hat{b} - m), \tag{19}$$

and substitution back into (18) yields

$$\tilde{b} = \hat{b} - GQ(Q'GQ)^{-1}(Q'\hat{b} - m). \tag{20}$$

For $\tilde{y} = X\tilde{b}$ the residual sum of squares after fitting the reduced model is

$$SSE \text{ (reduced model)} = (y - X\tilde{b})'(y - X\tilde{b}).$$

Substituting for $\tilde{b}$ from (20) this leads, after a little reduction, to

$$\begin{aligned} SSE \text{ (reduced model)} &= (y - X\hat{b})'(y - X\hat{b}) + t'Q'GQt \\ &= SSE \text{ (full model)} + t'Q'GQt. \end{aligned}$$

Hence the numerator of F is

$$\begin{aligned} SSE \text{ (reduced model)} &- SSE \text{ (full model)} \\ &= t'Q'GQt \\ &= (Q'\hat{b} - m)'(Q'GQ)^{-1}(Q'\hat{b} - m), \end{aligned}$$

this last result coming from the expression for t given in (19). Thus is the theorem proved.

This theorem means that the F-value for testing the hypothesis $Q'b = m$ can be calculated as

$$F = [(Q'\hat{b} - m)'(Q'GQ)^{-1}(Q'\hat{b} - m)]/s\hat{\sigma}^2. \tag{21}$$

With $Q'\hat{b}$ being the estimator of the estimable functions $Q'b$ in the full model it is apparent that once $\hat{b} = GX'y$ has been calculated F is readily obtainable.

A by-product of the theorem is the solution of the normal equations in the reduced model, given in (20), for which the variance-covariance matrix is

$$\mathrm{var}(\tilde{b}) = [G - GQ(Q'GQ)^{-1}Q'G]\sigma^2.$$

In situations where m is a null vector, $m = 0$, the expressions for F and $\tilde{b}$ reduce to the simpler forms

$$F = \hat{b}'Q(Q'GQ)^{-1}Q'\hat{b}/s\hat{\sigma}^2 \tag{22}$$

and

$$\tilde{b} = \hat{b} - GQ(Q'GQ)^{-1}Q'\hat{b}. \tag{23}$$

Example. In the example of the preceding Section

$$G = \begin{bmatrix} 0 & 0 & 0 & 0 \\ 0 & \frac{1}{3} & 0 & 0 \\ 0 & 0 & \frac{1}{2} & 0 \\ 0 & 0 & 0 & 1 \end{bmatrix} \quad \text{and} \quad \hat{b} = GX'y = \begin{bmatrix} 0 \\ 100 \\ 86 \\ 32 \end{bmatrix},$$

with the estimate of σ^2 being $\hat{\sigma}^2 = 70/3$ (Table 10). The difference $\alpha_1 - \alpha_2$ is estimable, and therefore a testable hypothesis is $\alpha_1 - \alpha_2 = 0$. Expressing this as $Q'b = 0$, Q' is a vector having rank $s = 1$: $Q' = [0 \quad 1 \quad -1 \quad 0]$. It will be found that $Q'GQ = 5/6$, $Q'\hat{b} = 14$, and hence from (22) the value of F for testing the hypothesis $\alpha_1 = \alpha_2$ is

$$F = \frac{14(5/6)^{-1}14}{1(70/3)} = \frac{1176}{5}\Big/\frac{70}{3} = 10.08.$$

A solution for b in the reduced model is, from (23),

$$\tilde{b} = \begin{bmatrix} 0 \\ 100 \\ 86 \\ 32 \end{bmatrix} - \begin{bmatrix} 0 \\ \frac{1}{3} \\ -\frac{1}{2} \\ 0 \end{bmatrix} (5/6)^{-1}14 = \begin{bmatrix} 0 \\ 94.4 \\ 94.4 \\ 32.0 \end{bmatrix}.$$

Calculations based on (22) and (23) are alternative to solving the normal equations for the reduced model, $X^{*\prime}X^*b^* = X^{*\prime}y$, which in this example are

$$\begin{bmatrix} 6 & 5 & 1 \\ 5 & 5 & 0 \\ 1 & 0 & 1 \end{bmatrix} \begin{bmatrix} \mu \\ \alpha_1 \\ \alpha_3 \end{bmatrix} = \begin{bmatrix} 504 \\ 472 \\ 32 \end{bmatrix}.$$

A solution is

$$\tilde{b}^{*\prime} = [\tilde{\mu} \quad \tilde{\alpha}_1 \quad \tilde{\alpha}_3] = [0 \quad 94.4 \quad 32.0]$$

and consequently the value of SSM is

$$SSM \text{ (reduced model)}$$
$$= \tilde{b}^{*\prime} X^{*\prime} y - n\bar{y}^2$$
$$= 94.4(472) + 32.0(32) - 504^2/6 = 3244.8.$$

Subtracting this from the sum of squares due to fitting the full model gives

$$3480 - 3244.8 = 235.2 = 1176/5$$

as the numerator of F, exactly as just given. These calculations may seem no more laborious in this example than using (22) and (23), but the advantages of the latter are clearly apparent in larger and more complex situations.

c. Fitting portions of a model.

In discussing regression we showed how to test the significance of fitting just some of the x-variables. This procedure can also be applied to the elements of a linear model, and it is particularly fitting to do so when these elements fall naturally into two or more groups and we want to test the significance of each group. For instance, in Examples *b* and *c* of Section 7, the α's represent row (i.e., variety) effects and the β's represent column (treatment) effects, and it is natural to want to test if the row effects are significant, and also if the column effects are significant. This can be done by developing an analysis of variance similar to Table 6 of Section 9.11. We illustrate the procedure in terms of an example.

Example. In Example c of Section 7 we were concerned with data consisting of yields from plants of three different varieties grown in two fertilizer treatments. To estimate variety and treatment effects the model

$$y_{ij} = \mu + \alpha_i + \beta_j + e_{ij} \tag{24}$$

was used, where α_i is the variety effect for variety i and β_j is the treatment effect for treatment j. Suppose that we now wish to test the significance of the contribution of the treatment effects (the β's) to the model given in (24). First, (24) would be fitted, the sum of squares due to fitting it being $SSM_{\alpha\beta}$, say. Then we would fit the model

$$y_{ij} = \mu + \alpha_i + e_{ij}, \tag{25}$$

with sum of squares SSM_{α}, say. Now the model (25) is equivalent to omitting the β's from (24), and so $SSM_{\alpha\beta} - SSM_{\alpha}$ represents the contribution of the β's to $SSM_{\alpha\beta}$. Hence this difference forms the basis for testing the significance of the β's in the model specified in (24).

In considering the two models, note that were the β's to be assumed equal in (24) it could be written as $y_{ij} = (\mu + \beta) + \alpha_i + e_{ij}$, which is equivalent to (25). Thus, although (25) is conceived as being (24) with the β's omitted, it is indistinguishable from (24) with the β's assumed equal. Therefore, with $\beta_1 - \beta_2$ being estimable (25) may be considered as (24) amended by the hypothesis $\beta_1 - \beta_2 = 0$, and consequently the difference $SSM_{\alpha\beta} - SSM_\alpha$ can be calculated exactly as discussed earlier in Section 8*b*; (24) corresponds to the full model and (25) corresponds to the reduced model.

For the sake of illustration let us suppose that in the example of Section 7*c* the vector of observations is

$$y' = [4 \quad 7 \quad 3 \quad 5 \quad 2 \quad 1]$$

so that the normal equations $X'Xb = X'y$ are

$$\begin{bmatrix} 6 & 3 & 2 & 1 & 4 & 2 \\ 3 & 3 & 0 & 0 & 2 & 1 \\ 2 & 0 & 2 & 0 & 1 & 1 \\ 1 & 0 & 0 & 1 & 1 & 0 \\ 4 & 2 & 1 & 1 & 4 & 0 \\ 2 & 1 & 1 & 0 & 0 & 2 \end{bmatrix} \begin{bmatrix} \mu \\ \alpha_1 \\ \alpha_2 \\ \alpha_3 \\ \beta_1 \\ \beta_2 \end{bmatrix} = \begin{bmatrix} y_{\cdot\cdot} \\ y_{1\cdot} \\ y_{2\cdot} \\ y_{3\cdot} \\ y_{\cdot 1} \\ y_{\cdot 2} \end{bmatrix} = \begin{bmatrix} 22 \\ 14 \\ 7 \\ 1 \\ 17 \\ 5 \end{bmatrix}.$$

Utilizing G of Section 7*c*, a solution is obtained as

$$\hat{b}' = (GX'y)' = (1/7)[0 \quad 20 \quad 15 \quad -12 \quad 19 \quad 0].$$

Hence

$$\begin{aligned} SSE &= y'y - y'XGX'y \\ &= 104 - 696/7 = 32/7 = 96/21 \end{aligned}$$

and therefore the sum of squares

$$\begin{aligned} SSM_{\alpha\beta} &= y'XGX'y - n\bar{y}^2 \\ &= 696/7 - 22^2/6 = 394/21. \end{aligned}$$

Thus, as in Table 9, the analysis of variance is

TABLE 11. ANALYSIS OF VARIANCE FOR FITTING THE FULL MODEL

Source of Variation	Degrees of Freedom	Sums of Squares
Fitting α's and β's	3	$SSM_{\alpha\beta} = 394/21$
Error	2	$SST - SSM_{\alpha\beta} = 96/21$
Total	5	$SST = 490/21$

To obtain SSM_α, the sum of squares due to fitting $y_{ij} = \mu + \alpha_i + e_{ij}$, we consider in the full model the hypothesis $\beta_1 = \beta_2$, namely $Q'b = 0$ for $Q' = [0 \quad 0 \quad 0 \quad 0 \quad 1 \quad -1]$. This is testable, the estimate of $\beta_1 - \beta_2$ is $Q'\hat{b} = 19/7$, and on computing $Q'GQ$ as $6/7$ the required difference between the two sums of squares is, as in (22).

$$SSM_{\alpha\beta} - SSM_\alpha = \hat{b}'Q(Q'GQ)^{-1}Q'\hat{b} = 19/7(7/6)19/7 = 361/42.$$

Thus

$$\begin{aligned} SSM_\alpha &= SSM_{\alpha\beta} - 361/42 = 394/21 - 361/42 \\ &= 427/42. \end{aligned}$$

In this way $SSM_{\alpha\beta}$ of Table 3 is split into two portions: $SSM_\alpha = 427/42$ due to fitting just the α's and $SSM_{\alpha\beta} - SSM_\alpha = 361/42$ due to fitting the β's in addition to the α's. The situation is summarized as follows in an analysis of variance table similar to that of Section 9.11,

TABLE 12. ANALYSIS OF VARIANCE FOR FITTING α-EFFECTS AND THEN β-EFFECTS

Source of Variation	D.F.	Sums of Squares
Fitting α's alone	2	$SSM_\alpha = 427/42$
Fitting β's, after fitting α's	1	$SSM_{\alpha\beta} - SSM_\alpha = 361/42$
Error	2	$SST - SSM_{\alpha\beta} = 192/42$
Total	5	$SST = 980/42$

As in the regression case, the value

$$F = \frac{2(SSM_{\alpha\beta} - SSM_\alpha)}{SST - SSM_{\alpha\beta}} = \frac{2(361)}{192} = 3.76$$

has an F-distribution with 1 and 2 degrees of freedom, so providing a test of the significance of the contribution of the β's to the full model.

A check on the calculations is provided by noting that the normal equations for the reduced model are

$$\begin{bmatrix} 6 & 3 & 2 & 1 \\ 3 & 3 & 0 & 0 \\ 2 & 0 & 2 & 0 \\ 1 & 0 & 0 & 1 \end{bmatrix} \begin{bmatrix} \mu \\ \alpha_1 \\ \alpha_2 \\ \alpha_3 \end{bmatrix} = \begin{bmatrix} 22 \\ 14 \\ 7 \\ 1 \end{bmatrix}$$

and the resulting sum of squares is

$$SSM_\alpha = 14^2/3 + 7^2/2 + 1^2/1 - 22^2/6 = 427/42,$$

as before.

Just as Table 12 has been developed to test the significance of the β's so can a table be established for testing the significance of the α's. This is done by noting that in the full model the hypothesis $\alpha_1 = \alpha_2 = \alpha_3$ can be represented as $Q'b = 0$ for

$$Q' = \begin{bmatrix} 0 & 1 & -1 & 0 & 0 & 0 \\ 0 & 1 & 0 & -1 & 0 & 0 \end{bmatrix}.$$

Since $Q'H = Q'$, the hypothesis is testable, $Q'\hat{b} = \begin{bmatrix} 5/7 \\ 32/7 \end{bmatrix}$, and $Q'GQ = (1/7)\begin{bmatrix} 6 & 2 \\ 2 & 10 \end{bmatrix}$ with $(Q'GQ)^{-1} = (1/8)\begin{bmatrix} 10 & -2 \\ -2 & 6 \end{bmatrix}$. Hence, with SSM_β denoting the sum of squares due to fitting the reduced model ($y_{ij} = \mu + \beta_j + e_{ij}$), we have

$$SSM_{\alpha\beta} - SSM_\beta = \hat{b}'Q(Q'GQ)^{-1}Q'\hat{b} = 411/28$$

which gives

$$SSM_\beta = 394/21 - 411/28 = 343/84.$$

The corresponding analysis of variance is shown in Table 13.

TABLE 13. ANALYSIS OF VARIANCE FOR FITTING β-EFFECTS AND THEN α-EFFECTS

Source of Variation	D.F.	Sums of Squares
Fitting β's alone	1	$SSM_\beta = 343/84$
Fitting α's after fitting β's	2	$SSM_{\alpha\beta} - SSM_\beta = 1233/84$
Error	2	$SST - SSM_{\alpha\beta} = 384/84$
Total	5	$SST = 1960/84$

The significance of the α's in the model is tested by comparing the ratio of the last two sums of squares against tabulated values of the F-distribution with 2 and 2 degrees of freedom.

Again the calculations can be checked from the normal equations, in this case

$$\begin{bmatrix} 6 & 4 & 2 \\ 4 & 4 & 0 \\ 2 & 0 & 2 \end{bmatrix}\begin{bmatrix} \mu \\ \beta_1 \\ \beta_2 \end{bmatrix} = \begin{bmatrix} 22 \\ 17 \\ 5 \end{bmatrix}.$$

From these,

$$SSM_\beta = 17^2/4 + 5^2/2 - 22^2/6 = 49/12 = 343/84,$$

as already obtained.

d. Balanced data

Tables 12 and 13 provide tests for fitting the α's on their own and for fitting them after fitting the β's and likewise for fitting the β's on their own and for fitting them after fitting the α's. Both tables apply whether data are balanced or unbalanced; we have considered a small example of unbalanced data. However, for balanced data it will be found that the two tables are algebraically the same, both reducing to the familiar expressions shown in Table 14. This reduction is due *solely* to the form of the various $X'X$ matrices in such cases.

TABLE 14. ANALYSIS OF VARIANCE FOR BALANCED DATA

Source of Variation	Degrees of Freedom	Sums of Squares
Rows (α's)	2	$\sum_i \sum_j (\bar{y}_{i\cdot} - \bar{y})^2$
Columns (β's)	1	$\sum_i \sum_j (\bar{y}_{\cdot j} - \bar{y})^2$
Error	2	$\sum_i \sum_j (y_{ij} - \bar{y}_{i\cdot} - \bar{y}_{\cdot j} + \bar{y})^2$
Total	5	$\sum_i \sum_j (y_{ij} - \bar{y})^2$

The reader should satisfy himself that for Example b in Section 7 both Tables 12 and 13 reduce to Table 14.

Although the expressions for the sums of squares in Table 14 have in themselves a more or less satisfactory empirical rationale, it is not a rationale that carries over to situations of unbalanced data. In that case the matrix development presented here will always lead to the appropriate results, and since it also applies to balanced data it seems more convenient to think of the analysis for balanced data as being a special case of the general matrix results and not as the general case itself.

This whole matter has, of course, been discussed in terms of the simplest of examples. But this is possibly as effective a way as any of presenting the general ideas involved. They can readily be extended to more complex situations, but to do so here would be beyond the scope of this book.

9. REGRESSION AS PART OF A LINEAR MODEL

The previous chapter on regression has been referred to frequently; and in this chapter we have so far dealt with what are usually called linear models. Sometimes, however, we may wish to use a combination of regression and a linear model. For example, in studying the weight gains of beef cattle or broiler chickens we may want to consider the effects of different diets taking into account the initial weights of the animals at the start of the study. A suitable model would be

$$y_{ij} = \mu + \alpha_i + \beta x_{ij} + e_{ij}$$

where y_{ij} is the weight gain of the jth animal receiving diet i, μ is a general mean, α_i the effect on gain due to diet i, x_{ij} the initial weight of the animal, β the regression of y_{ij} on x_{ij} and e_{ij} a random error term. The analysis procedure is exactly as before, based on normal equations $X'X\hat{b} = X'y$, only X is now a matrix involving both 0's and 1's and the x-observations, and b is the vector of parameters μ, α_i and β. We illustrate with a small hypothetical sample of data.

TABLE 15. WEIGHT GAINS OF FIVE ANIMALS

Diet i	Animal j	Weight gain y_{ij}	Initial weight x_{ij}
1	1	8	1
	2	3	2
	3	7	1
2	1	5	3
	2	4	2

Writing the observations in terms of the model as

$$\begin{aligned}
y_{11} &= 8 = \mu + \alpha_1 &&+ \beta + e_{11}\\
y_{12} &= 3 = \mu + \alpha_1 &&+ 2\beta + e_{12}\\
y_{13} &= 7 = \mu + \alpha_1 &&+ \beta + e_{13}\\
y_{21} &= 5 = \mu &&+ \alpha_2 + 3\beta + e_{21}\\
y_{22} &= 4 = \mu &&+ \alpha_2 + 2\beta + e_{22}
\end{aligned}$$

we have them in the familiar form $y = Xb + e$ where y and e are, as usual, the vectors of observations and error terms, and

$$X = \begin{bmatrix} 1 & 1 & 0 & 1 \\ 1 & 1 & 0 & 2 \\ 1 & 1 & 0 & 1 \\ 1 & 0 & 1 & 3 \\ 1 & 0 & 1 & 2 \end{bmatrix} \quad \text{and} \quad b = \begin{bmatrix} \mu \\ \alpha_1 \\ \alpha_2 \\ \beta \end{bmatrix}.$$

Thus

$$X'X = \begin{bmatrix} 5 & 3 & 2 & 9 \\ 3 & 3 & 0 & 4 \\ 2 & 0 & 2 & 5 \\ 9 & 4 & 5 & 19 \end{bmatrix}, \qquad X'y = \begin{bmatrix} 27 \\ 18 \\ 9 \\ 44 \end{bmatrix} = \begin{bmatrix} y_{..} \\ y_{1.} \\ y_{2.} \\ \Sigma xy \end{bmatrix},$$

and a generalized inverse of $X'X$ is

$$G = (1/7)\begin{bmatrix} 0 & 0 & 0 & 0 \\ 0 & 13 & 20 & -8 \\ 0 & 20 & 41 & -15 \\ 0 & -8 & -15 & 6 \end{bmatrix} \quad \text{with} \quad H = GX'X = \begin{bmatrix} 0 & 0 & 0 & 0 \\ 1 & 1 & 0 & 0 \\ 1 & 0 & 1 & 0 \\ 0 & 0 & 0 & 1 \end{bmatrix}.$$

Using $w' = [w_0 \quad w_1 \quad w_2 \quad w_3]$ as the arbitrary vector in equations (8) and (9) the estimable functions are

$$\begin{aligned} q'b &= w'Hb \\ &= (w_1 + w_2)\mu + w_1\alpha_1 + w_2\alpha_2 + w_3\beta, \end{aligned}$$

their estimators being

$$q'\hat{b} = w'GX'y = w'\hat{b} = (1/7)(62w_1 + 69w_2 - 15w_3).$$

At once we see that $\alpha_1 - \alpha_2$ and β are both estimable functions, their estimators being

$$\widehat{\alpha_1 - \alpha_2} = (1/7)(62 - 69) = -1$$

for $w_1 = 1$, $w_2 = -1$ and $w_3 = 0$, and

$$\hat{\beta} = (1/7)(-15) = -15/7,$$

for $w_1 = 0 = w_2$ and $w_3 = 1$. As before, in equation (10), the variance of the estimator of an estimable function is

$$w'Gw\sigma^2 = (1/7)(13w_1^2 + 40w_1w_2 - 16w_1w_3 + 41w_2^2 - 30w_2w_3 + 6w_3^2)\sigma^2,$$

which for $(\widehat{\alpha_1 - \alpha_2})$ is

$$\text{var}(\widehat{\alpha_1 - \alpha_2}) = (1/7)(13 - 40 + 41)\sigma^2 = 2\sigma^2$$

and for $\hat{\beta}$ is $\text{var}(\hat{\beta}) = 6\sigma^2/7$. The estimator of σ^2 will, from (11), be given by

$$(n - r)\hat{\sigma}^2 = y'y - y'XGX'y$$

i.e.,

$$\begin{aligned}(5 - 3)\hat{\sigma}^2 = 64 + 9 + 49 + 25 + 16 \\ - (1/7)[13(18^2) + 40(18)(9) - 16(18)(44) \\ + 41(9^2) - 30(9)(44) + 6(44^2)].\end{aligned}$$

Thus

$$2\hat{\sigma}^2 = 163 - 1077/7$$

which gives

$$\hat{\sigma}^2 = 32/7.$$

The analysis of variance is

TABLE 16. ANALYSIS OF VARIANCE FOR DIETS AND REGRESSION ON INITIAL WEIGHT

Source of Variation	Degrees of Freedom	Sums of Squares
Diets and regression	2	282/35
Error	2	320/35
Total	4	602/35

the total sum of squares being $163 - 27^2/5 = 602/35$.

Should we wish to test the significance of fitting the diet effects after allowing for regression on initial weights we must find the sum of squares due to fitting the model

$$y_{ij} = \mu + \beta x_{ij} + e_{ij}.$$

Since $\alpha_1 - \alpha_2$ is estimable the sum of squares for fitting this model, SSM_β, can be obtained by considering the hypothesis $\alpha_1 - \alpha_2 = 0$, for which $Q' = [0 \quad 1 \quad -1 \quad 0]$. A solution of the normal equations is

$$\hat{b} = GX'y = (1/7)\begin{bmatrix} 0 \\ 62 \\ 69 \\ -15 \end{bmatrix}$$

for which $Q'\hat{b} = -1$. Also, $Q'GQ = 2$, and hence

$$SSM_{\alpha\beta} - SSM_{\beta} = (-1)2^{-1}(-1) = \tfrac{1}{2}$$

which gives

$$SSM_{\beta} = 282/35 - \tfrac{1}{2} = 529/70.$$

The appropriate analysis of variance is therefore as follows.

TABLE 17. ANALYSIS OF VARIANCE FOR DIETS AND REGRESSION ON INITIAL WEIGHT

Source of Variation	Degrees of Freedom	Sums of Squares
Regression	1	529/70
Diets, after allowing for regression	1	35/70
Error	2	640/70
Total	4	1204/70

The observed value of the F-ratio for testing the significance of diets is therefore 35/320, which is not significant when tested against the tabulated F-distribution with 1 and 2 degrees of freedom. The reader should check the value of SSM_{β} either by noting that $y_{ij} = \mu + \beta x_{ij} + e_{ij}$ is a simple regression model, or by making use of the normal equations

$$\begin{bmatrix} 5 & 9 \\ 9 & 19 \end{bmatrix} \begin{bmatrix} \mu \\ \beta \end{bmatrix} = \begin{bmatrix} 27 \\ 44 \end{bmatrix}. \tag{26}$$

He might also show, from equation (26), that if the diet effects are neglected the estimate of the regression coefficient is $-23/14$. Testing the hypothesis that $\beta = 0$ could be considered too, and a model allowing for a different regression for each diet might also be fitted, $y_{ij} = \mu + \beta_i x_{ij} + e_{ij}$, and a test made of the hypothesis $\beta_1 = \beta_2$.

More complex cases of this sort can readily be envisaged—involving, for example, more than one regression variable. No matter how many such variables there are, nor how many elements in the purely linear model part of the complete model, the procedure remains the same. The matrix description given in equations (8) through (11) encompasses all possible situations.

10. SUMMARY OF CALCULATIONS

In line with the final section of the regression chapter we summarize here a list of formulae for deriving estimable functions, their estimators, variances of the estimators and associated analysis of variance table.

b = vector of parameters of model;
n = number of observations;
y = vector of observations;
$\bar{y}$ = mean of observations;
SST = corrected sum of squares of the observations;
X = design matrix, the matrix of coefficients of the elements of b in the model;
r = rank of X;
$X'X$ = matrix of coefficients of the elements of b in the normal equations $X'Xb = X'y$;
$X'y$ = vector of totals in the normal equations;
G = generalized inverse of $X'X$ (see Chapter 6);
$H = GX'X$;
w' = vector of arbitrary elements;
$q'b = w'Hb$: estimable function;
$q'\hat{b} = w'GX'y$: estimator of estimable function;
$\text{var}(q'\hat{b}) = w'Gw\sigma^2$: variance of estimator;
$\text{cov}(q_1'\hat{b}, q_2'\hat{b}) = w_1'Gw_2\sigma^2$: covariance between estimators of two estimable functions;
$SSM = y'XGX'y - n\bar{y}^2$: sum of squares due to fitting the model;
$SSE = SST - SSM$: error sum of squares;
$\hat{\sigma}^2 = SSE/(n - r)$: estimated error variance;
$F_{r-1,n-r} = \dfrac{SSM(n - r)}{SSE(r - 1)}$: F value for F-test;
$Q'b = 0$: testable hypothesis of s linearly in-dent estimable functions;
$F_{s,n-r} = \hat{b}'Q(Q'GQ)^{-1}Q'\hat{b}/s\hat{\sigma}^2$: F-test for hypothesis $Q'b = 0$.

REFERENCES

Anderson, T. W. (1958). *Introduction to Multivariate Statistical Analysis.* Wiley, New York.

Graybill, Franklin A. (1961). *An Introduction to Linear Statistical Models.* Vol. I, McGraw-Hill, New York.

Federer, W. T. (1955). *Experimental Design.* Macmillan, New York.

Kempthorne, Oscar (1952). *Design and Analysis of Experiments.* Wiley, New York.

Mood, A. M., and Franklin A. Graybill (1963). *Introduction to the Theory of Statistics.* McGraw-Hill, New York.

Rao, C. R. (1962). A note on a generalized inverse of a matrix with applications to problems in mathematical statistics. *J. Roy. Stat. Soc.* (*B*), **24,** 152–158.

Scheffé, H. (1959). *The Analysis of Variance.* Wiley, New York.

BOOK LIST

The following is a short list of books on matrix algebra which the reader might wish to consult for further reading.

Aitken, A. C. *Determinants and Matrices.* Fifth Edition. Oliver and Boyd, Edinburgh, 1948.

Browne, Edward T. *Introduction to the Theory of Determinants and Matrices.* University of North Carolina Press, Chapel Hill, 1958.

Ferrar, W. L. *Algebra, a Text-book of Determinants, Matrices and Algebraic Forms.* Oxford University Press, First Edition, London, 1941.

Frazer, R. A., W. J. Duncan, and A. R. Collar. *Elementary Matrices and Some Applications to Dynamics and Differential Equations.* Cambridge University Press, Cambridge, 1952.

Hohn, Franz E. *Elementary Matrix Algebra.* Macmillan, New York, 1958.

MacDuffee, C. C. *The Theory of Matrices.* Verlag von Julius Springer, Berlin, 1933. (Also Chelsea Publishing Company, New York, 1946, 1956.)

Parker, W. V. and J. C. Eaves, *Matrices.* Ronald Press, New York, 1960.

Perlis, S. *Theory of Matrices.* Addison-Wesley, Cambridge, Massachusetts, 1958.

Turnbull, H. W., and A. C. Aitken. *An Introduction to the Theory of Canonical Matrices.* Blackie and Son, London, 1932.

INDEX

Addition, 18
  conformability, 19
Adjoint matrix, 90
Adjugate matrix, 90
Age distributions, 94
Age distribution vectors, 163
  instability, 173
  stability, 164
Alternant, 69, 168
Analysis of variance, linear models, 273
  regression, 241
Augmented matrix, 142

Backcrossing, 109, 194
Bilinear form, 42
  with inverse matrix, 97
Biological Illustrations, age distribution of beetles, 166, 173
  age of wildlife, 94
  crop yields, 254
  dairy cow nutrition, 51, 225
  flour beetles, 12
  forest vegetation, 213
  genetics, autotetraploids, 172, 185, 217
    backcrossing, 109, 194
    brother-sister mating, 195
    cross-breeding, 38
    dwarfism in mice, 81
    genetic merit, 97
    genotype frequencies, 54
    genotypic values, 33, 108
    inbreeding, 26, 35
    selection index, 97, 208
    selfing, 54, 152, 194
    sib mating, 195
    zygotic frequencies, 78
  human physiology, 212, 244
  pecking order in poultry, 30
  population surveys, 18, 47, 50
  rabbit extermination, 185
  selection index, 97, 208
  tuberculosis diagnoses, 95
  weight gain in animals, 20, 51, 284

Canonical form, congruent, 130
  equivalent, 127
  orthogonal, 189
  similar, 168
Cayley-Hamilton theorem, 179
Characteristic equation, 165
  factorization, 186
Characteristic roots and vectors, 166; *see* Latent Roots; Vectors
Cofactor, 86
Complementary principal minor, 73
Computing, generalized inverse, 145
  inverse matrix, 98
  regression, *see* Regression
Conformability, 19
  addition, 19
  multiplication, 28
  necessity, 38
  subtraction, 21
Congruent reduction, 129
Consistency, 137
  tests for, 142
Consistent equations, 138
  generalized inverse, 145
  solution, 138; *see also* Solutions

Death rates, 163
Dependence (linear dependence), 113
Dependent variable, 226
Design matrix, 257
Determinant, 56
  addition, 70
  cofactors, 86
  diagonal expansion, 71
  direct product, 216

Determinant, direct sum, 214
elementary expansions, 63
expanding, 57
by diagonal elements, 73
by minors, 58
by row or column, 58
evaluating, 57
formal definition, 61
Laplace expansion, 74
matrix product, 75
multiplication, 75
n-order, 57
partitioned matrix, 95
reduction, 57
second-order, 57
subtraction, 70
third-order, 58
Diagonal, elements, 12
form, 127
matrix, 12
expansion, 71, 73
Differentiation, Hessian, 209
Jacobian, 209
quadratic forms, 205
vector of operators, 203
Dimensions, 12
Direct product, 215
Direct sum, 213
Dominance matrix, 30
Dot notation, 10

Eigenvalues and vectors, 166; *see* Latent Roots
Element, 4
diagonal, 12
*ij*th, 11
leading, 12
sub-diagonal, 164
Elementary operators, 121
determinants, 122
E-type, 121
inverses, 122
P-type, 122
post-multiplication by, 122
pre-multiplication by, 121
rank, 123
R-type, 121
transposed, 122
Equations, 136
consistent, 138; *see* Consistent equations
full rank, 142
more, or fewer, than unknowns, 143
normal, 257; *see* Linear models
not of full rank, 142
null right-hand sides, 153
solutions, *see* Solutions, 136
Equivalent matrices, 124, 125
Estimable functions, 262

Factorizing a matrix, 119
Fertility rates, 163
Full rank, 120

Gene effects, 81
General linear hypotheses, 275
Generalized inverse, 144
and diagonal form, 145
computing, 145
definitions, 144
linear models, 259
product with original matrix, 145
rectangular, 158
solving equations, 147
unique form, 144
weak, 145
Generation matrix, 26
Genes identical by descent, 54
Genetic merit, 97

Hessian, 209
Hypothesis testing, 275

Idempotent matrix, 145, 200
Illustrations, 15; *see* Biological and Statistical
Identity matrix, 37
Inbreeding, 26, 35
Independence (linear), 113
rows, and columns, of a matrix, 117
Independent variable, 226
Inner product, 29
Inverse matrix, 80
by partitioning, 210
computing, 98
direct product, 216
direct sum, 214
effective, 144
existence, 91

Inverse matrix, generalized, 144; *see* Generalized
  left, 92, 132, 145
  linear equations, 109
  partitioned, 210
  properties, 91
  pseudo, 144
  rank, 119, 120
  right, 92, 132, 145
  rounding error, 102
  weak generalized, 145

Jacobian, 209

Kronecker product, 215

Lambda roots and vectors, 166; *see* Latent
Laplace expansion, 74
Latent roots, 166
  all different, 167
    symmetric matrix, 189
  direct product, 217
  direct sum, 214
  dominant, 181
  multiple, 174
    symmetric matrix, 190
  non-zero, 179
  powers of, 177
  powers of a matrix, 167
  real, 191
  scalar product, 179
  sum and product, 178
Latent vectors, 170
Laws of algebra, 35
Leading term, 12
Least squares, 228
Left direct product, 215
Left inverse, 92, 132, 145
Linear combination, 110
  columns of a determinant, 113
Linear dependence, 110
  and determinants, 113
  vectors, 112
Linear equations, 136; *see* Equations
  inverse matrix, 109
  solutions, 108; *see* Solutions
Linear hypotheses, 275
Linear independence, 110
Linear models, 254
  analysis of variance, 273
  balanced data, 283
  convenient restraints, 259
  design matrix, 257
  error variance, 260
  estimable functions, 262
  examples, 264
  expected values, 260
  fitting parts of, 279
  full model, 275
  general description, 254
  generalized inverse, 259
  hypothesis testing, 275
  least squares estimation, 256
  no unique estimators, 255
  normal equations, 257
  obvious restriction, 259
  one-way classification, 255, 264
  reduced model, 275
  regression as part of, 284
  solving normal equations, 259
  summary of calculations, 288
  tests of hypotheses, 275
  variances, 260
Linear regression, 266; *see* Regression
Linear transformation, 33
Linearly dependent, 110
Linearly independent, 110
  vectors, number of, 114
LINN, 117
Lower triangular matrix, 12

Matrices, computing, 3
  data analysis, 1
  equivalent, 124, 125
Matrix, 11
  adjoint, 90
  adjugate, 90
  array, 4
  all elements equal, 197
  augmented, 142
  definition, 11
  design, 257; *see* Linear Models
  diagonal, 12
  direct product, 215
  direct sum, 213
  dominance, 30
  equality, 22
  generalized inverse, 144
  generation, 26
  idempotent, 145, 200

Matrix, identity, 37
  inverse, 80; *see* Inverse matrix
  Kronecker product, 215
  lower triangular, 12
  nil-potent, 202
  null, 22
  orthogonal, 91, 196
  partitioned, 47
  positive (semi-) definite, 45, 192
  probability transition, 13
  quasi-scalar, 32
  reciprocal, 83
  scalar, 37
  skew, 197
  sub-matrix, 47
  symmetric, 40
  transpose, 38
  triangular, 12
  uni-potent, 203
  unit, 37
  upper triangular, 12
  variance-covariance, 46
Matrix algebra, 4
Matrix functions, 212
  exponential, 212
  logarithmic, 213
Minor, 58
  complementary principal, 73
  principal, 72
  signed, 86
Multiple correlation, 243
Multiple regression, 226; *see* Regression
Multiplication, 22
  determinants, 76
  diagonal matrices, 32
  element-by-element, 32
  elementary operators, 121
  matrices, 26
  matrix by vector, 23
  products equal to identity, 84
  non-commutative, 36
  partitioned matrices, 49
  pre-multiplication, 30
  post-multiplication, 30
  rank, 132
  rank and elementary operators, 123
  scalar, 20
  term-by-term, 32
  vectors, 22, 31

Nil-potent matrix, 202
Normal equations, 257
Notation, 14
  LINN, 117
  matrices, 14
  null matrix, 165
  rank, 118
  regression, 228
Null matrix, 22

One-way classification, 255, 264
Order, 12
Orthogonal matrix, 91, 196

Partitioned matrix, 47
  determinant, 95
  inverse, 210
  multiplication, 49
Pecking order in poultry, 30
Positive (semi-) definitive matrix, 45, 192
Powers of a matrix, 169
  and Cayley-Hamilton theorem, 179
Principal minor, 72
Probability transition matrix, 13

Quadratic form, 42
  bilinear, 42
  differentiation of, 205
  distribution, 202
  expected value, 201
  positive definite, 45
  symmetric matrix, 44
  with inverse matrix, 96
Quasi-scalar matrix, 32

Rank, 118
  and elementary operators, 123
  and canonical form, 127
  and non-zero latent roots, 179
  augmented matrix, 142
  direct product, 215
  direct sum, 214
  factorizing a matrix, 119
  finding rank of matrix, 123
  full rank, 120
  inverse matrix, 119, 120
  non-singular minors, 119
  null matrix, 118
  product matrix, 132
  rectangular matrix, 119

Rank, square matrix, 119
Reciprocal matrix, 83; *see* Inverse
Reduction, 126
  canonical form, 126
  congruent form, 130
  determinants, 57
  orthogonal similarity, 189
  similar, 168
Regression, 225
  analysis of variance, 241
  computing, 236, 241, 247, 249
  dependent variable, 226
  deviations from means, 233
  error terms, 231
  error variance, 233, 239
  fitting some of the variables, 244
  fitting variables one at a time, 247
  general solution, 228
  independent variable, 226
  intercept model, 233
  linear, 226
  many variables, 230
  mathematical model, 231
  multiple, 226
  multiple correlation, 243
  no-intercept model, 231
  non-linear, 226
  notation, 228
  predicted y-values, 232
    variances of, 241
  summary of calculations, 252
  tests of significance, 243
  unbiased estimators, 232
  variances of estimators, 232, 236
Right direct product, 215
Right inverse, 92, 132, 145

Scalar, 13
Scalar matrix, 37
Selection index, 97, 208
Selfing, 54, 172, 194
Signature, 131
Skew matrix, 197
Solutions of equations, 136
  by partitioning, 139
  combinations of, 151
  combinations of elements, 154
  consistent equations, 138
  existence of, 138
  general solution, 148
  generalized inverse, 147
  independent, 150
  infinite number of, 141
  inverse matrix, 80
  LINN, 150
  many solutions, 137
Statistical Illustrations, chi-square distribution, 192, 199
  covariance, 284
  factorial experiments, 218
  least squares, 207
  linear models (Chapter 10), 254
  maximum likelihood, 209
  mean, 44
  multiple correlation, 243
  one-way classification, 255, 264
  quadratic form—distribution, 202
  quadratic form—expected value, 201
  regression (Chapter 9), 225
  tests of hypotheses (linear models), 275
  tests of significance (regression), 243
  two-way classification, balanced, 266
  two-way classification, unbalanced, 270
  variance, 44
  variance-covariance matrix, 46, 214, 223
Sub-matrix, 47
Subscripts, 5
  matrix elements, 5
  order of matrix, 30
Subtraction, 20
Summation notation, 6
Symmetric matrix, 40
  canonical form, 188
  congruent reduction, 130
  inverse, 91
  latent roots all different, 189
  multiple latent roots, 190
  positive definite, 192
  real latent roots, 191

Term, 4
  leading, 12
Tests of hypotheses (significance), 243
  regression, 243
  linear models, 275
Trace, 12, 73

Trace, of a product, 32
Transformation (linear), 33
  variance-covariance matrix, 46
Transition probabilities, 13, 50, 95
Transpose matrix, 38
  inverse, 91
Triangular matrix, 12
Two-way classification, no interaction, 266
  balanced data, 266
  unbalanced data, 270

Uni-potent matrix, 203
Unit matrix, 37
Upper triangular matrix, 12

Variance-covariance matrix, 46, 214, 223
Vectors, 13
  age distribution, 163
  characteristic, 166
  column, 13
  eigen, 166
  lambda, 166
  latent, 166
  linearly dependent, 112
  linearly independent, 114
  normalizing, 189
  of differential operators, 203
  row, 13